FUNDAMENTAL DESIGN AND AUTOMATION TECHNOLOGIES IN OFFSHORE ROBOTICS

FUNDAMENTAL DESIGN AND AUTOMATION TECHNOLOGIES IN OFFSHORE ROBOTICS

Edited by

HAMID REZA KARIMI

Series Editor

QUAN MIN ZHU

ELSEVIER

ACADEMIC PRESS

An imprint of Elsevier

Academic Press is an imprint of Elsevier
125 London Wall, London EC2Y 5AS, United Kingdom
525 B Street, Suite 1650, San Diego, CA 92101, United States
50 Hampshire Street, 5th Floor, Cambridge, MA 02139, United States
The Boulevard, Langford Lane, Kidlington, Oxford OX5 1GB, United Kingdom

Library of Congress Cataloging-in-Publication Data
A catalog record for this book is available from the Library of Congress

British Library Cataloguing-in-Publication Data
A catalogue record for this book is available from the British Library

ISBN: 978-0-12-820271-5

For information on all Academic Press publications
visit our website at https://www.elsevier.com/books-and-journals

Publisher: Mara Conner
Acquisitions Editor: Sonnini R. Yura
Editorial Project Manager: Fernanda A. Oliveira
Production Project Manager: Sojan P. Pazhayattil
Designer: Greg Harris

Typeset by VTeX

Contents

List of contributors

Andreas Birk
Jacobs University Bremen, Robotics Group, Computer Science & Electrical Engineering, Bremen, Germany

Zhenzhong Chu
Shanghai, China

Tobias Doernbach
Jacobs University Bremen, Robotics Group, Computer Science & Electrical Engineering, Bremen, Germany

Arturo Gomez Chavez
Jacobs University Bremen, Robotics Group, Computer Science & Electrical Engineering, Bremen, Germany

Nan Gu

Tao Jiang
School of Mechatronic Engineering and Automation, Shanghai University, Shanghai, China

Hamid Reza Karimi
Department of Mechanical Engineering, Politecnico di Milano, Milan, Italy

Daniel Köhntopp
Jacobs University Bremen, Robotics Group, Computer Science & Electrical Engineering, Bremen, Germany

Bo Li
Institute of Logistics Science and Engineering, Shanghai Maritime University, Shanghai, China

Tieshan Li

Lu Liu

Wen Liu
Institute of Logistics Science and Engineering, Shanghai Maritime University, Shanghai, China

Christian A. Mueller
Jacobs University Bremen, Robotics Group, Computer Science & Electrical Engineering, Bremen, Germany

Zhouhua Peng

Hongde Qin
College of Shipbuilding Engineering, Harbin Engineering University, Harbin, China

Ke Qin
Institute of Logistics Science and Engineering, Shanghai Maritime University, Shanghai, China

Yanchao Sun
College of Shipbuilding Engineering, Harbin Engineering University, Harbin, China

Dan Wang

Ning Wang
School of Marine Electrical Engineering, Dalian Maritime University, Dalian, China

Yueying Wang
School of Electrical Information and Electrical Engineering, Shanghai University, Shanghai, China
School of Mechatronic Engineering and Automation, Shanghai University, Shanghai, China

Bing Xiao
School of Automation, Northwestern Polytechnical University, Xi'an, China

Yongsheng Yang
Institute of Logistics Science and Engineering, Shanghai Maritime University, Shanghai, China

Guichen Zhang
Merchant Marine College, Shanghai Maritime University, Shanghai, PR China

Minjie Zheng
Navigation College, Jimei University, Xiamen, Fujian, China

Preface

With the rapid growth of offshore technology in various application fields such as oil and gas industry, wind energy, robotics, and logistics, many researchers in academia and industry have focused on technology-based challenges raised in offshore environment. To this aim, this book unifies existing and emerging concepts concerning advanced design and automation technologies of offshore robotics towards practical applications, such as guidance principles for motion control, autonomous underwater vehicles, autonomous surface vehicles, as well as measurement and fault detection. The book may be useful for graduate students and researchers in control systems, mechatronics, mathematics, mechanics and alike.

The book consists of one introductory chapter and 12 technical chapters, which are organized as separate contributions and listed according to the order of the list of contents as follows:

In Chapter 1, an introduction and some main characteristics are generally addressed for the design and automation technologies in offshore robotics. A continuous system integration scheme is proposed in Chapter 2, which is used for underwater perception applications. In Chapter 3, azimuthing thruster single-lever-type remote control system is designed to freely control the sailing direction and speed of a ship. Then, solutions for underwater perception of environment and object are proposed in Chapter 4, in particular, a joint framework for underwater imagery mosaicking is proposed for autonomous underwater vehicle (AUV) perception of underwater in a wider visual range. Afterwards, several autonomous control methods are proposed in Chapter 5 for trajectory tracking control of AUV under the influence of modeling uncertainties, ocean current, and thruster faults. In Chapter 6, a hybrid control architecture with the characteristics of hierarchical architecture and subsumption architecture is proposed for a small autonomous underwater vehicle. In addition, the trajectory tracking control problem of an underwater robot is presented in Chapter 7. In Chapter 8, thruster fault reconstruction is investigated for AUVs with thruster fault. Then, Chapter 9 investigates the robust stabilization problem for nonlinear sampled-data dynamic positioning of ships based on Takagi–Sugeno (T–S) fuzzy model. The problem of finite-time control of an autonomous surface vehicle (ASV) with complex unknowns is addressed in Chapter 10. Then, Chapter 11 deals with the problem of how to realize way-point tracking control for an underactuated unmanned surface vehicles (USVs). In Chapter 12, a guidance law is proposed for cooperative path maneuvering of multiple autonomous surface vehicles guided by one parameterized path, and *last but not least*, in Chapter 13, the problem of finite time fault-tolerant attitude stabilization control for unmanned aerial vehicles is studied without the angular velocity measurements, in the presence of external disturbances and actuator failures.

Finally, I would like to express appreciation to all contributors for their excellent contributions to this book.

Hamid Reza Karimi
Milan, Italy
17 August, 2020

CHAPTER ONE

Introduction to fundamental design and automation technologies in offshore robotics

Hamid Reza Karimi
Department of Mechanical Engineering, Politecnico di Milano, Milan, Italy

Contents

Abstract

Due to the rapid progress of offshore technology in energy and transportation sectors, the context of offshore robotics has obtained increasing attentions in recent years. To partly address some complexities in design and automation technologies associated with offshore robotics, this chapter highlights recent developments in motion modeling and control techniques applied to offshore robotics to leverage the performance of the system in the offshore environment. In particular, Guidance principles for motion control, Autonomous underwater vehicles, Autonomous surface vehicles, Measurement and fault detection are addressed in this chapter.

Keywords

Offshore robotics, Design, Automation technology, Motion control, Autonomous underwater vehicles, Autonomous surface vehicles, Measurement, Fault detection

1.1. Introduction

Offshore robotics technology has recently received increasing attention in both academic and industrial aspects due to the importance of offshore environment in the areas of energy, transportation and security. Examples of offshore robotics are remote operated vehicles (ROVs), teleoperation of unmanned drilling, autonomous underwater vehicles (AUVs),

Fundamental Design and Automation Technologies in Offshore Robotics
https://doi.org/10.1016/B978-0-12-820271-5.00006-7

unmanned surface vehicles (USVs), under-water welding, welding robots, for instance, see [1] and the references therein. The main motivation behind this robotization is the need for reducing operation and maintenance costs, simultaneously increasing the reliability of the system operation because of potentially harsh sea environments. Moreover, based on successful development of robotics systems for different applications in space, manufacturing, medical care, for instance, the development of robotics and manipulators for operation in offshore area has faced some complexities in terms of cost-effective design, safety and reliability in implementation.

On the other hand, advanced automation technology is deployed in robotic systems, especially those which require autonomous operations in industries. These systems are multidisciplinary and involve integrated systems engineering that combine various fields of study, including mechanical, electrical/electronic, control, and information disciplines. Although the automation of robotic systems has been developed greatly, there are still problems for what is a rapidly changing industry environment. These systems need to be more intelligent and more integrated within the industrial environment in which they operate. When it comes to offshore robotics, new solutions are required, particularly in the areas of advanced sensing and perception, motion planning, disturbance mitigation, intelligence, and control for robotics. These are the challenges facing the next generation of offshore robotics, so that these complexities can be incorporated in the future analysis and design of offshore robotic systems.

The main focus of this book is on the motion modeling and control techniques applied to offshore robotics to leverage the performance of offshore robotics with high efficiency.

This book provides a platform to facilitate interdisciplinary research and to share the most recent developments in various examples of offshore robotics. The specific areas represented include motion control, guidance, path tracking, advanced control techniques, and fault detection for offshore vehicles. In the following, we will divide the introduction into the following five parts, including guidance principles for motion control, autonomous underwater vehicles, autonomous surface vehicles, as well as measurement and fault detection.

1.1.1 Guidance principles for motion control

In [16], the authors present the distributed guidance law design for cooperative path maneuvering of autonomous surface vehicles with fully-actuated and underactuated configurations. Distributed guidance laws are based on the neighboring information for autonomous surface vehicles guided by a parameterized path. Cooperative path maneuvering guided by multiple parameterized paths is considered in [5] and [6] where the path variables are synchronized. With the partial knowledge of the path information, distributed path maneuvering of autonomous surface vehicles is addressed in [7] and [8].

Way-point tracking control for an underactuated Unmanned Surface Vehicle (USV) includes two aspects: path generation and path tracking for the generated path. In order to solve the problems of large tracking error and slow error convergence in the tracking process of underactuated asymmetric USVs in sharp turns and other extreme paths, an adaptive sliding mode control method based on a generalized predictive control (GPC) algorithm (LOS-GPC-SMC) is proposed in [20].

The robust stabilization problem for nonlinear sampled-data dynamic positioning (DP) ships based on Takagi–Sugeno (T–S) fuzzy model is addressed in [21]. Firstly, the T–S fuzzy model of the nonlinear sampled-data DP ship system is established. Next, based on Lyapunov method, a sufficient condition is given to guarantee the stability of the system and improve H_∞ performance. Then, the fuzzy sampled-data controller is designed by analyzing the stabilization condition. In [22], the authors propose an adaptive sliding mode control method based on the local recurrent neural network for trajectory tracking control of an underwater robot. In the current researches, most controllers are used to design the force/torque on each degree of freedom. Different from these, the authors use underwater robot as nonaffine nonlinear system for analysis, and then get an affine nonlinear system whose input is the control voltage of thrusters by affine transformation. According to the controlled system, the control voltage of thrusters is designed directly based on sliding mode control principle. For an unknown function term in the controller, an adaptive learning method is proposed based on the local recurrent neural network.

1.1.2 Autonomous underwater vehicles

In marine environment with high-pressure and poor-visibility, autonomous underwater vehicle (AUV) tasks cannot be achieved, when an accident happens in an AUV. As the important technology of AUVs, fault diagnosis has great significance on AUV's safety [3,4]. In [13], the authors investigate thruster fault reconstruction for an AUV with thruster fault. Since it is difficult to establish an accurate AUV dynamic model, a motion model with the affine form is developed based on an RBF neural network, whose input is the thruster control signal. Thus, the relationship between system input and output can be reflected clearly. The experiment results show that the developed motion model can well describe the AUV dynamics. For the AUV thruster fault reconstruction problem, a fault reconstruction method is developed based on a terminal sliding observer. Under the action of the developed method, the estimation errors of all states converge to zero in finite time and the thruster fault can be reconstructed quickly. The experiment results show the effectiveness of the developed method in terms of thruster fault reconstruction. In addition, in [14], the author designs an azimuthing thruster single-lever-type remote control system to freely control the sailing direction and speed of a ship by controlling azimuthing thrusters by using a lever mounted to the control stand.

In [18], the authors study motion control for underwater offshore vehicles due to its importance for multiple autonomous underwater vehicles (AUVs) and remotely operated vehicles (ROVs) working in the offshore regions [9–11]. Modern underwater vehicle control systems are based on a variety of design methods such as PID control, prescribed performance control, nonlinear neural networks control theory, and so on.

From the vertical perspective, in [19], the authors proposed a hybrid architecture consisting of management, function, and hardware layers for small autonomous underwater vehicles (SAUVs). Specifically, an overall design scheme of SAUV is introduced, and then the hybrid control architecture is described in detail.

1.1.3 Autonomous surface vehicles

In [17], the authors study finite-time control of an autonomous surface vehicle (ASV) with complex unknowns including unmodeled dynamics, uncertainties and/or unknown disturbances within a proposed homogeneity-based finite-time control (HFC) framework. Major contributions are as follows: (1) in the absence of external disturbances, a nominal HFC framework is established to achieve exact trajectory tracking control of an ASV, whereby global finite-time stability is ensured by combining homogeneous analysis and Lyapunov approach; (2) within the HFC scheme, a finite-time disturbance observer (FDO) is further nested to rapidly and accurately reject complex disturbances, and thereby contributing to an FDO-based HFC (FDO-HFC) scheme which can realize exactness of trajectory tracking and disturbance observation; and (3) aiming to exactly deal with complicated unknowns including unmodeled dynamics and/or disturbances, a finite-time unknown observer (FUO) is deployed as a patch for the nominal

HFC framework, and eventually results in an FUO-based HFC (FUO-HFC) scheme which guarantees that accurate trajectory tracking can be achieved for an ASV under harsh environments.

1.1.4 Measurement and fault detection

From technological aspect, an autonomous underwater vehicle (AUV) relies on an underwater camera and sonar for perception the surrounding environment including underwater image processing, object detecting and tracking. For underwater optical environment perception, the limited visual range of the camera severely limits the acceptance of detection information; in addition, underwater light is assimilated and scattered, which seriously deteriorates the underwater imaging, in particular decreases the image contrast. To address these issues, in [23], a joint framework based on a convolutional neural network (CNN) is proposed to improve underwater image quality and extract more matching feature points. Accurate registration between images, an underwater mosaicking technology, is also involved, and a fusion algorithm is implemented to mitigate artificial mosaicking traces. Experimental results show that the presented framework can not only keep underwater image detail information, but also exact more matching feature points for registration and mosaicking. For underwater acoustic image environment perception, mechanically scanned imaging sonars (MSIS) are usually equipped on AUV for avoiding obstacles and multiple-targets tracking. In [23], two multiple underwater objects tracking methods are presented. This proposed method is based on the cloud-like model data association.

In [12], the authors develop a continuous system integration framework (CSI) featuring a high fidelity simulator (SIL), and introduce it to the project development process of the DexROV project. Cost-intensive projects, such as DexROV, benefit from the proposed framework as it facilitates a quantitative assessment of the performance under various conditions before the roll-out of the system. Subsequently, such assessment allows reducing uncertainties and a more predictable project planning as required for and particularly challenging in multidisciplinary projects consisting of several workgroups and partners. The key contribution of the presented work is the synchronization of environmental and spatial conditions observed in real-world with the simulator of the CSI/SIL framework. The creation of a high fidelity simulator features the major challenge in the preparation process of such a continuous system integration system. The simulator fidelity is a product of several factors ranging from individual sensor models, CAD vehicle models to adaptation of external environment conditions within simulator and the introduction of algorithms and methods that accurately emulate real hardware and software components [2]. Efficient continuous system integration as presented in this chapter plays an important role as robotic systems become more complex and are applied to more challenging environment conditions performing more complex tasks. Therefore, a benchmark and validation framework, such as the proposed in this chapter, increasingly gains importance to quantify system performance, as well as to identify and analyze uncertainties, bottlenecks, limitations, etc.

Moreover, in [15], the authors study the finite-time fault-tolerant attitude stabilization control for unmanned aerial vehicles (UAVs) without the angular velocity measurements, in the presence of external disturbances and actuator failures.

References

[1] Amit Shukla, Hamad Karki, Application of robotics in offshore oil and gas industry—A review Part II, Robotics and Autonomous Systems 75 (B) (January 2016) 508–524.

[2] C.A. Mueller, T. Doernbach, A. Gomez Chavez, D. Koehntopp, A. Birk, Robust continuous system integration for critical deep-sea robot operations using knowledge-enabled simulation in the loop, in: International Conference on Intelligent Robots and Systems, 2018.

[3] M. Takai, T. Ura, Development of a system to diagnose autonomous underwater vehicles, International Journal of Systems Science 30 (1999) 981–988.

[4] Y.R. Xu, K. Xiao, Technology development of autonomous ocean vehicle, Acta Automatica Sinica 33 (2007) 518–521.

[5] R. Skjetne, T.I. Fossen, P.V. Kokotović, Robust output maneuvering for a class of nonlinear systems, Automatica 40 (Mar. 2004) 373–383.

[6] I.F. Ihle, M. Arcak, T.I. Fossen, Passivity-based designs for synchronized pathfollowing, Automatica 43 (9) (2007) 1508–1518.

[7] Z. Peng, J. Wang, D. Wang, Containment maneuvering of marine surface vehicles with multiple parameterized paths via spatial-temporal decoupling, IEEE/ASME Transactions on Mechatronics 22 (2) (2017) 1026–1036.

[8] Z. Peng, J. Wang, D. Wang, Distributed containment maneuvering of multiple marine vessels via neurodynamics-based output feedback, IEEE Transactions on Industrial Electronics 64 (5) (2017) 3831–3839.

[9] M. Bidoki, M. Mortazavi, M. Sabzehparvar, A new approach in system and tactic design optimization of an autonomous underwater vehicle by using Multidisciplinary Design Optimization, Ocean Engineering 147 (2018) 517–530.

[10] N. Wang, G. Xie, X. Pan, et al., Full state regulation control of asymmetric underactuated surface vehicles, IEEE Transactions on Industrial Electronics (2018), https://doi.org/10.1109/TIE.

[11] R.B. Wynn, V.A.I. Huvenne, T.P.L. Bas, et al., Autonomous underwater vehicles (AUVs): their past, present and future contributions to the advancement of marine geoscience, Marine Geology 352 (2) (2014) 451–468.

[12] C.A. Mueller, A. Gomez Chaveza, T. Doernbach, D. Köhntopp, A. Birk, Continuous system integration and validation for underwater perception in offshore inspection and intervention tasks, in: H.R. Karimi (Ed.), Fundamental Design and Automation Technologies in Offshore Robotics, 2020, Chapter 2.

[13] Z. Chu, Thruster fault reconstruction for autonomous underwater vehicle based on terminal sliding mode observer, in: H.R. Karimi (Ed.), Fundamental Design and Automation Technologies in Offshore Robotics, 2020, Chapter 8.

[14] G. Zhang, Azimuth thruster single lever type remote control system, in: H.R. Karimi (Ed.), Fundamental Design and Automation Technologies in Offshore Robotics, 2020, Chapter 3.

[15] B. Li, W. Liu, K. Qin, B. Xiao, Y. Yang, Finite-time extended state observer based fault tolerant output feedback control for UAV attitude stabilization under actuator failures and disturbances, in: H.R. Karimi (Ed.), Fundamental Design and Automation Technologies in Offshore Robotics, 2020, Chapter 13.

[16] Z. Peng, N. Gu, L. Liu, D. Wang, T. Li, ESO-based guidance law for distributed path maneuvering of multiple autonomous surface vehicles with a time-varying formation, in: H.R. Karimi (Ed.), Fundamental Design and Automation Technologies in Offshore Robotics, 2020, Chapter 12.

[17] N. Wang, Finite-time control of autonomous surface vehicles, in: H.R. Karimi (Ed.), Fundamental Design and Automation Technologies in Offshore Robotics, 2020, Chapter 10.

[18] H. Qin, Autonomous control of underwater offshore vehicles, in: H.R. Karimi (Ed.), Fundamental Design and Automation Technologies in Offshore Robotics, 2020, Chapter 5.

[19] Z. Chu, Development of hybrid control architecture for a small autonomous underwater vehicle, in: H.R. Karimi (Ed.), Fundamental Design and Automation Technologies in Offshore Robotics, 2020, Chapter 6.

[20] Y. Wang, T. Jiang, Way-point tracking control of underactuated USV based on GPC path planning, in: H.R. Karimi (Ed.), Fundamental Design and Automation Technologies in Offshore Robotics, 2020, Chapter 11.

[21] Y. Wang, M. Zheng, Robust sampled-data control for dynamic positioning ships based on T–S fuzzy model, in: H.R. Karimi (Ed.), Fundamental Design and Automation Technologies in Offshore Robotics, 2020, Chapter 9.

[22] Z. Chu, Adaptive sliding mode control based on local recurrent neural networks for an underwater robot, in: H.R. Karimi (Ed.), Fundamental Design and Automation Technologies in Offshore Robotics, 2020, Chapter 7.

[23] H. Qin, Autonomous environment and target perception of underwater offshore vehicles, in: H.R. Karimi (Ed.), Fundamental Design and Automation Technologies in Offshore Robotics, 2020, Chapter 4.

CHAPTER TWO

Continuous system integration and validation for underwater perception in offshore inspection and intervention tasks

Christian A. Mueller[a], Arturo Gomez Chavez[a], Tobias Doernbach, Daniel Köhntopp, Andreas Birk
Jacobs University Bremen, Robotics Group, Computer Science & Electrical Engineering, Bremen, Germany

Contents

[a] Authors share first-authorship.

Fundamental Design and Automation Technologies in Offshore Robotics
https://doi.org/10.1016/B978-0-12-820271-5.00007-9

Abstract

Continuous System Integration and Validation is an increasingly important factor for an efficient system development process. In particular, for underwater projects involving semi- to fully-autonomous robotic systems since they progressively become more complex, need to perform under more challenging environmental conditions and execute more intricate tasks. Therefore, a benchmark and validation framework become crucial to quantify the system performance, as well as to identify and analyze uncertainties, bottlenecks, limitations, etc., before field trials and missions.

We present a simulation-driven framework whose concepts ease the development of benchmark and validation tests that are of particular interest in interdisciplinary projects with multiple contributors at different locations. While considering aspects related to continuous system integration and validation, the presented framework facilitates a more efficient and subsequently effective project realization. Considering the preparation efforts of such system, these pay off as they allow for fast informed development cycles and qualified assessments, minimizing the integration labour on-site at field missions which usually involve high running costs and time constraints. The effectiveness of the mentioned framework is demonstrated through the development and analysis of the EU-H2020 DexROV project.

Keywords

Continuous System Integration, System Validation, Underwater System Development, Simulation-Driven System Validation

2.1. System development and integration in deep-sea robotics

Offshore environments represent a challenging scenario for the investigation and exploitation of automation technologies. Yet, the oceans are one of the forces driving commerce, employment, and scientific research growth, as well as one of the greatest sources of mineral, biological, and energy resources. Nonetheless, approximately two-thirds of the oceans are still unexplored, especially the deep layer (1000 m below the surface or more) which is subject to extremely poor illumination conditions and high pressure.

For these reasons, in the past decade there has been significant research and development of innovative Unmanned Underwater Vehicles (UUVs) including both Autonomous Underwater Vehicles (AUVs) and Remote Operating Vehicles (ROVs) required for inspection and intervention tasks in different application domains such as industrial, oceanographic, archaeological, and environmental scenarios. Nowadays marine vehicles offer higher bandwidths through acoustics, higher resolution in acoustic and camera imaging, and more energy-efficient sensors that enable longer periods of operation.

Nevertheless, required autonomy in complex missions, e.g., demanding advanced perceptual or dexterous manipulation skills, has not been achieved yet in order to perform safe, reliable, and fully unsupervised operations. Hence, human interventions are often necessary even when exposed to the latent risks found in offshore underwater mission scenarios. Already shallow waters (< 100 m) represent a perilous environment: the UK Health & Safety Executive's (HSE) 2016–2017 Offshore Safety Statistics [1] reports 39 major injuries and more than 400 dangerous occurrences, most of them performing maintenance and construction activities. On this account, both operation and development of marine technologies aim to reduce the personnel required on offshore facilities.

The development and continuous evaluation of such systems typically require an offshore crew consisting of coordinators, operators, and navigators, as well as engineers and researchers, responsible for the system's integration and maintenance. On the other hand, such endeavors require the deployment of valuable equipment, including a vessel enabled to deploy, operate, and retrieve UUVs at high seas. All of this rapidly increases the effort and cost of each development cycle. This, coupled with the fact that costs of failure are several orders of magnitude higher than in ground-robotics, calls for the need of specialized and efficient strategies to validate the precision, robustness, and reliability of the UUV developed capabilities.

As to alleviate these risks and costs, we introduce an approach based on a high fidelity simulator that embeds spatial and environmental conditions from recorded field trials data, we refer to it as **Simulation in the Loop (SIL)** (see Section 2.4). This methodology [2] interweaves simulated and real-world data at the development stage, effectively providing a platform to mitigate performance discrepancies between the development and application stage, which are caused by inconsistencies between simulated conditions during development and real-world conditions during the application stage, featured by uncertainties in underlying assumptions, e.g., sensor noise, illumination artifacts, system component interdependencies, etc. In turn, this platform serves to thoroughly investigate and benchmark behaviors of system components in real and simulated scenarios concurrently. This SIL methodology is complemented with a **Continuous System Integration (CSI)** scheme [3] (see Section 2.3) to facilitate the seamless integration of simulated system components with real roll-out ready ones into validation pipelines, beyond the traditional pre-deployment integration and testing cycles on real software and hardware components in complex integration scenarios.

While being constrained to particular system aspects, we primarily concentrate on using this scheme for underwater perception applications which is the core basis for various autonomous functionalities. Concretely, we describe the benefits arising from such a continuous integration and validation approach in tasks related to underwater perception, including object recognition and pose estimation, 3D mapping, and self-localization for navigation under challenging spatial and environmental conditions. As an example use case, we employ the proposed approach to the EU-H2020 DexROV [4,5] project in the context of deep-sea oil and gas industrial monitoring and maintenance as further introduced in the following Section 2.2.

2.2. Underwater perception for offshore inspection and intervention tasks

2.2.1 Importance and progress overview

Robotics plays an increasingly crucial role in underwater offshore technology development. By minimizing the risk caused by the exposure of humans to hazardous underwater offshore environments and increasing the automation of interventions, robotic advancements reach a state that particularly facilitates underwater inspection and intervention tasks. In particular, robotic inspection of vessels [6] and man-made structures [4], or monitoring of diver's condition [7] and subsea environment [8] are emerging application fields of offshore underwater robotics. Underwater perception featured in various tasks, such as data collection, collision avoidance, mapping, object detection and recognition, etc., represents a major pillar for these offshore robotic applications. Recent underwater perception advancements equip unmanned vehicles with capabilities that enable semi- to full-system autonomy to tackle complex tasks ranging from inspection and manipulation during surveying and maintenance to search and rescue missions. These vehicles are known in the field as I-AUVs (Intervention AUVs) representing the autonomous counterpart of the ROV.

Throughout the evolution of this research field, pioneer work started more than 20 years ago with the MBARI OTTER [9] and ODIN [10] systems using 1 degree of freedom (DOF) manipulators. But the first successful demonstration was made in 2003 as part of the ALIVE [11] project, where an I-AUV docked to a subsea panel and manipulated valves in a fixed-base manner. In 2009, within the SAUVIM project [12], the first floating-base intervention was performed by detecting, tracking, and retrieving a target object from the sea bottom using an imaging sonar. This technology was refined in the subsequent years in projects such as TRIDENT (2012) [13], PANDORA (2014) [14], and others, where capabilities were shown to plug and unplug subsea panel connectors and perform valve turning [15,16]. For a more comprehensive overview of I-AUV developments and progress we refer to [17].

In this chapter, we particularly focus and refer to a recent project, the EU-H2020 DexROV – *effective dexterous ROV operations in presence of communication latencies* project [4,5]. DexROV addresses one of the current most demanding domains: deep sea intervention and manipulation tasks in oil and gas industry through long distance teleoperation. The project represents a heterogeneous system composed of multidisciplinary fields: robot vision, navigation, manipulation, mapping, acoustic communication, latencies compensation, among many others. Such a complex system requires a well-designed continuous system development and integration regime to guarantee an efficient and effective exploitation of human and material resources, as well as time. Therefore, DexROV is taken as an application scenario to describe and apply the concepts of the proposed **Continuous**

System Integration (CSI) and **Simulation in the Loop (SIL)** methodologies.

2.2.2 Application scenario: remote dexterous ROV interventions under communication latencies (EU-H2020 DexROV)

DexROV [4,5] deals with the challenge that the state-of-the-art for underwater manipulation is dominated by costly ROV operations. These offshore operations require significant financial and human resources to control and maintain the robotic platform under challenging environmental conditions at sea. Furthermore, oil and gas routine interventions are still performed by experts through low-level, manual control of the manipulators and the vehicle itself. These experts are part of the offshore crew at a surface vessel commonly affected by hazardous weather, high currents, or low-visibility above and below sea level.

The DexROV core idea (Fig. 2.1) is to mitigate these costs and risks by enabling operations from an onshore control center, i.e., the ROV can be teleoperated via satellite communication from a distant on-land facility. This also includes the process to shift from the mentioned low-level teleoperation to a semi-autonomous scheme. In a concise manner, operators in the onshore control center interact with a real-time simulation, providing access to the ROV's real condition and made observations about the environment during operations. A cognitive engine analyzes operator's requests, and transmits them to the ROV as motion primitives to be executed on the field (Fig. 2.1).

During the field trial for this project, the onshore center was located in Zaventem, Belgium, and the offshore operations were performed off the coast of Marseille, France, with the Janus II vessel of COMEX.SA (see Fig. 2.2A) hosting an Apache ROV (see Fig. 2.2B). For testing and validation, a test panel (Fig. 2.2C) was constructed to emulate different perception and manipulation scenarios (Fig. 2.2D). The panel features three sides respectively consisting of a set of valves and wet-mate connectors from oil and gas industry interfaces (ISO 13628 standard), a mockup of corals for biological sampling tasks, and an archaeological box with ceramic mockups.

Based on this setup, the DexROV strategy aims to accomplish several objectives as shown in Fig. 2.1. Note that in Fig. 2.1 also the location (ROV, vessel, or onshore control center) is indicated, at which a task is primarily performed to accomplish the respective objective:

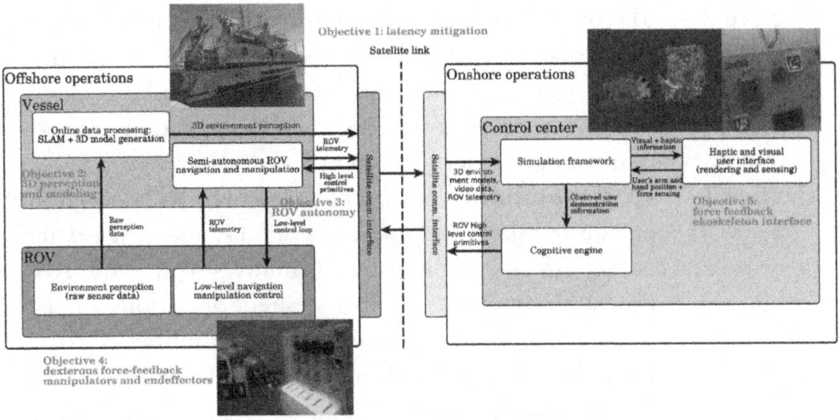

Figure 2.1 Illustration of the distributed DexROV architecture [4].

(A) COMEX Janus II vessel

(B) ROV ready for deployment

(C) Oil & gas panel side

(D) ROV and test panel during field trials

Figure 2.2 DexROV field trials setup at Marseille, France [4].

Objective 1 – Latencies mitigation. Establishment of a reliable data flow management between offshore and onshore systems while mitigating communication latencies.

Objective 2 – 3D perception and modeling. Based on localization and visual sensory data streams, a model of the environment is created and transmitted to the onshore control center for its analysis. This includes detection and pose estimation of objects of interest, e.g., panel valves.

Objective 3 – ROV autonomy. A machine learning based cognitive engine translates the operator's actions and requests into navigation and manipulation primitives so that the ROV can autonomously perform desired operations, despite possible communication delays or interruptions.

Objective 4 – Dexterous force feedback manipulators. To enable complex manipulation tasks and a more intuitive user control interface, the ROV is equipped with a pair of force sensing manipulators and three-finger end-effectors.

Objective 5 – Exoskeleton interface. The motion primitives for the mentioned ROV manipulator are generated from the onshore control center through an exoskeleton with haptic feedback.

In order to efficiently achieve each of these objectives and for their successful integration in the overall project, a continuous system integration and validation regime over the entire development process is a necessity. For this purpose, we introduce the proposed simulation-based continuous system integration and validation framework (Sections 2.3 and 2.4), describe its core concepts, general use cases, and implementation aspects in the context of the DexROV project.

2.3. CSI: continuous system integration and validation

The traditional product development and validation workflow is commonly represented using the Vee model [18], which describes a sequential approach to follow during the life cycle of a product or system. The Vee model first defines concurrent *verification* (left-side of V) and *validation tests* (right-side of V) for system baselines. As time and system maturity advances, the development team can either focus on the baseline *verification* (top-down) – from requirement analysis to low-level source code specifications – or the team focuses on system *validation tests* (bottom-up) – from source code module testing to baseline acceptance tests. This process allows performing opportunity and risk management investigations and constantly verifying that the defined baselines are acceptable to the final users. The strengths of such sequential method are its predictability, stability, repeatability, and high assurance. The system improvement focuses on standardization, measurement, and control, all of which rely on a "master plan" to anchor processes, unit tests, and communication amongst all workgroup members, from developers to users.

Essentially, in following this traditional scheme, the development workgroup must wait for a complete product design, build, and validation cycle in order to know if the overall system, with both hardware and software components, fulfilled the requirements. This approach is neither suitable nor efficient for our application of interest (DexROV, see Section 2.2.2) or for most other offshore deep sea intervention systems. As mentioned in previous sections, the design and organization of validation tests on the field represent large expenses, let alone starting a new production cycle. In order to surmount this challenge, *hardware-in-the-loop* (HIL) techniques are gaining widespread adoption as reflected in the inclusion of HIL standards in the rules of the *Det Norske Veritas* (DNV-GL) maritime classification society [19]. HIL methods aim to support system integration and verification by modeling components and environmental inputs, mathematically or through simulation, and running a diverse number of scenarios that emulate real-world situations. Then, selected and modeled components can be replaced by their physical counterpart to incrementally verify that the system requirements are satisfied.

Hardware-in-the-loop proves to be a useful approach. However, application examples are tailored to the development of very specific low-level or system-level components and behaviors, e.g., a power management unit (PMU) microcontroller and the current generation of an electrical grid, respectively. An offshore underwater intervention operation comprises several modules from different disciplines, i.e., vehicle navigation and localization, visual perception, object manipulation, etc. And each module is developed and tested by specialized workgroups, often from different institutions that have to integrate their work progress into the overall system at later stages in the development cycle. It would be an extremely daunting task to coordinate and define static development and integration stages among all these workgroups; for this reason, we propose an adoption of a *Continuous System Integration* (CSI) approach that provides a usable software platform at all times, so that all workgroups are able to develop, integrate, and validate their methods and algorithms, regardless of the development state of other components led by different groups.

We initially described *Continuous System Integration* (CSI) in our previous work [3], which follows the idea of the software engineering concept of *Continuous Integration* (CI) [20]. Within CI, members of a workgroup develop and very frequently integrate their local work, e.g., new software components to the system framework which subsequently triggers automated testing pipelines. In turn, integration problems are mitigated, which are often the result of long-term divergence between the local efforts of each workgroup member. In this chapter, we extend this concept to the integration of system-level distributed components from different robotic disciplines. Therefore, ***Continuous System Integration*** can be defined as a **system life cycle approach for distributed component development and validation that follows an underlying system structure and interface, all of which preserves the functionality and consistency of exchanged data between components at all times.** Additionally, any module can be replaced by its simulated counterpart anytime without jeopardizing the overall system behavior. Based on this, all business logic is developed as mockup components first, then replaced with high-fidelity simulation components, and finally with components providing the desired final functionality.

Continuity, in the CSI sense, refers to the concept of seamless, hybrid simulation/real-world component integration. This means that, whenever an expected roll-out ready component is not yet available due to underestimated resource allocations or setbacks during the development, validation, or integration stages, such component can be replaced by its simulated counterpart to allow for continuous testing of the complete system. Thus, a major priority in a project employing CSI is the setup of a fully-functional and modular simulation environment which allows for seamless replacement of simulated components with roll-out ready ones. To achieve this and follow the proposed paradigm, we introduce in Section 2.4 a pre-deployment simulation environment to test and evaluate single components, as well as the entire system, in an efficient way.

In summary, CSI can significantly reduce the costs of verification and benchmark tests for offshore underwater technologies in comparison with projects following traditional schemes such as the Vee model. The high costs of integration and deployment campaigns can be substantially mitigated through extensive pre-deployment validation tests with a combination of physical and simulated components. To explain this in more detail, next we describe the use cases for which the CSI methodology can be applied, and the advantages it introduces into complex software project structures over conventional methods (see Section 2.3.1). Then, although presenting itself as a continuous concept, we discretize the CSI approach into several stages to highlight at which point during project runtime a particular validation step has been performed (see Section 2.3.2).

2.3.1 CSI validation concepts and use cases

Not only in deep sea offshore operations, but especially in the presence of harsh conditions and limited data transmission capabilities, efficient integration and testing are crucial during the development of complex systems. However, in projects demanding for workgroups with particular expertise divided amongst disciplines, departments, and institutions that are often spatially distributed around the world, system integration becomes an exponential effort with increasing project size.

In the DexROV project, which comprises seven partners from all across Europe and providing different engineering disciplines, we identified several requisites and introduced them as design concepts for the proposed continuous system integration approach in which simulation plays a crucial role. Similar concepts are considered and described in the literature [21] as advantages of modularized simulation-based approaches against conventional monolithic testing, and here we extend them to the CSI approach presented.

Concept I – **Interface/pipeline testing.** *Before even involving physical hardware, interfaces and processing pipelines have to be thoroughly tested using simulated data.*
Especially when hardware interfaces still have to be designed, implemented and tested, developers from each workgroup can use simulation to optimize and validate their internal processing pipelines, as well as the agreed external interfaces. This generally is performed before concrete methods and algorithms are fully implemented in order to facilitate field testing in terms of "time to market", i.e., time until field data collection and/or validation missions can be launched.

Concept II – **Regression/degradation testing.** *Specific constraints such as data bandwidth, communication delays, processing power, environmental conditions, sensor noise, among others, are individually taken into account in the system design and verification.*

Since only a selection of the mentioned constraints may occur for certain components and conditions, particular constraint profiles can be defined in order to test each component. Especially stochastic noise in sensor measurements and algorithms can be accounted for when performing baseline validations tests by using switchable noise plug-ins in the simulation. In distributed systems where intercomponent connections are critical, bandwidth constraints, network degradation, and physical mediums are major factors to consider.

Concept III – **Distributed deployment.** *Developers from various workgroups collaborate together to develop the complete system, individually replacing specific simulated components with real ones according to validation purposes or due to progression of project milestones.*

This is relevant in integration settings where each group has to deploy their own components in the processing pipeline and there is only one physical platform. Preferably, simulated components are used at first to verify inputs/outputs and data flow, as well carry out proof-of-concept tests. As soon as each algorithmic method and physical component reaches maturity, they can gradually replace their simulated counterpart. In this manner, each workgroup can determine their own development timeline with the certainty of being able to benchmark their components at any given time. Distributed Deployment is the main and most widely used concept within the DexROV scenario as it is described in Section 2.5.

Concept IV – **Parallelized testing.** *Developers are able to evaluate their components in parallel using several instances of the complete system. These instances can be purely simulated or run using local resources instead of onboard devices, including software and hardware.*

Especially when one or a few hardware instances exist or are available due to maintenance and calibration procedures, simulated components can compensate this unavailability of resources. Using packaging and virtualization tools (see Section 2.4.3) makes it easy to deploy the entire system multiple times in parallel among different local hardware, not necessarily the one required in the final tests, which also may increase the computational power available, and subsequently reduces testing times. Parallelized testing offers many other advantages such as verifying the functionality of different component subsets concurrently, e.g., perception (cameras) and localization (motion sensors), or perception (cameras) and manipulation (robot arms). Even multiple robot missions can be simulated and executed automatically at the same time, either using the complete simulation or recorded real-world sensor data as explained in Section 2.4.3.

Concept V – Fault recovery/safety testing. *Instead of bearing the risk of damaging or losing hardware equipment (e.g., sensory, manipulator, or even UUV/ROV) at offshore underwater locations, many foreseeable situations, as well as extreme or unusual circumstances, have been verified in simulation prior to deployment under real conditions.*

Hardware equipment generally poses limitations in borderline/extreme situations which may damage the equipment. Test driving comprehensive and fast platforms in simulation can help prevent software errors and assessing the system behavior in circumstances which would be hard to reproduce in the real world. Since simulated components can be restored with no costs and efforts, hardware can come into play only in the final overall-system verification steps.

2.3.2 System life cycle stages

In order to showcase all advantages of the presented methodology with respect to the validation concepts described above, a number of tasks and corresponding experiments within the DexROV project have been conducted as described in detail in Section 2.5. It is required to identify in which stage of the system life cycle each of these task experiments are performed because this determines the number of simulated and roll-out ready components involved. Therefore, the following presented CSI approach is discretized into three high level stages to distinguish and classify the component state and conducted experiments evaluating project tasks.

Stage I *Development stage*

The first stage of the proposed CSI method initially consists of simulation-only components. These do not necessarily have to provide all functionalities of the roll-out ready components envisioned for the final stages of the system life cycle. However, the goal of this first stage is to mainly *design, develop and verify interfaces, data formats, and communication protocols* in order to provide a reliable software platform to all workgroups contributing to the project. Essentially, the groundwork for the usage of CSI concepts I and II is built. This framework can be easily run and validated in a single-site office setting, providing all necessary infrastructure required for developing the software business logic. Workgroups can independently develop their components while consuming expected (simulated) data according to the designed software platform.

Stage II *Integration stage*

This intermediate stage provides a continuous transition from stage I to stage III using a mix of simulated and roll-out ready components in the setup. The ratio between these type of components varies depending on the respective software development phase, project milestones and schedule. Therefore, this stage is important as it allows *integrating* and *benchmarking* algorithms, sensors, hardware, or sequencing and orchestration procedures at different development phases. Typically, most of the project runtime is spent in this stage because workgroups can work on their own components while using or interacting with simulated or real components from other workgroups. Hence, the CSI concepts introduced in the previous section are going to be applied and exploited mainly at this stage.

Stage III *Validation stage*

The final project stage presents a runnable, integrated platform with all required components and submodules. Eventually, the goal of this stage is to *validate the complete project setup in the application scenario with real world data* such as from field trials of the DexROV project. Thus, no simulated components are used at this stage since all hardware and software functionalities need to be tested and benchmarked under real conditions, i.e., the system application is deployed in the desired final setting (offshore deep sea oil and gas intervention scenario). At this point, the fulfillment of project requirements is analyzed and eventually validated; note that, depending on the validation results, further stage cycles (I to III) could be conducted in order to satisfy the required levels of performance. The successful completion of this *validation stage* and the project itself depends on how well workgroups exploited the CSI concepts in an efficient and strategic manner, e.g., concept V (fault recovery and safety testing) in order to prevent costly hardware damages or losses.

2.4. SIL: simulation in the loop – synchronization of real-world observations with simulation

As described in the previous sections, a simulator that reflects the environment with high fidelity and offers enough flexibility to modularize the system components and behaviors is one of the main building blocks of the proposed CSI methodology. It is an indispensable factor for the exploitation of the CSI validation concepts (Section 2.3.1) such as parallel testing and distributed deployment, and hence for the evaluation and integration of all system components. However, such a simulator faces challenging tasks: creating models of the world with sufficient detail to reproduce physics and photo-realistic graphics, and running these models while not exceeding the available computational processing power. Thus, a trade-off has to be decided between model accuracy and runtime performance of the simulator.

To account for all these factors, in this section the following key aspects are described: (1) criteria to determine the simulation platform choice and its adaptation (Section 2.4.1), (2) packaging and virtualization of the system and its components (Section 2.4.2), and (3) simulation augmentation through the incorporation and synchronization of real-world sensor data into the simulated workflow (Section 2.4.3), referred as *simulation in the loop* (SIL). In the literature the term *simulation in the loop* has been introduced in different contexts, for instance, with simulation as a sampling and verification step for motion planning approaches [22]. It has also been used as a synonym for *continuous system integration* [23] as per our definition in this work. Other authors even use this term to merely indicate the substitution of real hardware with a simulated representation [24]. However, all of these definitions treat and integrate the simulation with real components as an open loop system, here it is treated as a closed loop to enhance the simulator based on real-world data and provide a test bed for a Continuous System Integration and Validation.

2.4.1 Simulation platform

The selection of a simulation platform featuring realistic and efficient simulations while providing modularity and extensibility is particularly important for the development of underwater intervention systems such as in the DexROV project. The following base criteria were defined to select an appropriate simulation platform. They were ordered from the most applicable to any underwater offshore project to those more necessary to follow a *Continuous System Integration* (CSI) methodology.

Criteria 1 – **Visual fidelity (graphics engine).** Visualization of realistic renderings of underwater environments and the availability to modify dynamically physical settings, e.g., floating particles, light backscattering, etc.

Criteria 2 – **Physics fidelity for rigid-bodies (physics engine).** The physics engine is capable of accurately simulating collisions, object manipulation, force control, position and velocity of objects, external dynamics.

Criteria 3 – **Sensor modeling.** Availability of required sensors for navigation, perception, etc., tasks, as well as required efforts to modify existing sensor suites or to create new ones.

Criteria 4 – **Adequate documentation and regular maintenance.** The platform must have a well-defined roadmap for updates and new releases, as well as a clear documentation or support for the development of new features. The objective is built upon a well-known platform to allow others to implement the CSI methodology with minimal effort.

Criteria 5 – **Interface with (robotic) middleware.** In order to have more efficient and fast-paced development cycles, it is fundamental to use an appropriate middleware that manages hardware abstraction, message-passing between processes, low-level device control, package management, among others. And as such, the simulation framework of choice must share compatible interfaces as the physical hardware with the middleware. For the DexROV project, *Robot Operating System* (ROS) [25] is used since it is one of the most standard and widely used middleware for distributed systems and robots featuring multiple functionalities such as related to perception, manipulation, and navigation tasks.

Criteria 6 – **Modularity and extensibility.** Ease of integration of additional modules and verification of existing ones. This criterion goes hand in hand with CSI concepts I and III – interface testing and distributed deployment.

Criteria 7 – Capability of multirobot simulation. Features available in the platform to create several instances of the same or different robots (system high-level components) and emulate their communication through a specified network.

As it was pointed out before, the selection of a simulator engine is one of the first steps in the CSI approach because it is the common platform on which all workgroups from a multidisciplinary project will build upon. For this reason, it is important to mention that the following simulator descriptions and comparisons are made based on their status until 2017, when the DexROV project effectively moved from the initial development stage to a phase of several verification, validation, and benchmark cycles. We specifically analyze two simulation environments affine with DexROV: (1) *UWSim v1.4.1* [26] and (2) *Gazebo v7.0* [27]. There are not as many options dedicated to underwater robotic systems as for their ground or aerial counterparts; the latter has a wider range of options to choose from, e.g., Unreal Engine [28], OpenRave [29], USARSim [30], WeBots [31], V-rep [32], among others.

Simulator 1 – UWSim. It is a simulator for marine robotics developed in the scope of the TRIDENT project (2010–2013) [13] that has been used and adapted in several subsequent research endeavors [33–35]. It offers a wide range of sensor models (DVL (Doppler Velocity Log), IMU (Inertial Measurement Unit), GPS (Global Positioning System), Sonar, etc.), and it is possible to configure different visualizations of underwater environments and water surfaces (reflection/refraction, waves, etc.) through its graphics engine OpenSceneGraph and plug-in osgOcean [36]. The Bullet physics engine underlying *UWSim* provides collision detection and rigid body dynamics, for example, manipulators can interact with objects but the simulator does not provide force feedback. The only compatible middleware is Robot Operating System (ROS) [25].

It supports simulation of vehicle dynamics and multiple AUVs instances and scenarios; however, its configuration might be cumbersome and laborious as the project scales because the settings for the environment, vehicles, objects, dynamics, and middleware interfaces are defined in a single monolithic XML file. Likewise, in order to add new sensor types, the software must be modified, there is no ad hoc functionality for this purpose.

Simulator 2 – **Gazebo + UUV simulator.** It is a general-purpose open-source robotics simulator maintained by the Open-Source Robotics Foundation (OSRF) that has been featured extensively in previous academic research [37–39] and is used to run the DARPA Virtual Robotics Challenge. Similar to *UWSim*, it offers a variety of sensor models except for an off-the-shelf DVL implementation; but it has a well-documented API to create new sensors as plug-ins. In the same way, *Gazebo* has no built-in functionality to create a water plane or underwater effects; nonetheless, the research community has presented open-source solutions such as *freefloating-gazebo* [40] and *Unmanned Underwater Vehicle (UUV) simulator* [41]. The first one bridges *Gazebo* with *UWSim* to transmit the effect of hydrodynamic forces on objects to the *UWSim* visualization environment; unfortunately, this limits runtime *Gazebo* capabilities as the settings of the complete simulation have to still be defined a priori in an XML file for *UWSim* to be initialized. The latter, UUV simulator, is a collection of dedicated *Gazebo* plug-ins that add core functionalities and sensors from underwater systems to the simulation engine, i.e., flow velocity, DVL, camera model including light backscatter, etc.

Gazebo can be compiled against one of several physics engines, each offering different advantages: ODE, Bullet, Simbody, and DART. Through these, all the mentioned physics phenomena simulated in *UWSim* can be achieved, plus force feedback and control, which are useful for designing complex manipulation schemes. The software is tightly coupled with ROS as middleware, but third-party wrappers can be used to interface with other middleware options such as YARP. It also supports vehicle dynamics simulation and multivehicle operations. Robots and world models can be added or modified at runtime using the graphical interface or through a programmatic approach.

Table 2.1 Comparison of simulators based on CSI criteria.

	Visual fidelity	Physical fidelity	Sensor modeling	Adequate documentation	Middleware interface	Modularity & extensibility	Multi-robot simulation
UWSim	High	Medium	Low‡	Low	Medium	Low‡	Medium
Gazebo	Medium†	High	High†	High	High	High†	Medium

† Extensible through APIs/plug-ins.
‡ Extensible by modifying source code only.

Based on this brief summary and Table 2.1 that compares each simulator with respect to the seven CSI criteria previously described, *Gazebo* was considered as the best option. Although *UWSim* offers an easy environment configuration to obtain realistic underwater visualizations, the *Gazebo* interface allows for high control of light sources and its API enables quick development of plug-ins to achieve similar visual effects, e.g., the *UUV simulator* plug-in suite. This last feature of *Gazebo* is what makes the simulator suitable for modular and incremental project development and permits the user to implement any missing functionality without going through a steep learning curve. Besides that, the *Gazebo* community is very large and active, constantly improving and documenting the software.

Fig. 2.3 shows an example scenario of the simulation environment. A terrain mockup was created from a recorded point cloud in underwater trials using a surface reconstruction method [42]. The scenario also contains a functional ROV and a model of the test panel (Fig. 2.2C) that features different types of valves, specifically designed to test the ROV manipulation capabilities (see Section 2.2.2). To correctly simulate kinematics and dynamics, detailed CAD models and physical properties of the ROV, payload skid, manipulator arm, and end-effector are integrated (see Section 2.5.1), along with low and high-level controllers for navigation [3].

Not last in importance, **CSI Criterion 7**: *capability of multi-robot simulation* has to be addressed. Table 2.1 shows that none of the simulators offer optimal functionalities to instantiate several robot components and subsequently model their intercommunication characteristics, which is a crucial step to validate and benchmark the components developed by multidisciplinary workgroups. On this ground, we introduce and describe a packaging, virtualization, and networking strategy based on software containers, i.e., Docker [43], in the next Section 2.4.2.

(A) Example simulator scene containing ROV, testing panel and terrain

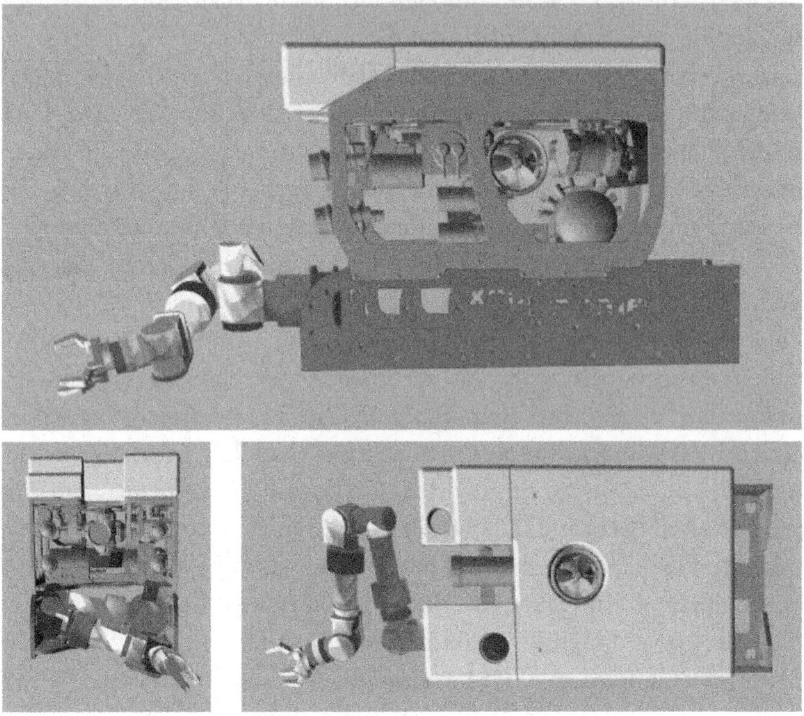

(B) ROV model

Figure 2.3 Example simulation environment with ROV model [4].

2.4.2 Packaging, virtualization, and networking

As a tool for packaging, virtualization, and easy deployment, Docker containers [43] are introduced. These containers are stand-alone executable packages of software that have all the requirements to run a specific applications, i.e., source code, system environment settings, system tools and library dependencies, etc. Note that, since these containers share the host machine's operating system (OS) kernel, opposite to virtual machines (VMs) which instantiate a full copy of the OS, they are very lightweight to deploy. A further important advantage of packaging each individual system into a Docker container is that the user can replace real machines by emulated ones on demand, and vice versa. A multidisciplinary team of multiple workgroups, spatially located at different places, can then perform separate experiments to test their components by instantiating containers from other workgroups.

In the DexROV scenario, three physical machines are defined: the ROV, Janus II vessel, and the onshore control center, each represented by a Docker container counterpart. Additionally, another container is defined to run the Gazebo+UUV simulator, this is extremely useful to run tests under simulated conditions similar to the expected ones during field trials. Specifically, these containers are created to emulate the architecture of Fig. 2.1:

Simulation – virtual underwater environment with CAD model of an oil and gas panel (see Fig. 2.3), plus ROV physics and sensor/actuator simulation which can be replaced by the real ROV and sensor inputs.

ROV – low-level processes onboard the ROV that execute microfunctions to manage the input/output of sensors and actuators, e.g., stereo image acquisition, propeller control, and localization.

Vessel – autonomous high-level capabilities running on the support vessel, e.g., motion planning, navigation, mapping, visual detection, and recognition.

Control center – high-level operator interface in an onshore control center which receives ROV status, maps, localization and object information over the satellite link in order to decide the next action to take, e.g., which valve in the panel to turn.

In addition to a logical separation of tasks and processes, intermachine networking plays an important role in a multirobot/component scenario (CSI criterion 7). Especially in open-sea robotic systems, it is common to have lower bandwidths, delays, and lost packages if acoustics or long cables are used for data transmission. To address this, we developed a network simulator [44] for Docker containers based on the Linux tool package NetEm [45] to define communication link parameters and constraints. As described in Section 2.2.2, in DexROV there are two main communication links: *ROV–vessel* and *vessel–control center*. The former uses a fiber-optic cable to transmit data from the ROV to the vessel, and the latter uses a satellite link between the vessel and the onshore control center. In all experiments related to DexROV, the network simulator has therefore been configured with the respective limitations of each link in order to create a realistic intercomponent communication scenario during all development stages.

2.4.3 Incorporation and synchronization with real-world observations

The proposed Continuous System Integration (CSI) approach that incorporates a simulation of the application scenario provides a powerful tool to decrease the project costs by minimizing integration efforts at field trials or by increasing the system reliability through rigorously conducting system benchmarks under controlled conditions without exposing valuable equipment to risks. Eventually, such a development process provides a reliable and predictable planning, as well as accelerates the preparation and execution of field trials and operations.

The concept of *simulation in the loop* (SIL), as introduced in our previous work [2,46], goes one step further with respect to general CSI concepts. Instead of conducting work cycles in a sequential development–integration–validation process with data generated from simulation and later from real-world field trials, SIL aims to combine both steps. Therein, our particular focus is set on closing the discrepancy between simulated and real-world data in order to achieve high fidelity in simulation with respect to the real-world conditions. To achieve this, recorded real-world data is projected into the simulation environment to provide close to actual conditions such as the robot configuration in space, perceived sensor data, and environmental constraints (see Fig. 2.4).

As a result, our SIL framework fulfills the following main key points:

Synchronizes simulated and real-world data by incorporating environmental and spatial feedback captured from field trials.

Provides an augmented virtual environment reflecting environmental/spatial conditions from real-world missions to continuously and exhaustively test, benchmark, and compare behaviors of system modules.

Preserves the benefits of continuous system integration to perform such benchmarks using real or simulated components or a combination of both.

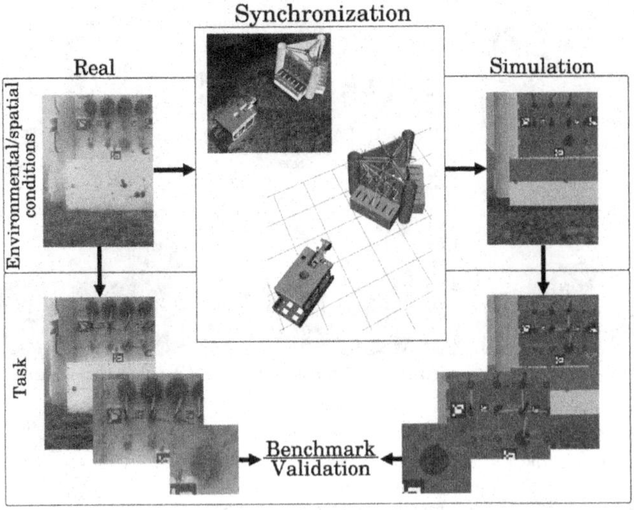

Figure 2.4 *Simulation in the loop* illustration [2] of a DexROV-related perception task (panel component pose estimation of valves and levers) to benchmark and validate its performance.

Facilitates the execution of evaluations in various integration scenarios, e.g., distributed deployment, system under communication regression/degradation, or fault recovery/safety instances.

In this synchronization process, we particularly consider the projection of *spatial* and *environmental conditions* from recorded field trial data to simulation. To reliably achieve *spatial* synchronization, we introduce in this work reference landmarks which are present in both worlds – simulation and real – and whose 3D pose can be accurately estimated. Generally, reliable landmarks are objects whose appearance, e.g., spatial dimensions or texture, is known beforehand, such as pipelines, ship wrecks, tailor-made markers/patterns or other man-made objects. In the DexROV scenario, the mockup oil and gas panel is used as such a reference since its accurate detection is a major project objective (see Section 2.5.2) and generally the first step in many of the task pipelines. For example, based on its 3D pose, spatial relationships between ROV and panel can be modeled (Fig. 2.5D) and incorporated for ROV localization purposes. Thus, as to guarantee robust pose estimation and high accuracy for the *spatial* synchronization, offline knowledge such as its CAD model (Section 2.5.1.3) and predefined visual fiducial markers and their locations on the panel are exploited (see Fig. 2.5).

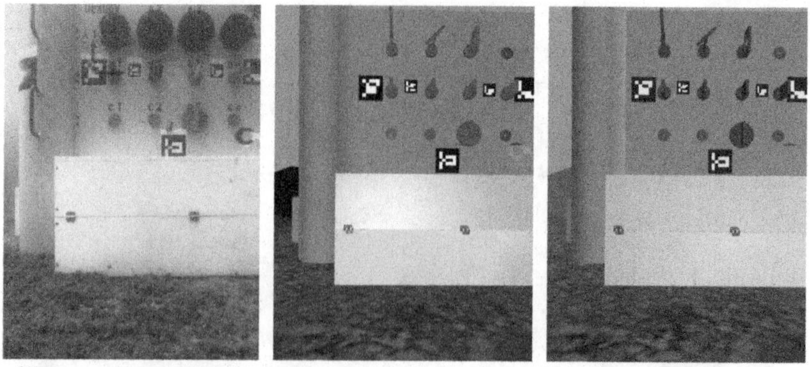

(A) Real-world camera image   **(B)** Simulated camera image   **(C)** Adapted simulated camera image

robot in camera frame: $^C_R\mathbf{T}$

camera in marker frame: $^M_C\mathbf{T}$

marker in panel frame: $^P_M\mathbf{T}$

panel in odometry frame: $^O_P\mathbf{T}$

robot in marker frame: $^M_R\mathbf{T} = {}^M_C\mathbf{T}\,{}^C_R\mathbf{T}$

marker in odometry frame: $^O_M\mathbf{T} = {}^O_P\mathbf{T}\,{}^P_M\mathbf{T}$

robot in odometry frame: $^O_R\mathbf{T} = {}^O_M\mathbf{T}\,{}^M_R\mathbf{T}$

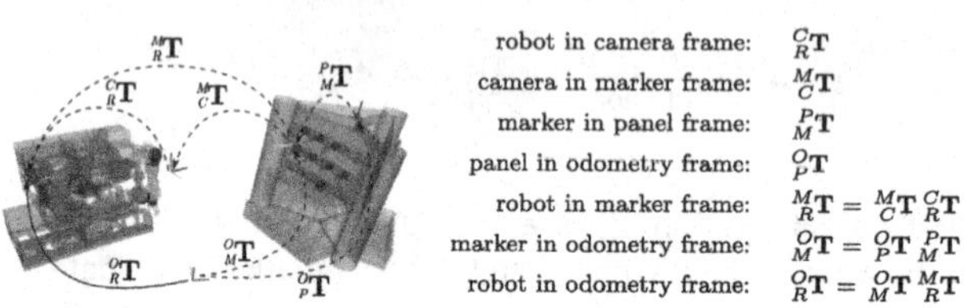

(D) Robot–panel space transformations and notations   **(E)** Panel with projected kinematic model

Figure 2.5 Illustration of the spatial and environment synchronization in SIL [2]. Synchronization of *spatial conditions*: robot pose (D) is inferred from real-world test panel observation (A) and accordingly projected into the simulation environment. Captured images in simulation (B) are accordingly adapted to the *environment conditions* (C) featured in (A). Given the panel pose (D), the kinematic state of test panel components (E) is projected in simulation.

Given the panel model augmented with ArUco markers [47], a reliable 3D panel pose can be computed by aggregating and cross–validating each marker pose and, as mentioned, projecting the relative *spatial conditions* between panel and ROV into simulation (see Figs. 2.4 and 2.5). Likewise, states of panel components (valves, switches, or wheels) from real observations can accordingly be projected to their simulated counterpart considering the a priori known kinematic model (Fig. 2.5E). Furthermore, real observed *environmental conditions,* such as camera image noise, haze, and illumination, can be estimated from recorded data and introduced into the simulation, i.e., visual fidelity of simulated data (Fig. 2.5B) can be enhanced from real-world samples (Fig. 2.5A) to benchmark components in more realistic environments (Fig. 2.5C). Effectively, we are boosting the visual engine capabilities of underwater simulators which often fall short from the

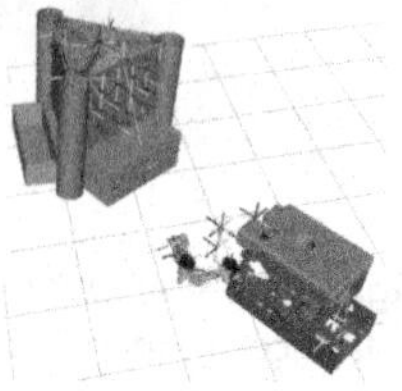

Algorithm 1 Simulation in the Loop (SIL).

Input: real-world sensor data $\mathcal{R}(\mathcal{E})$, task $\mathcal{T}$
  initialize knowledge base (see Section 2.5.1.3)
  infer environment conditions $\mathcal{E}$ from sensor data $\mathcal{R}(\mathcal{E})$
  initialize simulation environment under conditions $\mathcal{E}^* \approx \mathcal{E}$
  spawn test panel model
  for all real-world samples $r(t) \in \mathcal{R}(\mathcal{E})$ at time steps t **do**
    detect visual markers in $r(t)$ (see Fig. 2.5A)
    estimate panel pose in odometry frame ${}^{O}_{P}\mathbf{T}$ from marker poses (Section 2.5.2)
    infer robot pose in odometry frame ${}^{O}_{R}\mathbf{T}$ (see Fig. 2.5D)
    set robot pose in simulation according to ${}^{O}_{R}\mathbf{T}$ (see Fig. 2.5B)
    generate simulated sensor data $s(t) \in \mathcal{S}(\mathcal{E}^*)$
    create benchmarking sensor data pair $b(t) \leftarrow \langle r(t), s(t) \rangle$ (Figs. 2.5A and 2.5C)
    calculate measure $m(\mathcal{T}, b(t))$, (Sections 2.5.2–2.5.4)
    $\mathcal{B} \leftarrow \mathcal{B} \cup b(t)$, $M \leftarrow M \cup m(\mathcal{T}, b(t))$
  end for
Output: synchronized data sequence $\mathcal{B} \leftarrow \{\langle r(t), s(t) \rangle\}$,
sequence of measures $M(\mathcal{T})$ for task $\mathcal{T}$ as function of $\mathcal{B}$

necessary requirements to develop robust offshore systems (see Section 2.4.1 and Table 2.1).

In SIL, *spatial* and *environmental conditions* perceived from real observations are *continuously* reflected in simulation in a cyclic manner, as illustrated in Figs. 2.4 and 2.5. This *processing loop* of observations acquired from real data and their projection into simulation is shown in Algorithm 1.

Through this loop, various optimization and benchmarking tasks $\mathcal{T}$ can be iteratively performed on simulated $\mathcal{S}$ and real-world data $\mathcal{R}$, such as object recognition, manipulation, 3D modeling, localization, etc. This yields for each task a sequence $M(\mathcal{T})$ of individual measurements $m(\mathcal{T}) \in M$, defined according to the specific characteristics of a task $\mathcal{T}$, available ground truth and the project objectives required to benchmark.

The proposed *simulation in the loop* is extremely valuable for the continuous system development of projects which are expected to robustly perform under challenging and dynamic conditions like in deep-sea scenarios shown in DexROV. In the next Section 2.5, we evaluate in detail a series of tasks $\mathcal{T}$ which are paramount to the successful completion of

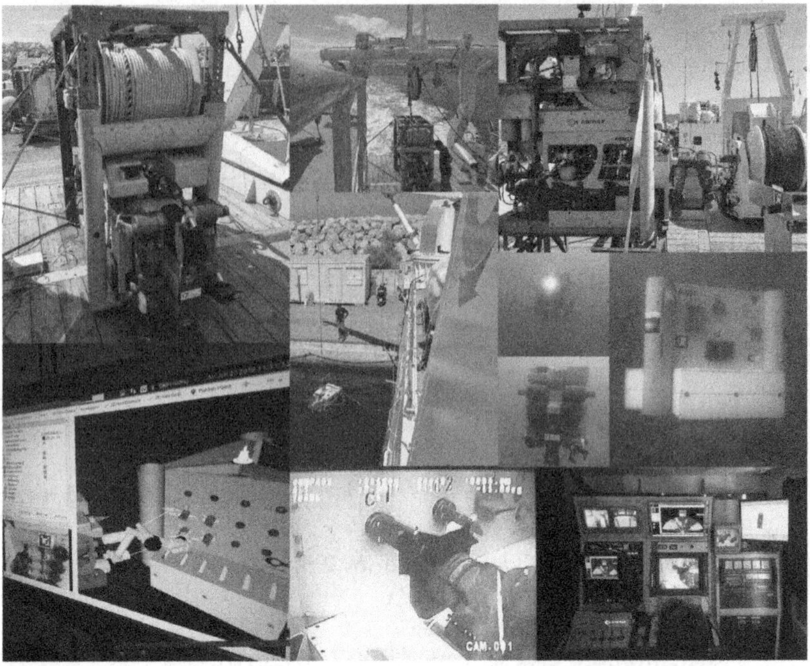

Figure 2.6 Impressions of sea trials in Marseille, showing ROV and the DexROV test panel.

DexROV offshore intervention missions. We further emphasize how the proposed CSI/SIL framework enables the seamless integration of system components and show the rapid adjustment and parametrization of methods and algorithms in order to enhance the performance of the evaluated tasks.

2.5. System benchmark and validation

The benchmark and validation process is demonstrated in the context of the DexROV scenario. The experimental setup is accordingly built and based on two pillars: (i) real-world data collection from sea field trials (Fig. 2.6) and (ii) the CSI/SIL framework (Section 2.4). Our goal is to combine both pillars to prepare a test bed that allows to evaluate and benchmark DexROV-relevant tasks under controlled simulated and real field trial conditions. In this process collected sea field trial data (sensor readings captured from camera to navigation sensors) are exploited to emulate realistic conditions in simulation. During the course of the DexROV project, in

(A) Recorded panel from ROV perspective.

(B) Simulated ROV placed at recorded position.

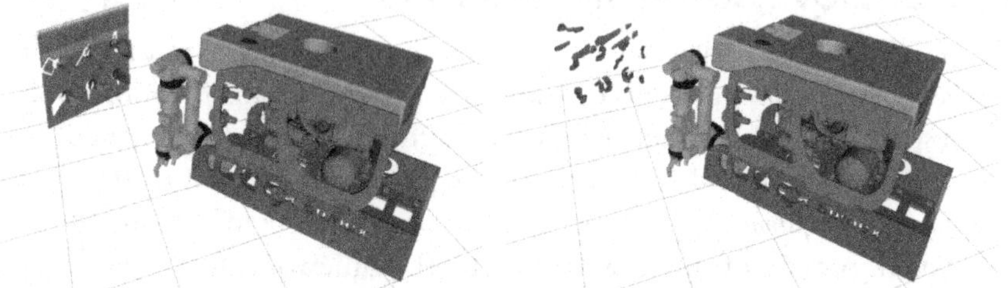

(C) Simulated point cloud (depth camera).

(D) Stereo point cloud generated from simulated images. (stereo point cloud sparsity is caused due to homogeneous textures captured from simulated noisy-free sensor-feed.)

Figure 2.7 Illustration of the synchronization between real and simulation.

two field trials sensory data have been collected under various conditions caused by different sea depths as it affects the illumination, intensity of marine snow, etc.; note that, for later reference, we denote these data collection campaigns as *field trial I* and *II*. Subsequently, in the field trials, captured circumstances are synchronized with the simulation as proposed by the CSI/SIL framework (Section 2.4.3). In Fig. 2.7 a sample of such synchronization is shown where the ROV is aligned to the test panel, i.e., real and simulated sensor feed are coaligned.

In this test bed, various experiments can be conducted in order to benchmark tasks under various conditions ranging from noisy-free simulated over noise-level observed in field trial data to extremely noisy conditions for system failure testing. Besides the enormous benefit during the development of system components, such benchmark process allows ultimately investigating bottlenecks, constraints, and anticipation of expected performance under certain environmental conditions or specific configurations of an integrated system such as DexROV represents, before introducing the system to real field trial conditions.

For the following benchmark and validation, the *perception* and *navigation* tasks of the DexROV system are evaluated using the proposed CSI/SIL framework. Initially in Section 2.5.1 an overview is provided about the related design and hardware setup of DexROV that is exploited as basis to perform these tasks. In Sections 2.5.2 and 2.5.3, perception benchmarks for pose estimations tasks are conducted regarding the test panel and its components. Eventually, in Section 2.5.4 the ROV localization task is benchmarked.

2.5.1 DexROV system design and setup

We provide a brief overview about design and hardware setup of the DexROV system components developed to accomplish the mentioned objectives, see Fig. 2.1; further details can be found in [4,5,48]. Therein, we describe key design and hardware aspects particularly related to the *vision* and *navigation system* as these system components are used to illustrate the benchmark process of the proposed CSI/SIL framework. In the following a brief summary is presented of the related system components:

Underwater Manipulator Setup. Two manipulators have been constructed based on an Underwater Modular Arm (UMA) design. Each arm consists of a 2-DOF hand (for flexion and abduction) and 6-DOF arm. This manipulation system is integrated within a skid that is mounted below of the Apache ROV. The overall length of the arm is 1 m when stretched. Furthermore, the manipulation system is foldable to prevent adverse effects on ROV navigation.

Vision and Navigation System Setup. These two components are described in Sections 2.5.1.1 and 2.5.1.2 in more detail. Additionally, in Section 2.5.1.3, we introduce a knowledge database containing valuable mission-relevant information in order to particularly facilitate and support reliability of applied perception and navigation methods.

Satellite Communication Setup. A maritime very-small-aperture terminal (VSAT) system is deployed. It includes a Ku-band Cobham Sailor 800 tracking antenna with its controllers and related modems. The nominal data bandwidths for uplink and downlink channels is 768 and 256 kb/s, respectively, with a round trip delay of 620 ms. In this context, the data encoding of the status of the ROV and the environment requires greater bandwidth (uplink) than the action primitives send from the control center (downlink). Finally, an ROS–DDS proxy (bridge) binds the two sides for managing data flows with dynamic adjustment of QoS (Quality of Service).

Control Center and Exoskeleton Setup. The onshore center located in Zaventem (near Brussels, Belgium) consists of a monitoring and control room, along with an exoskeleton consisting of two 7-DOF arms with 6-DOF hands each. It features force feedback sensors and a passive gravity compensation system to counteract the exoskeleton's and the operator's mass – the operator is given the impression that he operates in neutral buoyancy as a diver would experience. The design is partially based on the work for the European Space Agency in [49] and further improved in the EU-FP7 ICARUS project [50].

Note that the *vision* and *navigation system* along with the underwater manipulator setup are part of the *skid* that can be attached or detached from the ROV. This enables existing ROVs to be equipped in a modular fashion with variable capabilities required for specific vision, navigation, or manipulation tasks.

2.5.1.1 Vision system setup

The perception – in particular, the vision stream – of the environment is a crucial capability to facilitate the generation of an internal belief state about the ROV's surroundings in which it is operating in.

The perception framework consists of several components such as object detection, recognition and scene modeling components, or a semantic map that reflects the observations made including observed objects. Besides navigation data, these components mainly consume vision data captured from the stereo camera system. Particularly, the panel and panel component detection consume vision data as described and benchmarked in Sections 2.5.2 and 2.5.3.

Stereo camera system

Commercial ROVs such as the Apache (Fig. 2.6) used in DexROV typically rely on *analog* camera systems. However, the image quality is strongly influenced by the connection to the support vessel. Due to long cables used in deep-sea operations, captured images are prone to noise and interrupted streams, which eventually prevents reliable processing of such data, for example, in detection, collision avoidance, or scene modeling tasks. For this

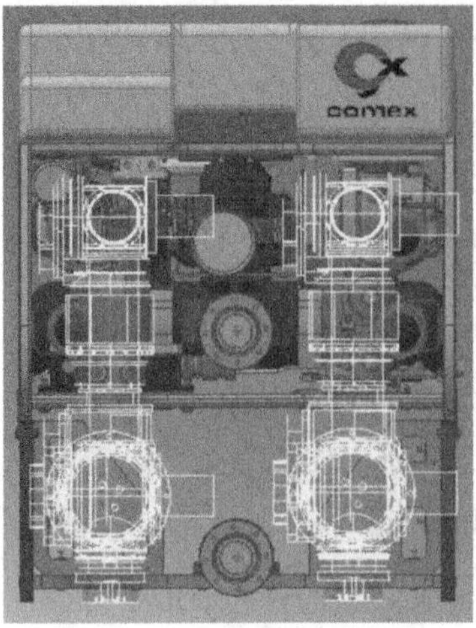

(A) Position of the perception system (orange), aligned vertically between manipulators (yellow).

(B) Single Camera of the Stereo Vision System

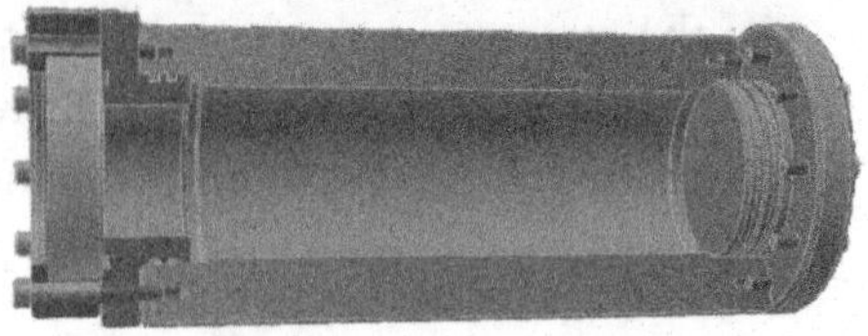

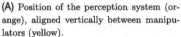

(C) Custom designed pressure housing, cut-away view

Figure 2.8 DexROV stereo camera system design [4].

reason, it was opted to use digital camera systems, which provide high-resolution images while allowing to control and maintain image quality under transmission constraints.

A stereo vision system was designed to meet the demanding requirements needed in DexROV. Point Grey Grasshoppers 2 cameras were selected which utilize Sony ICX285 CCD sensors, known for good performance in low light conditions. These cameras are further equipped with FireWire interfaces to ensure accurate synchronization for stereo reconstruction; the FireWire interface provides hardware-level synchronized capturing for all daisy-chained cameras.

A total of three cameras were originally built to provide two different baselines to maximize both range and accuracy; however, due to buoyancy and design constraints, a two-camera setup was eventually deployed with a baseline of 30 cm that reflects a compromise between stereo reconstruction quality and desired manipulation workspace (1–2 m). Fig. 2.8A shows the two-camera stereo vision system position between the two manipulators. Furthermore, in Fig. 2.8B, a single camera of the stereo vision system is shown featuring a flat sapphire glass panel (Fig. 2.8C) and deep-sea housing that is pressure tested for 2000 msw.

Underwater stereo image rectification

Before being able to consume image data for perception tasks, the stereo camera system requires to be calibrated (i) *intrinsically* against each other and (ii) *extrinsically* against the ROV platform. For intrinsic calibration, our previous work [51] was applied, which allows for calibrating the stereo system *in-air*, prior to underwater deployment with no need of further in-water calibration. Camera parameters estimated by our calibration method take into account salinity, temperature, and pressure. Subsequently varying depth can easily be taken care of. For further details on the intrinsic calibration, we refer to [51].

Using the calibrated camera model, images from both cameras are rectified to remove distortions which stem from refraction caused by the water and the protective glass panel in front of the cameras. After receiving rectified images, several preprocessing steps are executed to enhance the image quality and speed up further processing. Particularly haze, introduced by artificial light, is mitigated as described in [52] and [46]. The resolution of images may also be lowered to increase processing speed. Thereafter, the stereo image disparity is calculated, and subsequently point clouds are generated using the *Triclops Stereo Vision SDK* library by the camera manufacturer, PointGrey Research, Inc.

2.5.1.2 Navigation system setup

In order to ensure reliable navigation capabilities, well-established sensors in marine robotics are deployed such as a Doppler Velocity Log (DVL) *Navquest 600P micro* and an Altitude and Heading Reference System (AHRS) *Xsens MTi-300*.

The DVL provides altitude and velocities (speed over ground) in x, y, and z for translation movement computation. Furthermore, the AHRS is deployed mainly for the navigation and localization system in order to capture ROV orientation and acceleration. The localization estimate is further enhanced through visual fiducial marker pose estimates and computer vision algorithms. These inputs, along with the stereo camera images, are processed in a pressure-tested computer bottle equipped with an Intel NUC Kit 4th-Gen Intel Core i5-4250U.

Sensory feed captured through this navigation setup is mainly consumed by the ROV localization system as described and benchmarked in Section 2.5.4.

2.5.1.3 Knowledge database

Deep-sea missions are cost-intensive and bear a risk to life and equipment. Therefore, prior knowledge about the mission objectives decreases risk of failures and increase safety. Particularly in visual inspection or manipulation tasks of man-made structures, the incorporation of prior knowledge can be exploited to increase efficiency and effectiveness of conducted missions. Therefore, a database is built offline which contains properties of mission objectives, including information about target objects such as the deployed test panel in DexROV. Along with basic information, like their 3D shapes and CAD models of target objects, it also contains task-specific static data, such as weights, orientations, or dimensions of, for instance, levers, valves, and switches of the test panel.

2.5.2 Panel pose estimation task ($\mathcal{T}_P$)

The 3D pose estimation of the panel $\mathcal{T}_P$ is one of the core tasks since its accurate computation facilitates reliable manipulation of valves and levers located on the panel, not to mention that it is the basis for projecting the panel model and its kinematic properties into simulation as illustrated in Fig. 2.5E and enabling the SIL methodology (see Section 2.4.3). Therefore, $\mathcal{T}_P$ has been selected as a main benchmark task.

2.5.2.1 Method: accurate fiducial marker-based panel detection

Estimating an accurate panel pose is a challenging task given noisy sensor data, e.g., low quality camera images, spurious localization readings, etc. Common approaches can be fed with the available visual streams in form of RGB or RGB-augmented depth information (RGB-D) to estimate an accurate 3D (6-DOF) object pose. To minimize the risk of failures and increase safety, the proposed detection methodology incorporates prior-offline knowledge such as the CAD panel model and visual landmarks (ArUco markers [47]) which are attached at predefined locations on the panel (Fig. 2.9).

In Fig. 2.9D a sample detection of a marker is shown. Note that, even though the panel is partially observed, the panel pose can be accurately estimated through a chain of space transformations (Fig. 2.5D) given the panel model (Fig. 2.10D). In this process, the panel pose in odometry frame ${}^{O}_{P}\mathbf{T}$ can be estimated using the computed pose of a given detected marker w.r.t. the camera frame ${}^{C}_{M}\mathbf{T}$, the camera pose on the robot frame ${}^{R}_{C}\mathbf{T}$, the panel pose in marker frame ${}^{M}_{P}\mathbf{T}$, and the current robot pose in odometry

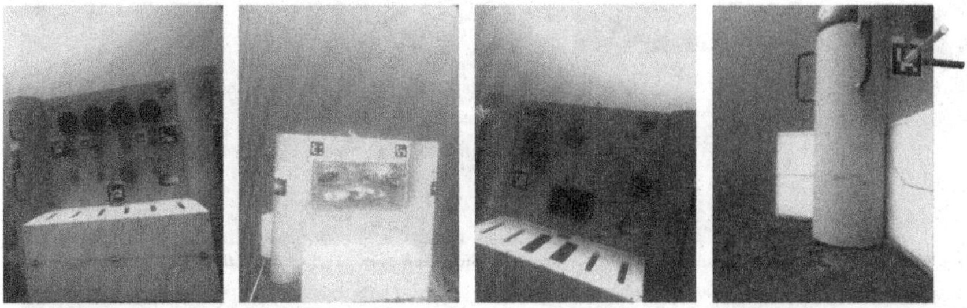

Figure 2.9 RGB panel images with ArUco markers captured at sea trials are shown in (A)–(C). A sample ArUco marker detection under sea trial conditions is shown in (D).

frame ${}^{O}_{R}\mathbf{T}$ (see Fig. 2.5D), as follows:

$$ {}^{O}_{P}\mathbf{T} = {}^{O}_{R}\mathbf{T}\,{}^{R}_{C}\mathbf{T}\,{}^{C}_{M}\mathbf{T}\,{}^{M}_{P}\mathbf{T} \tag{2.1} $$

Consequently, n marker observations lead to n panel pose estimates ${}^{O}_{P}\mathbf{T}$ that eventually allow for a more reliable pose estimate by computing over all marker observations the mean pose that consists of a mean position and orientation determined by *spherical linear interpolation (Slerp)* [53].

Given the offline available panel model and a reliable panel pose, the SIL framework is used to project the panel model and its kinematic properties into the simulation at the determined pose w.r.t. the ROV as illustrated in Figs. 2.5E and 2.10D. It is important to note that this process can also be done entirely in simulation to validate the algorithmic behavior of the markers detection and the spatial transformation to obtain the 6-DOF panel pose, as it is usually done in the CSI/SIL development stage. Furthermore, this task $\mathcal{T}_P$ can be rigorously benchmarked under various controlled spatial and environmental conditions, as well as similar to the real conditions, in order to quantify the accuracy of the proposed methods and their limitation, e.g., maximum viewpoint angle and distance to the panel/markers. In the following Section 2.5.2.2 we show this benchmark and validation process for the proposed panel detection method $\mathcal{T}_P$.

2.5.2.2 Benchmark and validation task

This benchmark task $\mathcal{T}_P$ evaluates the accuracy of the panel pose estimation, and inherently validates the accuracy of detected fiducial markers under simulated and real underwater conditions. For this we define the *SIL synchronization* and *measurement* $m(\mathcal{T}_P)$ as follows.

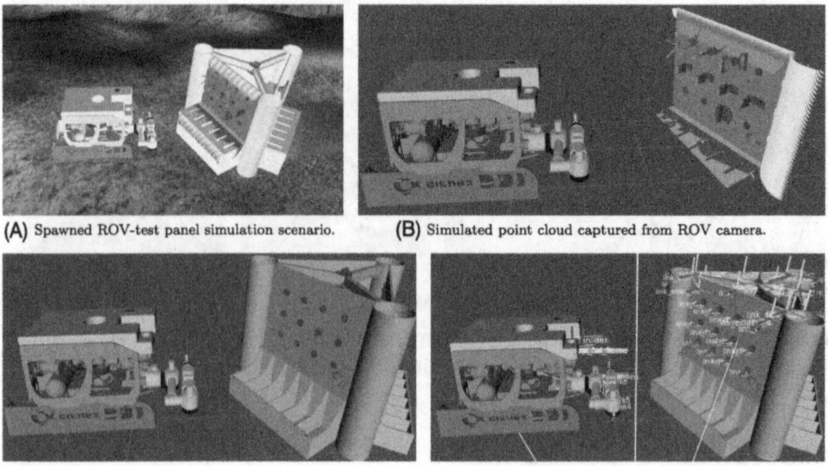

(A) Spawned ROV-test panel simulation scenario. (B) Simulated point cloud captured from ROV camera.

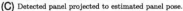

(C) Detected panel projected to estimated panel pose. (D) Detected panel with kinematic model.

Figure 2.10 Illustration of the panel detection processing: from a spawned simulation scenario (A) to the detection of the test panel and the projection of its kinematic model (D).

SIL synchronization is performed for both spatial and environmental conditions for this task. First, to compute a $\mathcal{T}_P$ error under realistic conditions, an ROV trajectory around the test panel is replicated from field trial observations in which routine trajectories are performed by the ROV operators or by the navigation system. Relative ROV poses w.r.t. the panel are inferred from detected panel markers observed in the field trial recordings. These 3D poses are exploited as waypoints which the ROV follows in the simulation environment to achieve realistic visuospatial properties regarding ROV viewpoint on the panel, ROV-panel distances, etc.; such a sample trajectory is shown Fig. 2.21A. As for the environmental properties such as illumination and camera noise, these can be adjusted heuristically through the simulator interface or more advance processes as described in Section 2.5.3.3.

The **measurement** $\boldsymbol{m(\mathcal{T}_P)}$ is the difference between the ground-truth panel pose in simulation ${}^O_P\mathbf{T}_S$ and the panel pose determined from the detection of simulated markers ${}^O_P\mathbf{T}_M$ (Eq. (2.2)) while the ROV follows the given trajectory determined by SIL, i.e.,

$$m(\mathcal{T}_P) = d\left({}^O_P\mathbf{T}_S, {}^O_P\mathbf{T}_M\right) = \langle d\left({}^O_P\overline{\mathbf{p}}_S, {}^O_P\overline{\mathbf{p}}_M\right), d\left({}^O_P\overline{\mathbf{q}}_S, {}^O_P\overline{\mathbf{q}}_M\right)\rangle \quad (2.2)$$

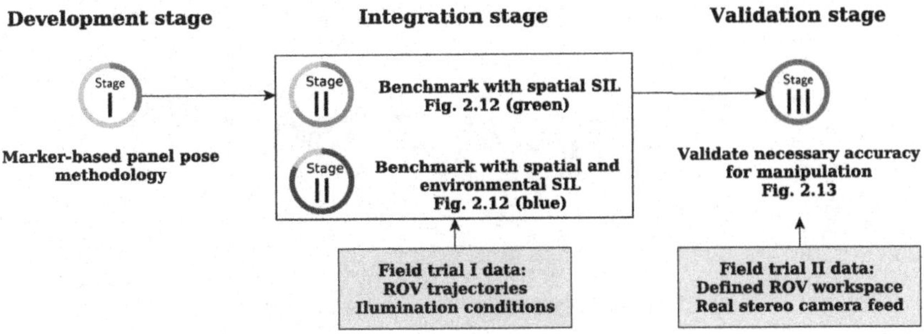

Figure 2.11 Panel pose estimation task $\mathcal{T}_P$ life cycle.

where $d\left({}^{O}_{P}\overline{\mathbf{p}}_S, {}^{O}_{P}\overline{\mathbf{p}}_M\right)$ is the Euclidean distance between positions and $d\left({}^{O}_{P}\overline{\mathbf{q}}_S, {}^{O}_{P}\overline{\mathbf{q}}_M\right)$ is the minimal geodesic distance between orientations [54].

Panel pose estimation task $\mathcal{T}_P$ life cycle is shown in Fig. 2.11. At the development stage, the methodology explained in Section 2.5.2.1 is implemented. Then, during

Integration stage

The mentioned SIL spatial synchronization is performed to replicate ROV trajectories from *field trial I* in simulation, and the accuracy is computed under a noise-free environment (standard in-air illumination and colors), denoted as $\mathcal{E}^0$. The translation and orientation mean error measures $\bar{M}(\mathcal{T}_P)$ are 0.02 m and 1.2°, respectively, as shown in Fig. 2.12. Note that one may interpret these values as the error that is mainly caused by the algorithm itself and not by external factors such sensor deficiencies.

After spatially synchronizing the simulator with ROV poses from real recordings, the simulated environment condition is heuristically adapted to reflect similar underwater illumination and haze conditions as the real camera input from *field trial I* data. Under these conditions which we denote as $\mathcal{E}^*$, a panel pose mean error of 0.118 m and 4.2° for translation and orientation, respectively, can be expected during the validation stage. It can be assumed that this error is close to the maximum expected one since the ROV cameras capture the test panel from various viewpoints and distances while following the evaluated trajectories, whereas in the validation stage, the panel is captured under constraint conditions (generally near frontal panel viewpoints) stem from the limited panel component manipulation workspace.

Technical insight. Since the panel pose task $\mathcal{T}_P$ is an initial task to be performed in various project pipelines, at this point of the integration stage,

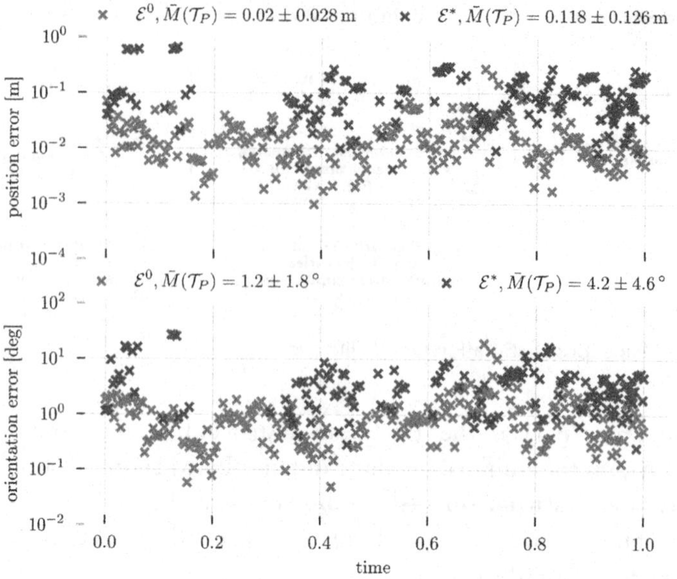

Figure 2.12 $\mathcal{T}_P$ **performance results: panel pose estimation errors** $M(\mathcal{T}_P)$ **for noise-free** $\mathcal{E}^0$ **and underwater** $\mathcal{E}^*$ **conditions, based on spatial and environmental data from** ***field trial I*****. Time is normalized.**

obtained results can be exploited to enhance future tasks. For example, the variance $\sigma^2(M(\mathcal{T}_P))$ of detected panel poses w.r.t. the ROV can be used to fine-tune the robot pose covariance matrix ${}^O_R\mathbf{C}$ for ROV localization (Section 2.5.4), as it is an estimation of relative pose precision (Section 2.5.4). As for manipulation tasks, applying a moving average over consecutive panel detections would eliminate outliers (inaccurate marker detections) in order to achieve more reliable results (Section 2.5.3).

Stage III

Validation stage

As mentioned, in the validation stage, the measurements $M(\mathcal{T}_P)$ are computed from the *field trial II* (real-world stereo images), while the ROV was located within a workspace of 2.5 m from the panel and a within viewing angle of 45° relative to the panel. This panel detection workspace has been defined to provide a reliable panel pose for inspection and maintenance purposes but also to facilitate reliable dexterous manipulation of the panel considering manipulator arm specifications and constraints. For a safe manipulation, concerning the avoidance of collisions including self-collisions, the project requirement states a maximum panel translation and orientation error of 0.05 m and 5°, respectively. Fig. 2.13

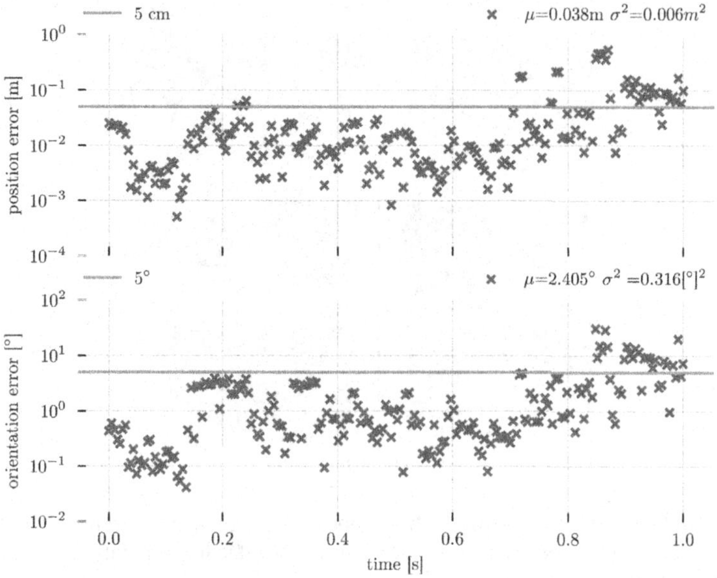

Figure 2.13 $\mathcal{T}_P$ validation results: panel pose error on the defined 2.5 m workspace with viewpoints not greater than 45°. Cyan horizontal lines indicate the DexROV project requirements to achieve robust manipulation. (For interpretation of the colors in the figure(s), the reader is referred to the web version of this chapter.)

shows that the average position error of 0.038 m and 2.4° fulfills this requirement, while the measurements generally show a lower error except for few outliers.

2.5.3 Panel component pose estimation ($\mathcal{T}_C$)

Given an estimated pose of the panel ($\mathcal{T}_P$), it is crucial for the DexROV application scenario to reliably estimate the pose of valves and levers located on the panel as they represent potential manipulation candidates. Therefore, $\mathcal{T}_C$ represents another main task to benchmark and validate – panel component pose estimation.

2.5.3.1 Method: superellipse-guided active contours-based pose estimation

Since the targeted application is the dexterous manipulation of underwater structures, accurate pose estimation of the panel components is necessary to guarantee their reliable manipulation. Once the position and orientation of the panel are estimated, the panel model provided by the knowledge

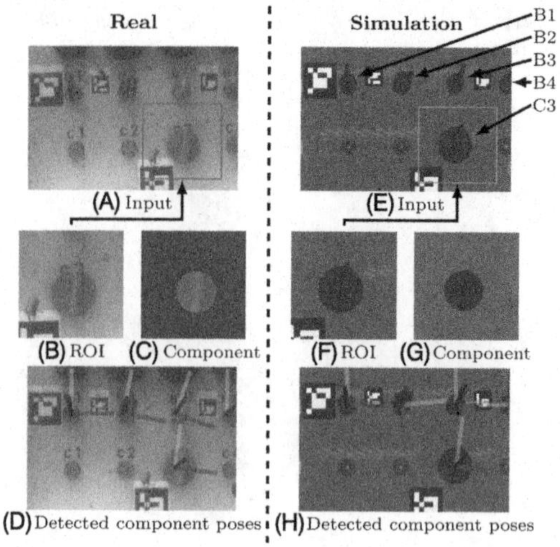

Figure 2.14 Computation stages of the panel component pose estimation method which is evaluated on identical viewpoints for real (A)–(D) and simulation data (E)–(H) within the SIL framework (see Section 2.4.3 [2]).

database (Section 2.5.1.3) can be exploited to detect and estimate the position and orientation of panel components (levers, wheels, valves, etc.) as illustrated in Fig. 2.14.

Given the estimated panel pose from task $\mathcal{T}_P$ and the kinematic model from the mentioned knowledge database, the relative model position of a component on the panel is accordingly computed and a 3D region of interest (ROI) is extracted in form of an oriented bounding box with the dimensions of the targeted panel component. At this point, an accurate estimation of the 3D panel component pose is known, but its orientation (roll) with respect to the panel needs to be estimated for manipulation actions, e.g., the grasp configuration or desired manipulation task are directly dependent on the estimated orientation of the panel component.

Various methods can generally be applied to estimate this orientation of the component model within the ROI. We propose an image-based method that estimates the component orientation using the extracted 3D ROI projected into the RGB-image 2D space. Fig. 2.14A shows the input image captured from one of monocular cameras of the real stereo camera, and Fig. 2.14B shows the computed ROI of the panel component labeled as C3 using the pose from $\mathcal{T}_P$ and panel model from our knowledge base. For a precise localization of the panel components within the image, the *superellipse-guided active contours segmentation* algorithm [55] is applied to each image patch representing an ROI. This algorithm is a particularly good fit for the presented use case because the panel components such as levers are round or ellipsoid-shaped when observed from approximately frontal viewpoints by the ROV. In Fig. 2.14C the result is shown of using the ROI (Fig. 2.14B) as an input for the superellipse-guided active contours segmentation. Subsequently, the panel component orientation can be inferred from the most prominent straight edge in the region extracted by the superellipse-based method. A Canny edge detector is used to detect edge points, followed by a Hough transformation for the respective detection of lines. In Fig. 2.14D the estimated component poses are overlaid on the panel. Based on these estimated orientations, the overall state of the panel is accordingly updated as shown in Fig. 2.5E.

Given the proposed CSI/SIL framework, the algorithm can be validated as described in the next Section 2.5.3.2 using identical camera viewpoints in simulation as in real-world experiments for comparison (Fig. 2.14). Note that panel component pose estimates retrieved from real (Fig. 2.14D) and simulated (Fig. 2.14H) data may deviate due to different signal-to-noise ratios featured in the respective data sources which lead to different image qualities. To further enhance the robustness of the algorithm when used in the envisioned DexROV scenario, a moving average of the detected panel component orientations from consecutive frames is introduced in order to mitigate the effects of inaccurate estimations on single captured camera images. Moreover, each image from the stereo camera is processed separately to cross-validate the component pose. This provides two detection streams from different perspectives at each instance of time which enhances the overall pose estimation accuracy when fused.

2.5.3.2 Benchmark and validation task

An accurate pose estimation of the panel components (valves, levers, etc.) is of particular interest as it enables dexterous manipulation in confined spaces under challenging underwater conditions. SIL provides a powerful tool to assess the panel component detection performance prior to the field trial application at sea. Particular deviations in component pose estimations may lead to a degradation of successful manipulations, hence, in the following experiments, we evaluate the absolute error of the estimated panel component poses.

SIL synchronization is performed for both spatial and environmental conditions for this task, and the experiments are performed once these factors are synchronized. As mentioned, by the design of the test panel, the

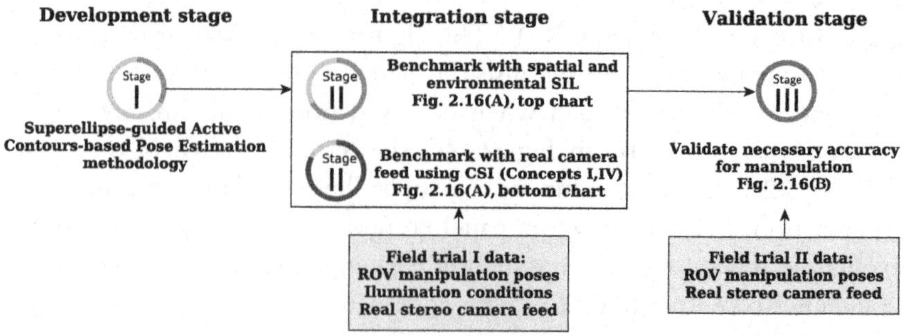

Figure 2.15 Panel component pose estimation task $\mathcal{T}_C$ life cycle.

position of panel components can be inferred from the CAD panel model given a detected panel pose from $\mathcal{T}_P$. First, the panel is projected to the respective pose in the real/simulated environment and component positions can be subsequently inferred according to the panel model. Note that component positions have been validated by the position of the panel in task $\mathcal{T}_P$ (Section 2.5.2).

Therefore, the **measurement $m(\mathcal{T}_C)$** particularly represents the difference between the estimated and ground truth component *orientation*, i.e., roll or rotation around the front-to-back axis, in degrees computed through minimal geodesic distance (Eq. (2.2), [54]). The ground truth is retrieved by human inspection of the valves, levers, etc., prior the experiments.

Panel component pose estimation task $\mathcal{T}_C$ life cycle is shown in Fig. 2.15. At the development stage, the methodology explained in Section 2.5.3.1 is implemented. Then, during

Integration stage

The mentioned SIL spatial and environmental synchronization is performed using the data from *field trial I*. In this case, it was an engineering decision to set the simulated environmental conditions $\mathcal{E}^*$ with lower illumination than the one observed in the real data, as it can be seen in Fig. 2.14E, in order to observe if the method would be robust enough in a more adversarial environment. It is also important to note that only robot poses within a predefined workspace for manipulation were synchronized through SIL, i.e., approximately within 2.5 m from the test panel and within 60° viewpoint with respect to the panel components. This is performed based on the panel detection methodology $\mathcal{T}_P$, see Section 2.5.2, and subsequently the panel component detection is executed.

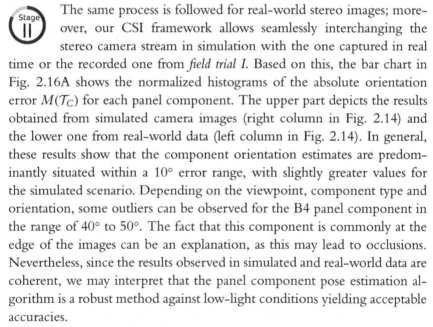

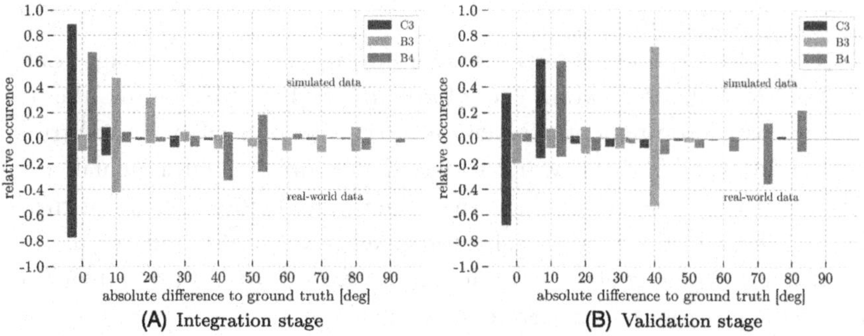

Figure 2.16 $\mathcal{T}_C$ **results: normalized histograms of panel component orientation error** $M(\mathcal{T}_C)$ **for simulated and real data.**

Stage II

The same process is followed for real-world stereo images; moreover, our CSI framework allows seamlessly interchanging the stereo camera stream in simulation with the one captured in real time or the recorded one from *field trial I*. Based on this, the bar chart in Fig. 2.16A shows the normalized histograms of the absolute orientation error $M(\mathcal{T}_C)$ for each panel component. The upper part depicts the results obtained from simulated camera images (right column in Fig. 2.14) and the lower one from real-world data (left column in Fig. 2.14). In general, these results show that the component orientation estimates are predominantly situated within a 10° error range, with slightly greater values for the simulated scenario. Depending on the viewpoint, component type and orientation, some outliers can be observed for the B4 panel component in the range of 40° to 50°. The fact that this component is commonly at the edge of the images can be an explanation, as this may lead to occlusions. Nevertheless, since the results observed in simulated and real-world data are coherent, we may interpret that the panel component pose estimation algorithm is a robust method against low-light conditions yielding acceptable accuracies.

Technical insight. In the course of the integration stage and during mock-up tests on-land, panel component B1 and B2 have been modified to enhance their orientation estimation accuracy by adding black coating to them. This modification has been made to investigate contrast benefits for edge-based image algorithms like the used one (see Figs. 2.14A and E). However, this modification causes a discrepancy between the simulated and real data, therefore B1 and B2 are disregarded in the results shown in Fig. 2.16.

Validation stage

The experiment was repeated during the validation stage using recorded real-world data of *field trial II*. Fig. 2.16B shows the respective panel component orientation error with respect to ground truth for the simulated and real-world images under similar environmental conditions. For C3 the orientation error is generally below 10°, and it turns out that B3 and B4 have several error spikes near the 40° and 70° marks, respectively. These effects were not present (for B3) or less prominent (for B4) in the previous integration stage results shown in Fig. 2.16A.

This outcome shows that the presented panel component pose estimation method strongly depends on the viewpoint and distance to the panel, as generally expected for monocular image-based perception methods. The recorded *field trial II* data used in this validation stage is indeed biased by viewing angles largely deviating from the panel front normal, which explains the accuracy decrease with respect to the more suitable integration stage with real-world data from *field trial I*.

Nevertheless, this benchmark task is an illustrative example of how to drive Continuous System Integration (CSI) in several iterations, also when not achieving the desired performance levels. The proposed test bed allows investigating, under controlled conditions, which circumstances are prone to estimation errors. This goes hand in hand with the CSI concept V (Safety testing) from Section 2.3.1. In Fig. 2.17A, a panel detection sequence is analyzed of over 350 consecutive frames – the sequence is normalized to a [0, 1] time interval. Therein, the panel component pose error with respect to ROV–panel distance is particularly analyzed. The observed estimation errors allow quantifying the impact of this distance to the orientation error. Subsequently, the uncertainty of the estimation can be anticipated depending on the distance which, for instance, allows preventing the execution of panel component manipulation actions if the uncertainty is high. For this purpose, an uncertainty threshold is chosen in Fig. 2.17B, i.e., in order to provide reliable estimates, the panel component detection is only performed if the panel is detected in a distance below the chosen threshold. Such informed design enhancements by considering the individual method's behavior allow improving the reliability of estimations and eventually increasing the robustness and safety of the overall system.

2.5.3.3 Further system development cycle

The benchmark and validation process for the panel component detection revealed insights about the method's behavior under different environ-

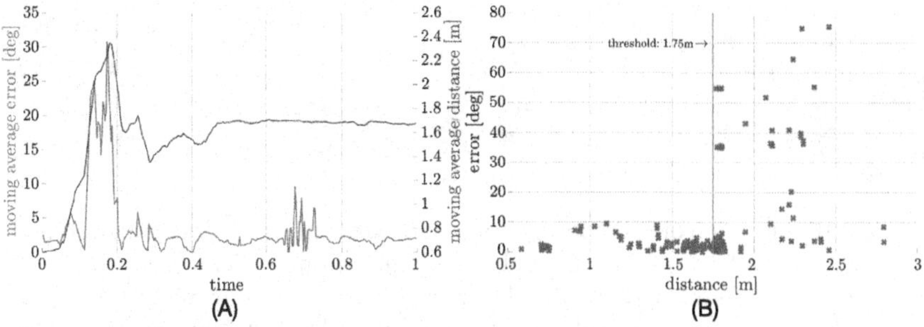

Figure 2.17 Panel component orientation error as the ROV changes its distance with respect to the panel. In (A), ROV-to-panel distance is shown in blue and the corresponding orientation error in red, time is normalized. Further, test (B) verified that a distance threshold of 1.75 m, red line, can be used to perform panel component pose estimation $\mathcal{T}_C$ when the ROV is located below this value.

mental and spatial conditions (viewpoint to the test panel) which allow quantifying the performance and subsequently conducting an informed enhancement step of the method. In this section we focus on two enhancement steps for our integration steps that could potentially help boosting the accuracy of the panel component pose estimation task $\mathcal{T}_C$. One addresses the stereo camera fidelity in simulation, whereas the other focuses on image enhancement (underwater dehazing).

Stereo camera fidelity enhancement

In order to further diminish the discrepancy of the method's behavior under simulated and real-world conditions, we aim to increase the fidelity of the visual sensor stream in simulation with respect to the real-world. Particularly, we focus on the adaptation of the simulated stereo camera image formation to the light behavior underwater [46], specifically for color attenuation. An exponential damping effect on the pixel intensity is applied to the simulated camera [41]:

$$i_c^* = i_c e^{-za_c} + (1 - e^{-za_c})b_c \quad \forall c \in \{R, G, B\} \tag{2.3}$$

where i_c and b_c correspond to the pixel and background intensity value for color channel c, a_c is a color-dependent attenuation factor, and i_c^* is the attenuated color value. The attenuation depends on the distance z between camera and object which can be directly extracted from the simulator. Additionally, the fog functionality provided by the Gazebo simulator [27] is introduced to further increase the image fidelity to deep-sea scenarios that are usually prone to light backscattering. Instead of adapting the attenuation

Figure 2.18 Illustration of PSO fidelity optimization w.r.t. the real image (left) over multiple iteration (*t*). From Tobias Doernbach, Arturo Gomez Chavez, Christian A. Mueller, Andreas Birk, High-Fidelity Deep-Sea Perception Using Simulation in the Loop, IFAC-PapersOnLine, Volume 51, Issue 29, 2018, Pages 32–37. Used with permission from IFAC.

factor a_c manually, it is adapted using Particle Swarm Optimization [56] and the Feature Similarity Index FSIM [57]; the result of the iterative optimization is illustrated in Fig. 2.18.

Through this proposed fidelity enhancement of the simulated stereo camera, more coherent benchmark results in simulation w.r.t. real-world conditions can be achieved for tasks that rely on the camera's vision data stream, such as in case of the panel component pose estimation $\mathcal{T}_C$.

Image enhancement using dark-channel prior

The degradation of underwater images is often caused by backscattering [59,60] leading to a haze effect. The dark-channel prior (DCP) method [58,61] allows mitigating this effect. Generally, the first step is to estimate the dark channel in order to compute the image atmospheric light and transmission map which are denoted as e^{-za_c} and b_c, see Eq. (2.3).

Given the synchronized spatial condition between the simulated and real world through SIL framework, the ground-truth *transmission map* $t(z) = e^{-za_c}$ (Eq. (2.3)) can be inferred using the known model of the DexROV test panel and the simulator's camera, i.e., for each image pixel the depth (distance camera and panel) can be accurately retrieved. This allows constructing the dehazed image by solving for i_c, see Eq. (2.3). Fig. 2.19 illustrates dehazing results made feasible by the SIL framework which offer greater image quality than traditional DCP-based methods. Further details can be found in [46].

2.5.4 ROV localization ($\mathcal{T}_L$)

An accurate ROV self-localization, task $\mathcal{T}_L$, is a necessity for various offshore intervention missions, as it enables important tasks related to 3D mapping, object pose estimation (oil and gas panel in Section 2.5.2) or floating-base manipulation control (Section 2.5.3). In order to achieve an

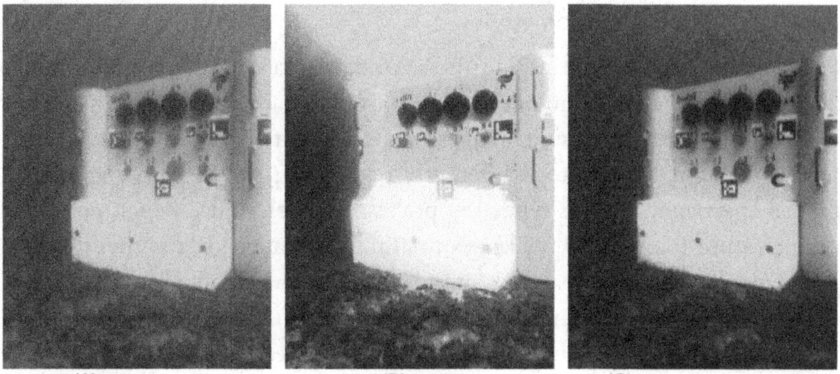

(A) Real world (B) DCP (C) SIL-based method

Figure 2.19 Illustration of image enhancement based on DCP [58] and proposed SIL-based approach. From Tobias Doernbach, Arturo Gomez Chavez, Christian A. Mueller, Andreas Birk, High-Fidelity Deep-Sea Perception Using Simulation in the Loop, IFAC-PapersOnLine, Volume 51, Issue 29, 2018, Pages 32–37. Used with permission from IFAC.

accurate self-localization, we propose a multimodal approach to enhance the robustness and precision of $\mathcal{T}_L$ by considering different methods and sensory feeds.

2.5.4.1 Method: multimodal ROV localization

ROV localization is a challenging task, particularly in underwater conditions, due to noisy sensor readings typically captured from acoustic devices like Ultra-Short Baseline (USBL) systems, single-beam or multibeam sonars, Doppler Velocity Log (DVL), or relative readings provided by Inertial Navigation Systems (INSs). Consequently, localization methods rely on the fusion of multiple modalities to increase reliability [62,63]. In the context of self-localization, a typical and well-established approach to deal with sensor fusion is the Extended Kalman filter (EKF) [64] which allows incorporating these modalities while considering their individual uncertainty. In order increase the ROV pose estimation accuracy, we exploit the test panel as a visual landmark due to its static pose on the seafloor and its visual augmentation with multiple fiducial markers (ArUco markers [47]), see Section 2.5.2.1 and Fig. 2.9. Once the panel pose is computed, the robot pose can be inferred and used as an additional EKF input modality.

In the following, the EKF-based localization system that incorporates multiple sensor readings and visual landmark pose estimates is described. Afterwards, in Section 2.5.4.2 the benchmark and validation of the method are accordingly conducted.

Visual landmark-based odometry

Vision-based self-localization in sub-sea or deep-sea scenarios is particularly challenging as the environmental conditions drastically deteriorate the quality of captured stereo-vision images, e.g., sunlight intensity reduces with depth and light backscatters differently depending on the viewing angle. Typical approaches of this type are optical-flow techniques which perform image feature tracking or tracking of natural landmarks, but as mentioned, they depend on quality features, abundant texture and stable camera movements which are not common in ROV operations.

As there is no guarantee to observe distinctive natural landmarks underwater, we make use of the fiducial markers located on the DexROV test panel (Fig. 2.5A). The patterns on these markers are designed with the goal of being robust to highly skewed viewpoints, low illumination, and partial occlusions. Hence, the ROV localization benefits from these stable distinctive references. Since the panel is utilized as a static landmark (see Section 2.5.2), the robot pose ${}^{O}_{R}\mathbf{T}$ can be subsequently estimated through a series of space transformations as follows:

$$ {}^{O}_{R}\mathbf{T} = {}^{O}_{P}\mathbf{T}\,{}^{P}_{M}\mathbf{T}\,{}^{M}_{C}\mathbf{T}\,{}^{C}_{R}\mathbf{T} \tag{2.4} $$

where ${}^{O}_{P}\mathbf{T}$ is the panel pose in odometry frame, ${}^{P}_{M}\mathbf{T}$ is one marker pose in panel frame, ${}^{M}_{C}\mathbf{T}$ is the camera pose w.r.t. the marker, and ${}^{C}_{R}\mathbf{T}$ is the robot fixed pose w.r.t. the camera. Further on, mean position ${}^{O}_{R}\overline{\mathbf{p}}$ and orientation ${}^{O}_{R}\overline{\mathbf{q}}$ w.r.t. the odometry frame can be inferred from multiple marker detections for a particular time instance. In addition, a covariance matrix ${}^{O}_{R}\mathbf{C}$ for the robot pose is computed (Eq. (2.5)).

$$ {}^{O}_{R}\mathbf{C} = \mathrm{diag}(\sigma^2_{\mathbf{p}_x}, \sigma^2_{\mathbf{p}_y}, \sigma^2_{\mathbf{p}_z}, \sigma^2_{\mathbf{q}_\phi}, \sigma^2_{\mathbf{q}_\theta}, \sigma^2_{\mathbf{q}_\psi}). \tag{2.5} $$

The full robot pose estimate ${}^{O}_{R}\mathbf{T} = \langle\, {}^{O}_{R}\overline{\mathbf{p}},\ {}^{O}_{R}\overline{\mathbf{q}} \rangle$, along with the respective covariance matrix ${}^{O}_{R}\mathbf{C}$, is then incorporated as an input for the localization filter in the final setup (see below – Extended Kalman Filter); $\{x, y, z\}$ represent the robot position in Euclidean space, and $\{\phi, \theta, \psi\}$ orientation as Euler angles.

Navigation sensor feed odometry

Since a visual feedback of the landmarks cannot always be guaranteed throughout the course of the DexROV vehicle, additional navigational feedback is captured from relative measurements of onboard DVL and INS.

These relative pose measurements such as linear/angular acceleration or velocity are also incorporated into the localization filter in the final setup (see below – Extended Kalman Filter). In this manner, even when not having a visual reference within the field of view, a rough estimate of the ROV pose can be computed.

Extended Kalman filter

The EKF filter [64] is applied to estimate the robot pose over time considering a state space consisting of position x, y, z, orientation ϕ, θ, ψ, translational $\dot{x}$, $\dot{y}$, $\dot{z}$, and angular velocities $\dot{\phi}$, $\dot{\theta}$, $\dot{\psi}$, as well as translational accelerations $\ddot{x}$, $\ddot{y}$, $\ddot{z}$. We only incorporate direct sensor measurements to the EKF, and no integrated or differentiated values. The onboard INS produces angular and linear accelerations, the DVL provides position outputs in form of altitude readings and linear velocities, and the mentioned landmarks are incorporated as absolute pose readings. To increase the localization filter robustness, obvious outliers from sensor readings or visual odometry are rejected based on heuristically tuned thresholds obtained in experiments. Moreover, it is important to note that the EKF also uses the covariance matrix from each input as a method to determine the uncertainty of the information they provide, and in turn, to provide a covariance matrix for the ROV localization estimate itself.

2.5.4.2 Benchmark and validation task

The proposed localization task $\mathcal{T}_L$ (Section 2.5.4) is benchmarked with our SIL methodology in the same fashion as the panel pose $\mathcal{T}_P$ and panel component pose $\mathcal{T}_C$ estimation tasks.

SIL synchronization is performed for both environmental and spatial conditions in one step. Thus, benchmarks are only conducted after both are synchronized since the algorithmic behavior (in noise-free environment) of the marker detection has been validated in task $\mathcal{T}_P$ (Section 2.5.2). The spatial conditions, trajectories from *field trial I* and *II* are retrieved as described in Section 2.5.2.2. Additionally, the environmental conditions (e.g., illumination) of the simulator are accordingly adjusted as in previous experiments.

The **measurement** $\boldsymbol{m(\mathcal{T}_L)}$ is defined to provide meaningful quantitative results to assess the ROV localization error. In this benchmark $m(\mathcal{T}_L)$ is a measure to compute the robot pose estimate error, stemming from either simulation or real-world field trial data:

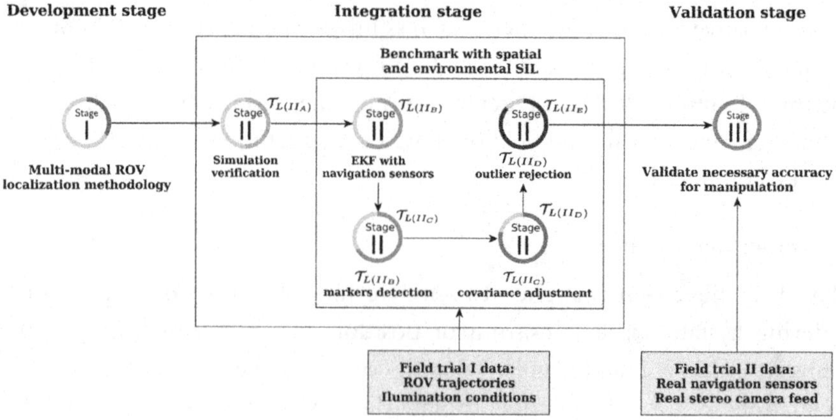

Figure 2.20 ROV localization task $\mathcal{T}_L$ life cycle.

Simulation. The robot pose estimate based on the visual marker detection ${}^{O}_{R}\mathbf{T}_M$ and the EKF output ${}^{O}_{R}\mathbf{T}_F$ are benchmarked against the simulation ground truth ${}^{O}_{R}\mathbf{T}_S$, denoted as $m_{S,M}(\mathcal{T}_{L1}) = d\left({}^{O}_{R}\mathbf{T}_S, {}^{O}_{R}\mathbf{T}_M\right)$ and $m_{S,F}(\mathcal{T}_{L1}) = d\left({}^{O}_{R}\mathbf{T}_S, {}^{O}_{R}\mathbf{T}_F\right)$ respectively, as per Eq. (2.2).

Real-world. To evaluate benchmarking tasks on real-world data, we use the robot pose estimate given by the marker ${}^{O}_{R}\mathbf{T}_M$ as reference ground truth, and compute the mean and standard deviation of the measure $m_{M,F}(\mathcal{T}_{Li}) = d\left({}^{O}_{R}\mathbf{T}_M, {}^{O}_{R}\mathbf{T}_F\right)$ plus the *lag-one autocorrelation* $m_A(\mathcal{T}_{Li}) = \sum_t {}^{O}_{R}\mathbf{T}_F(t)\, {}^{O}_{R}\mathbf{T}_F(t-1)$ on the EKF-predicted poses. Another measure $m_A(\mathcal{T}_{Li})$ is introduced to provide insights about the trajectory smoothness. Smoothness of a trajectory is particularly important in order to analyze and subsequently prevent sudden robot motions which can interfere with the manipulation reliability.

ROV localization task $\mathcal{T}_L$ life cycle is shown Fig. 2.20. As in the previous tasks, the development stage encompasses the software and hardware implementation of the methodology (Section 2.5.4.1). Note that this task in particular has several integration steps to ensure adequate accuracy which consist of fine-tuning processes of the available parameters. Furthermore, after the first validation stage, another integration loop is performed to make our methodology more robust and boost the performance of the $\mathcal{T}_L$ task as described in Section 2.5.4.3. Based on this procedure, the integration and validation stages are described as follows:

Table 2.2 Description of the localization task $\mathcal{T}_L$ integration steps.

Task	Description
$\mathcal{T}_{L(II_A)}$	EKF using (simulated) navigation sensors and (simulated) visual marker detections
$\mathcal{T}_{L(II_B)}$	EKF using only (real) navigation sensors
$\mathcal{T}_{L(II_C)}$	EKF using (real) navigation sensors and (real) visual markers detection with default parameters
$\mathcal{T}_{L(II_D)}$	$\mathcal{T}_{L(II_C)}$, plus covariance adjustment of the robot pose estimates from marker detections ${}^{O}_{R}\mathbf{C}$ based on the results from task $\mathcal{T}_P$, i.e., using $(0.126\ \mathrm{m})^2$ and $(4.6°)^2$ (see Fig. 2.12) as diagonal values for single marker detections
$\mathcal{T}_{L(II_E)}$	$\mathcal{T}_{L(II_D)}$, plus rejection of marker pose estimates whose distance $d\left({}^{O}_{R}\mathbf{T}_M, {}^{O}_{R}\mathbf{T}_F\right)$ to the current prediction are greater than 1 m and 12°; determined from $\mathcal{T}_{L(II_A)}$ and the results in Fig. 2.21

Integration stage

As shown in Fig. 2.20, the integration stage consists of several intermediate steps or milestones which are described in detail here, a summarized description is presented in Table 2.2.

$\mathcal{T}_{L(II_A)}$ **– Simulation verification.** First of all, the correct algorithmic behavior is evaluated of the proposed localization filter as a new modality is integrated into the EKF. Initially, an ROV absolute pose estimation is derived from visual markers detection. In order to retrieve the ROV pose, the correct chain of spatial transformations needs to be verified.

Given the SIL framework, the spatial synchronization for this task $\mathcal{T}_{L(II_A)}$ is performed in which a simulated robot follows a realistic trajectory around the panel. The trajectory is directly extracted from real trajectories captured in *field trial I* data by inferring way points from detected fiducial markers as described in $\mathcal{T}_P$ (Section 2.5.2).

In simulation, we can directly use as ground truth the robot pose given by Gazebo ${}^{O}_{R}\mathbf{T}_S$ to compare it against the robot pose determined through the detected simulated markers ${}^{O}_{R}\mathbf{T}_M$ and the robot pose computed by the EKF filter ${}^{O}_{R}\mathbf{T}_F$. In this setup, EKF only uses the pose estimated from the fiducial markers because at the time of testing a DVL plug-in for the simulator was not available, and its velocity readings are necessary to prevent the filter from drifting. Nevertheless, this benchmark has the goal of proving quantitatively that the EKF converges to the ground truth pose when markers are detected.

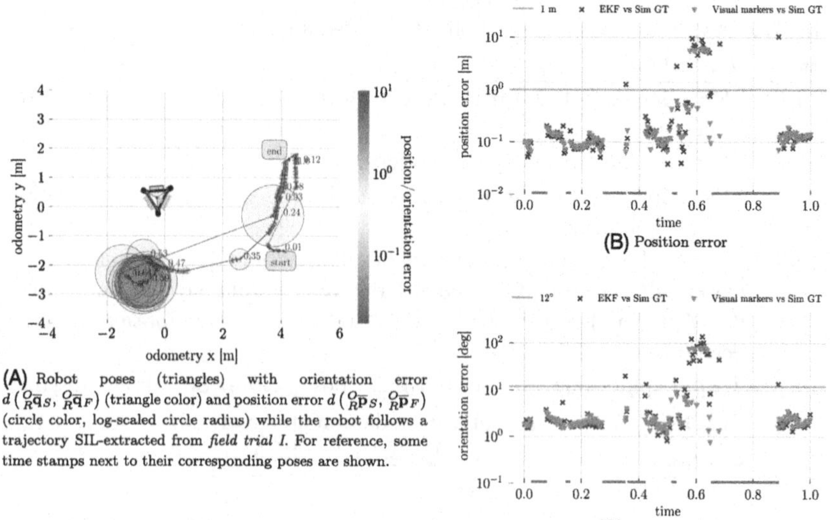

(A) Robot poses (triangles) with orientation error $d\left({}^{O}_{R}\overline{\mathbf{q}}_{S},\, {}^{O}_{R}\overline{\mathbf{q}}_{F}\right)$ (triangle color) and position error $d\left({}^{O}_{R}\overline{\mathbf{p}}_{S},\, {}^{O}_{R}\overline{\mathbf{p}}_{F}\right)$ (circle color, log-scaled circle radius) while the robot follows a trajectory SIL-extracted from *field trial I*. For reference, some time stamps next to their corresponding poses are shown.

(B) Position error

(C) Orientation error

Figure 2.21 $\mathcal{T}_{L(II_A)}$ results: (A) ROV computed trajectory around test panel (B)–(C) position errors $d\left({}^{O}_{R}\overline{\mathbf{p}}_{S},\, {}^{O}_{R}\overline{\mathbf{p}}_{M}\right)$ / $d\left({}^{O}_{R}\overline{\mathbf{p}}_{S},\, {}^{O}_{R}\overline{\mathbf{p}}_{F}\right)$ and orientation errors $d\left({}^{O}_{R}\overline{\mathbf{q}}_{S},\, {}^{O}_{R}\overline{\mathbf{q}}_{M}\right)$ / $d\left({}^{O}_{R}\overline{\mathbf{q}}_{S},\, {}^{O}_{R}\overline{\mathbf{q}}_{F}\right)$ between ground-truth robot pose from Gazebo simulation and marker-based / EKF-based robot pose estimates; no marker detected for sampling times marked green .

The pose estimate errors of ${}^{O}_{R}\mathbf{T}_{M}$ and ${}^{O}_{R}\mathbf{T}_{F}$ with respect to simulation ground truth are shown in Fig. 2.21, denoted as $m_{S,M}(\mathcal{T}_{L1}) = d\left({}^{O}_{R}\mathbf{T}_{S},\, {}^{O}_{R}\mathbf{T}_{M}\right)$ and $m_{S,F}(\mathcal{T}_{L1}) = d\left({}^{O}_{R}\mathbf{T}_{S},\, {}^{O}_{R}\mathbf{T}_{F}\right)$ as in Eq. (2.2). The trajectory in Fig. 2.21A and the detailed error breakdown in Figs. 2.21B and 2.21C show that if markers have not been detected consecutively during a certain time period, the EKF error increases significantly, but then it quickly reconverges towards ground truth on the next reading. On parts of the trajectory where markers are constantly visible, the localization error decreases satisfactorily below 0.3 m/3° respectively, e.g., between time marks 0.1 and 0.25. In this way, it can be observed that poses computed from visual markers facilitate the convergence of the EKF to ground truth which validates the proposed methodology. Likewise, from the results, it can be noticed that marker pose detections become unreliable at distances beyond 2.5 m approximately, e.g., concentration of red circle errors in Fig. 2.21A; and that pose changes of more than 1 m/12° indicate the presence of outlier detections as shown in the magenta lines in Figs. 2.21B and 2.21C.

Stage II

$\mathcal{T}_{L(II_B)}$ – **Only real navigation sensors.** In order to prepare a baseline to compare the performance of the localization filter when integrating visual landmarks, in this integration stage only DVL and INS measurements are used as inputs. This also allows for fine-tuning of sensor parameters such as sampling rate, sensor bias, and their extrinsic calibration.

Technical insight. For example, a sampling rate of 2 and 1 Hz for the INS and DVL, respectively, were selected due to low ROV velocities facilitating the execution of careful panel monitoring and manipulation actions. It was found that high sampling frequencies caused the EKF to drift because of sudden spurious readings and sensor noise. Likewise, when the ROV is close to the panel (< 1.0 m) when performing manipulation tasks, the DVL responds with a large number of spurious readings due to its acoustic beams colliding with the bottom structure of the test panel. For this reason, a module was added that enables/disables the use of DVL inputs when manipulating valves from the test panel. Furthermore, to ensure safety and robustness, it was decided to use one arm for fixing the ROV position by grasping a dedicated bracket of the panel, while the other arm is carrying out manipulation actions according to the given task.

As discussed, from this integration step forward, the used ground truth of the robot pose is inferred from detected marker poses as these estimates are reliable even under these challenging underwater conditions. Note that both pose error and autocorrelation measures are accordingly computed.

Stage II

$\mathcal{T}_{L(II_C)}$ – **Real navigation sensors and visual markers.** Starting from this stage, the pose obtained from the fiducial visual markers is included in the localization filter. At this point, the navigation sensors have been optimized as described in the previous integration stage. The default parameters for the marker detection are used, i.e., covariance matrix given by the *Slerp* algorithm and minimum number of different markers detected to compute a robot pose (1 marker). The objective is to check that the localization accuracy improved the baseline approach $\mathcal{T}_{L(II_B)}$ that uses only the navigation sensors.

Stage II

$\mathcal{T}_{L(II_D)}$ – **Robot pose covariance adjustment.** In this step, we build upon the previous $\mathcal{T}_{L(II_C)}$ and aim to optimize the covariance matrix ${}^{O}_{R}\mathbf{C}$ given by the robot pose computation from the marker detections. By default, the covariance matrix is obtained from the *Slerp* algorithm [53]; however, marker misdetections or incorrect pose estimations yield large covariance values than directly affect the EKF accuracy. Thus, it was opted to use fixed covariance values obtained from the ex-

Table 2.3 Measure results for task $\mathcal{T}_L$ during all steps of the integration stage $\mathcal{T}_{L(II_i)}$* and validation stage $\mathcal{T}_{L(III)}$.

Task measures	Integration stages				Validation stage	
	● $\mathcal{T}_{L(II_B)}$	▽ $\mathcal{T}_{L(II_C)}$	◇ $\mathcal{T}_{L(II_D)}$	✖ $\mathcal{T}_{L(II_E)}$	$\mathcal{T}_{L(III)}$	$\mathcal{T}_{L(III)}$ (CSI/SIL)
$\bar{m}_{M,F}(\mathcal{T}_{Li}\langle\bar{\mathbf{p}}\rangle)$[m]	2.11±0.94	0.26±0.39	0.29 ±0.32	0.28 ±0.35	1.43±1.20	0.65±0.58
$\bar{m}_{M,F}(\mathcal{T}_{Li}\langle\bar{\mathbf{q}}\rangle)$[deg]	15.59±7.33	10.24±7.57	8.82 ±5.17	8.86 ±5.19	17.65±9.29	14.65±8.42
$m_A(\mathcal{T}_{Li})$	0.95	0.72	0.91	0.94	0.79	0.88
	0.95	0.72	0.91	0.94	0.79	0.88

*$\mathcal{T}_{L(II_A)}$ was not included since it is entirely based on simulated data.

periments performed for the panel pose estimation task $\mathcal{T}_P$ (Section 2.5.2). This means that we use the average position and orientation variance of the marker pose detection, i.e., $(0.126\text{ m})^2$ and $(4.6°)^2$ (see Fig. 2.12) as diagonal values for ${}^O_R\mathbf{C}$. In this way, although the uncertainty of the EKF does not reflect the quality of the current sensor input data, many outliers that cause the EKF to drift or not converge are filtered out. This option was preferred because sudden motions in the localization output make control algorithms for navigation and manipulation difficult to design – it is a trade-off among high frequency outputs and precision.

$\mathcal{T}_{L(II_E)}$ – **Outlier rejection.** The last step in this series of integration stages for localization optimization is the inclusion of an outlier rejection module that filters out EKF outputs that substantially deviate from the previous outputs. This is done as a preemptive step in order not feed other system modules (manipulation, 3D mapping, etc.) with data that can adversely affect their functionality. For example, a sudden change in the robot pose during manipulation can trigger the robot arms to perform movements which are physically not possible and that can damage the hardware. Based on this and on the results obtained from $\mathcal{T}_{L(II_A)}$ integration stage in simulation only, EKF values are rejected that represent pose changes of more than 1 m/12° indicated by the magenta lines in Figs. 2.21B and 2.21C.

Results for each of these steps of the integration stage are shown in Table 2.3 and Fig. 2.22. As expected, the EKF using only navigation sensor data (● $\mathcal{T}_{L(II_B)}$) has the greatest error, but since its sampling rate is higher and continuous it also has the highest autocorrelation value (smoother trajectories) – fiducial marker detection depends highly on the viewpoint, image quality, and having the test panel in the field of view. Integrating the visual makers into the EKF (▽ $\mathcal{T}_{L(II_C)}$) reduces the pose error substantially; however, it reduces the autocorrelation measure as well, which can cause several jerky motions while navigating. Then, we have shown that based on

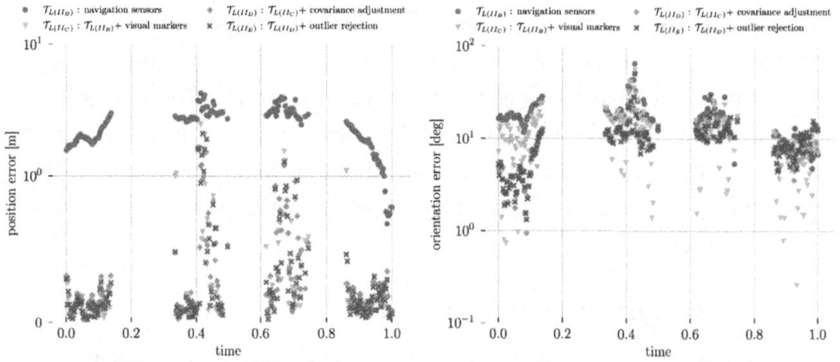

(A) Position error $d({}^{O}_{R}\bar{\mathbf{p}}_M, {}^{O}_{R}\bar{\mathbf{p}}_F)$ **(B)** Orientation error $d({}^{O}_{R}\bar{\mathbf{q}}_M, {}^{O}_{R}\bar{\mathbf{q}}_F)$

Figure 2.22 Benchmark for the localization task integration steps $\mathcal{T}_{L(II_i)}$. Time is normalized and *y*-axis is logarithmically scaled for better visualization purposes.

previous tasks and/or integration stages developed through our CSI/SIL framework, i.e., $\mathcal{T}_P$ (Section 2.5.2, Fig. 2.13) and $\mathcal{T}_{L(II_A)}$ (Fig. 2.21), the localization performance can be further optimized by adjusting the pose estimate covariances as in ◇ $\mathcal{T}_{L(II_B)}$, and rejecting outliers as in ✖ $\mathcal{T}_{L(II_E)}$. Note that, in these last two integration stages, the pose error slightly increases, but the autocorrelation value significantly improves, which, as explained before, is a preferred outcome for the workgroup designing the control algorithms for manipulation and localization.

Stage III

Validation stage

As shown in the task $\mathcal{T}_L$ life cycle in Fig. 2.20, the same measurements $M(\mathcal{T}_L)$ are computed for an unknown ROV trajectory followed during the *field trial II*. It is important to stress that no parameter fine-tuning was performed beforehand, the parameter and threshold values obtained in last integration stage $\mathcal{T}_{L(II_E)}$ were directly used in order to verify that the CSI/SIL methodology offers competent results without executing extensive experiments on the field. For comparison, we also benchmark the localization filter with default parameters, as if no integration stages beyond $\mathcal{T}_{L(II_C)}$ had been followed – which cannot be executed without CSI/SIL. The results are shown in Table 2.3 as $\mathcal{T}_{L(III)}$. It is expected to observe an increase in the pose error and a reduction in the autocorrelation measure as the environmental conditions, location, and trajectory are completely different from *field trial I*. We can see that the validation experiment using the parameters optimized through the presented CSI/SIL framework (last column in Table 2.3) provide better results than solely conducting parameter optimization experiments under uncontrolled conditions, directly on field trials (next to last column in Table 2.3). The obtained measures are common and reliable for navigation and monitoring tasks, also competitive for 3D mapping [65]. Nonetheless, the required precision of 5 cm/5° for manipulation tasks set in the DexROV project has still not being achieved. For this reason, another iteration through our integration and validation stage is performed to improve the algorithms chosen for localization, as described in the next Section 2.5.4.3.

2.5.4.3 Further system development cycle

The initially proposed ROV localization method relies on navigation sensor feed and visual landmarks-based odometry. The benchmark and validation cycles in Section 2.5.4.2 reveal a satisfying accuracy for the DexROV application scenario; however, in order to achieve higher reliably for safety-critical system, we present a further development cycle of the ROV localization system component. Therein, informed enhancements are incorporated based on the validation results made from the initially proposed method.

A multimodal two-stage navigation scheme [66] is proposed that in the first-stage generates a coarse probabilistic map of the workspace which is used to filter noise from computed point clouds. In the second-stage, it uses vision-based modalities (feature tracking and 3D plane registration) depending on the image quality of the inputs. A complete overview is shown in Fig. 2.23. Initially the test panel is approached by the ROV until an accurate panel pose is estimated through our panel pose detection method $\mathcal{T}_P$ (Section 2.5.2). Given this fixed panel pose, the ROV closely navigates around the target area, i.e., the test panel, using odometry estimates from navigation sensors and visual landmarks, i.e., the method described and benchmarked in the previous Section 2.5.4.2. While navigating with odometry uncertainty, a probabilistic map in form of a 3D occupancy grid is generated from the stereo input. This navigation stage is called *workspace definition* as depicted in Fig. 2.23A–C.

Furthermore, in the second stage of the scheme, an adaptive decision-making approach is introduced that determines which additional perception cues to incorporate into the localization filter in order to optimize accuracy and computation performance. In addition to the previously introduced odometry inputs in Section 2.5.4.1 (visual landmark-based and navigation sensor feed odometry), two additional visual odometry (VO) inputs are proposed in order to enhance the localization accuracy:

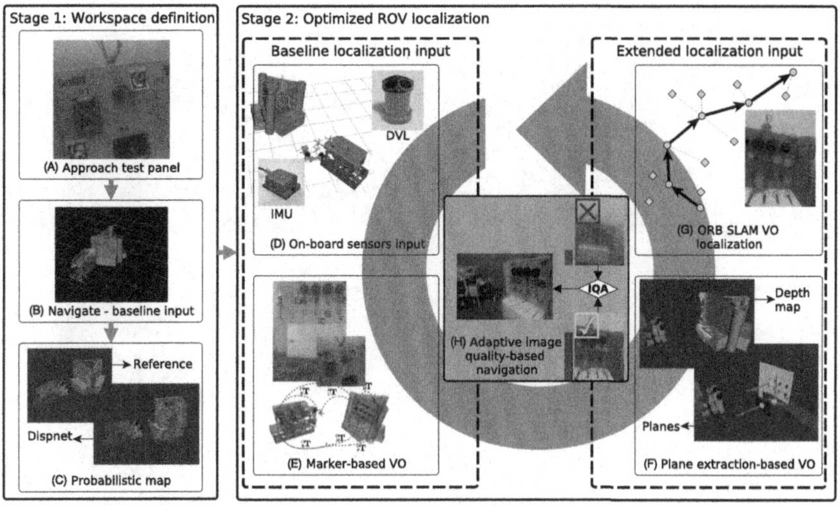

Figure 2.23 Illustration of the proposed two-stage navigation scheme [66]. **First stage** - *Workspace definition*: (A) detect the test panel and compute its pose based on visual markers (Fig. 2.4A), (B) navigate close to the panel based on navigational sensors and visual markers, i.e., baseline localization shown in (D) and (E), (C) generate probabilistic map with stereo imagery and Dispnet to define the workspace; RGB-D camera based probabilistic map displayed for reference. **Second stage** - *Optimized localization*: (F)–(H) show multimodal localization inputs which are incorporated to a final Kalman filter-based localization estimate. An image quality assessment (IQA) is introduced (H) to validate reliability of the extended localization inputs to boost the accuracy of the estimates given by the baseline localization inputs.

VO-planes Extracted planes from computed dense point clouds (Dispnet [67]) which are filtered using the probabilistic map that is computed in the first stage to prevent large drifts and noise artifacts. These planes are exploited as features for registration.

VO-ORB Extracted and tracked robust 2D features from imagery using ORB-SLAM [68].

In both cases, additional relative pose estimates are inferred and introduced to the given EKF (Section 2.5.4.1). However, these VO inputs require additional computational resources from the limited resources provided from the ROV and vessel, but they will enhance the overall localization performance if the visual conditions allow. Therefore, to increase the efficiency of available resource usage, an image quality assessment (IQA) (see Fig. 2.23H) is proposed to decide when these additional visual odometry inputs are introduced, i.e., if the visual quality suggests a reliable estimate either from

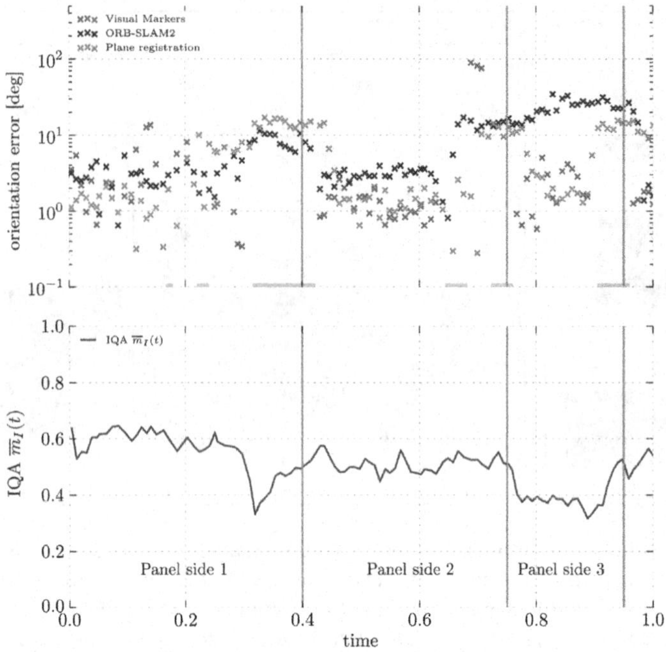

Figure 2.24 Orientation error for different visual odometry methods (top). No markers detected for sampling times marked orange, and changes of panel side with a red line. Normalized image quality measurement $\overline{m}_I(t)$ per stereo pair (bottom).

the plane registration or ORB feature tracking only then these inputs are activated to eventually benefit the localization performance.

The image quality is assessed by aggregating a non-reference image quality measure based on Minkowski Distance Measure (MDM) [69] and the number of tracked ORB features between consecutive frames. The MDM provides three values in [0, 1] interval describing the contrast distortion in the image; thus, the number of ORB features is normalized based on the predefined maximum number of features to track. Subsequently, the average of all three values presents the final IQA measurement.

In Fig. 2.24 the visual odometry inputs are evaluated according to the orientation error and IQA measurement over a sequence in which the ROV maneuvers around the test panel in our CSI/SIL framework, in simulation only. The observed correlation between error and visual quality can be utilized to perform an informed activation of the individual VO-inputs. Using SIL, an IQA threshold (0.45) has been determined to trigger the computationally expensive plane registration only when the image qual-

Table 2.4 Image quality based navigation performance.

	EKF-all	**EKF-adaptive**
position error [m]	0.73 ±0.38	0.61 ±0.14
orientation error [deg]	8.93 ±4.22	3.02 ±1.06
trajectory autocorrelation	0.92	0.95

Table 2.5 Tests measure results for position/orientation error and trajectory autocorrelation in *field trial II*.

	$\mathcal{T}_{L(III)}$ **(Section 2.5.4.2)**	$\mathcal{T}_{L(III)}{}^{*}$	$\mathcal{T}_{L(III)}{}^{**}$
$\bar{m}_{M,F}(\mathcal{T}_{Li}\langle\bar{\mathbf{p}}\rangle)$ [m]	0.65 ± 0.58	0.31 ± 0.11	0.85 ± 0.22
$\bar{m}_{M,F}(\mathcal{T}_{Li}\langle\bar{\mathbf{q}}\rangle)$ [deg]	14.65 ± 8.42	7.21 ± 2.10	11.89 ± 4.55
$m_A(\mathcal{T}_{Li})$	0.88	0.94	0.91

* IQA navigation scheme + navigation sensors + visual markers.
** IQA navigation scheme + only navigation sensors.

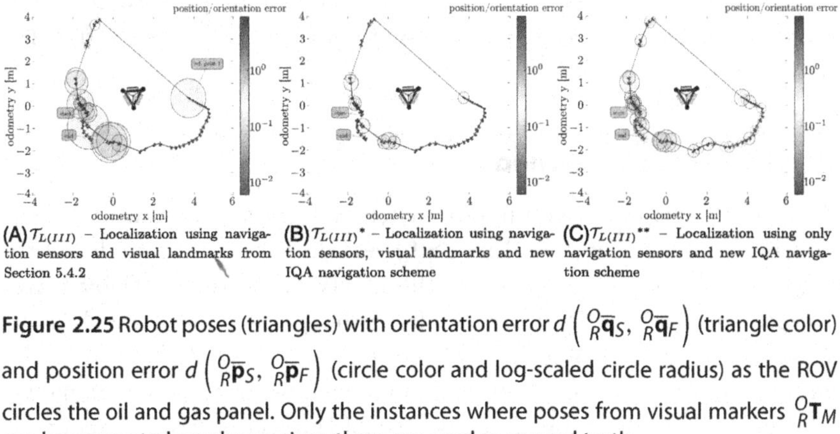

(A) $\mathcal{T}_{L(III)}$ – Localization using navigation sensors and visual landmarks from Section 5.4.2
(B) $\mathcal{T}_{L(III)}{}^{*}$ – Localization using navigation sensors, visual landmarks and new IQA navigation scheme
(C) $\mathcal{T}_{L(III)}{}^{**}$ – Localization using only navigation sensors and new IQA navigation scheme

Figure 2.25 Robot poses (triangles) with orientation error $d\left({}^{O}_{R}\bar{\mathbf{q}}_S, {}^{O}_{R}\bar{\mathbf{q}}_F\right)$ (triangle color) and position error $d\left({}^{O}_{R}\bar{\mathbf{p}}_S, {}^{O}_{R}\bar{\mathbf{p}}_F\right)$ (circle color and log-scaled circle radius) as the ROV circles the oil and gas panel. Only the instances where poses from visual markers ${}^{O}_{R}\mathbf{T}_M$ can be computed are shown since these are used as ground truth.

ity is poor and there are scarce features to track. When using the IQA to decide which inputs to integrate into the localization filter, we reduce the pose error and increase the smoothness of the followed trajectory, see Table 2.4.

Our observation is that integrating all inputs to the filter does not boost performance as the Kalman filter does not reason about the quality of the sensor data except for examining the input covariance matrices. On the contrary, the proposed two-stage scheme allows an informed decision-making when to activate VO-input for the benefit of accuracy and uncertainty reduction.

This enhancement of the navigation method, illustrates the utility of the proposed CSI/SIL framework as it supports the development of the proposed multimodal two-stage navigation scheme by providing a test bed to validate the effectiveness of improvements, as well as to support the parameterization process; further details about the two-stage navigation scheme can be found in [66]. We show the final results of this new methodology in Table 2.5 and Fig. 2.25, tested with the data acquired from *field trial II*, the same that was used in our previous validation stage in Section 2.5.4.2 for comparison purposes. Based on this, it can be seen quantitatively and visually that the integration of the new scheme, tested in simulation first (Fig. 2.24), boosts the localization performance ($\mathcal{T}_{L(III)}{}^{*}$) to levels required for manipulation tasks. Finally in test $\mathcal{T}_{L(III)}{}^{**}$, we analyze the performance of our method without the use of visual landmarks. The objective is to strive towards a more general localization filter that can function without fiducial landmarks. Table 2.5 shows that although the position and orientation error increase, they are not far from $\mathcal{T}_{L(III)}$ results where fiducial landmarks have been used.

2.6. Lessons learned

In this chapter, a continuous system integration framework (CSI) featuring a high fidelity simulator (SIL) was described and introduced to the project development process of the DexROV project. Cost intensive projects, such as DexROV, benefit from the proposed framework as it facilitates quantitative assessment of the performance under various conditions before roll-out of the system. Subsequently, such assessment allows reducing uncertainties and a more predictable project planning as required for particularly challenging multidisciplinary projects consisting of several workgroups and partners.

In the following, we reflect about our main experiences and lessons learned during the development process with the proposed CSI/SIL framework in the DexROV project.

Incremental simulator fidelity improvement. The key contribution of the presented work is the synchronization of environmental and spatial conditions observed in real-world with the simulator of the CSI/SIL framework. The creation of a high fidelity simulator features the major challenge in the preparation process of such Continuous System Integration system. The simulator fidelity is a product of several factors ranging from individual sensor models, CAD vehicle models to adaptation of external environment conditions within simulator or the introduction of algorithms and methods that accurately emulate real hardware and software components.

Due to constraint resources which such undertakings generally feature, regarding workforce, time, or project milestone schedule, our practice has shown success by incrementally equipping the simulator according to necessity and priority with new functionalities and components such as additional simulated sensors, actuators, or other simulation capabilities; note that the preparation of an accurate sensor model is cumbersome, i.e., a trade-off has to be made between the desired fidelity and required effort to reach the desired fidelity considering designed behavior and noise model, etc. Furthermore, in this context a successful practice was the incremental addition of simulated system components required to perform particular tasks for benchmarking and validation, e.g., the simulated stereo vision and panel detection were a priori introduced to enable the benchmark of vision-based systems such as the panel component detection.

Validated system architecture. For a successful development process within the CSI/SIL framework, the system architecture is desired to be identical during development and deployment regarding, e.g., software components introducing program logic (ROV localization, object recognition, etc.), interfaces and data flow. At the beginning of the DexROV project an architecture has been introduced including sensor, actuator drivers, software components, communication layers, etc. Particularly, software interfaces and component structure were initially defined. In this initial phase, the program logic reflecting the function of software components is mimicked according to their desired behavior, i.e., components may directly use information from the simulator such as the ROV pose to mimic the ROV localization component. However, in the development process components have been step by step replaced with methods and algorithms using respective real sensor feed. This architecture design and procedure is fundamental and a core aspect of the CSI/SIL concept. From the very beginning of the project, it facilitates a continuous verification of interfaces, dataflow, and system behavior, which is a not to be an underestimated factor for the system integration in a multidisciplinary project in which workgroups contribute from different institutions. Particularly in the DexROV project, costly harbor, ROV, and ship time during field operations were dedicated to conduct valuable missions instead of tackling common integration issues when *not* using continuous system integration mechanisms, for example, solving issues regarding communication difficulties among components caused by interface discrepancies or solving pipelining difficulties caused by data-flow bottlenecks. Consequently, the CSI/SIL framework reduces integration efforts and allows permanent evaluation of the system's state which is necessary for an efficient execution of the project.

Distributed development and validation. An important aspect of the proposed CSI/SIL framework is that it facilitates a distributed development among workgroups which are responsible for different tasks and developments of the project while they are geographically located at different sites. In the DexROV project, workgroups of seven institutions could independently develop their components and also continuously validated their progress with respect to the fully-integrated system as for each workgroup the CSI/SIL test bed was locally made available. Note that this portability of the test bed has immensely benefited from virtualization and deployment mechanisms introduced through, e.g., Docker [43]. This test bed allowed in the DexROV project fast (*validated*) development cycles prior to field operations, eliminating of architectural deficiencies, data provider–consumer optimizations, failure detection and recovery improvements, etc.

The task validation process relies on individually designed benchmarks and measures. These measures, which usually require ground truth, have a direct effect on the validation of particular tasks. However, defining ground truth to estimate deviations is particularly cumbersome in the underwater domain. In DexROV, ground truth was established through landmarks in the form of fiducial makers, which provided highly accurate estimates. Due to the accuracy, these landmarks were also exploited for the synchronization of the simulator with real field trial data (see Section 2.4). Therefore, landmarks used for referencing purposes is a particular crucial aspect. Depending on the application scenario, reference alternatives have to be evaluated, ranging from artificial landmarks like man-made objects to natural landmarks extracted from the environment.

Informed enhancement cycles. In the process of development and integration, the benchmark procedures are needed to be individually designed according to the task. In additional to the benchmark, measures are required to be defined regarding the individual tasks. These measures have a direct effect on the assessment of particular tasks.

Therefore, tasks and their validation measures have to be carefully designed and defined. Given the ease of integration enabled through the CSI/SIL framework and despite the effort of designing a benchmark including measures to quantify the performance, CSI/SIL have shown enormous benefits on the thorough evaluation and validation which led to an informed optimization and verification of modifications to enhance methods and algorithms of particular tasks. In the DexROV context, we could identify limitations of methods during the development phase of the CSI/SIL framework without exposing the system to risks present in the real world scenario. Consequently, informed enhancement cycles were performed prior to roll-out of the system which allowed predicting an expected system performance and increasing system reliability.

Efficient Continuous System Integration – as presented in this chapter – plays a more and more important role, as robotic systems become more complex and are applied to more challenging environment conditions performing more complex tasks. Therefore, a benchmark and validation framework, such as the proposed one, increasingly gains importance to quantify system performance, as well as to identify and analyze uncertainties, bottlenecks, limitations, etc. Through such measures, informed and qualified assessments can be conducted for an effective continuous system development. Considering the preparation effort of such a framework, efforts pay-off as they allow fast informed development cycles and minimizing the integration efforts on-site at field missions where usually high running costs and time constraints are present. Consequently, the proposed framework eases the development of interdisciplinary projects with multiple contributing workgroups located at different locations. It facilitates a more efficient and effective project realization.

2.7. Future directions and beyond continuous system integration with simulation-in-the-loop

Ground truth accuracy. Generally, a benchmark requires ground truth for a valuable system validation; furthermore, the validation and its assessment can only be as accurate as the accuracy of the – often approximated – ground truth. In our work, visual landmarks have been introduced to provide an accurate ground truth regarding position and orientation of objects-of-interest. As these landmarks represent an artificial modification of the environment, such modifications may not generally be applicable depending on the application scenario. Therefore, these landmarks could be replaced by other methods providing accurate estimates, possibly relying on other artificial or even natural landmarks. Such landmarks have to be carefully selected, since they may introduce estimation errors regarding their pose which will eventually lead to a degradation of the benchmark quality.

Data collection quality. The development process and the subsequent benchmark can only be as meaningful as the fidelity of the characteristics featured in the recorded real-world data used for evaluation. It is crucial that the sampling distribution of the dataset encodes realistic circumstances of the application scenario in order to achieve coherent results during development in simulation, as well as during operations in reality. A measure indicating the dataset quality has to be introduced in order to guarantee a useful assessment of the development progress. Such measure shall reveal deficiencies (e.g., redundancies, outliers, etc.) of the data collection process and facilitate the quantification of the gain and eventually classification of the collected data used for benchmarking. The design of the measure could be inspired from areas related to Information Theory that allow quantitatively assessing the gain of collected data and ultimately of the data collection method.

SIL enabling internal representation. Going beyond the Continuous System Integration use case, the proposed SIL approach can be exploited as a mechanism to create an internal representation about the current external state, i.e., conditions and affordance of the environment, spatial relationships among observed objects, etc. A system that is aware about such external circumstances through a simulation environment that is driven by SIL is a powerful tool that opens new possibilities beyond conservative sense–plan–act system architectures. Possibilities, such as anticipation of external changes or continuous planing and optimization in simulation fed with external observations and circumstances that are currently being made.

Acknowledgment

The research leading to the presented results has received funding from the European Union's Horizon 2020 program within the project "Effective Dexterous ROV Operations in Presence of Communication Latencies" (DexROV).

References

[1] UK Health & Safety Executive (HSE), Offshore safety statistics bulletin, http://www.hse.gov.uk/offshore/statistics/hsr2017.pdf, 2017. (Accessed 1 August 2019).

[2] C.A. Mueller, T. Doernbach, A. Gomez Chavez, D. Koehntopp, A. Birk, Robust continuous system integration for critical deep-sea robot operations using knowledge-enabled simulation in the loop, in: 2018 IEEE/RSJ International Conference on Intelligent Robots and Systems (IROS), 2018, pp. 1892–1899.

[3] T. Fromm, C.A. Mueller, M. Pfingsthorn, A. Birk, P. Di Lillo, Efficient continuous system integration and validation for deep-sea robotics applications, in: OCEANS 2017 - Aberdeen, 2017, pp. 1–6.

[4] J. Gancet, P. Weiss, G. Antonelli, M. Pfingsthorn, S. Calinon, A. Turetta, C. Walen, D. Urbina, S. Govindaraj, P. Letier, X. Martinez, J. Salini, B. Chemisky, G. Indiveri, G. Casalino, P. di Lillo, E. Simetti, D. de Palma, A. Birk, T. Fromm, C. Mueller, A. Tanwani, I. Havoutis, A. Caffaz, L. Guilpain, Dexterous undersea interventions with far distance onshore supervision: the DexROV project, in: IFAC-PapersOnLine – 10th IFAC Conference on Control Applications in Marine Systems CAMS, 2016, pp. 414–419.

[5] A. Birk, T. Doernbach, C. Mueller, T. Łuczynski, A. Gomez Chavez, D. Koehntopp, A. Kupcsik, S. Calinon, A.K. Tanwani, G. Antonelli, P. Di Lillo, E. Simetti, G. Casalino, G. Indiveri, L. Ostuni, A. Turetta, A. Caffaz, P. Weiss, T. Gobert, B. Chemisky, J. Gancet, T. Siedel, S. Govindaraj, X. Martinez, P. Letier, Dexterous underwater manipulation from onshore locations: streamlining efficiencies for remotely operated underwater vehicles, IEEE Robotics & Automation Magazine 25 (4) (2018) 24–33, https://doi.org/10.1109/MRA.2018.2869523.

[6] M. Eich, F. Bonnin-Pascual, E. Garcia-Fidalgo, A. Ortiz, G. Bruzzone, Y. Koveos, F. Kirchner, A robot application for marine vessel inspection, Journal of Field Robotics 31 (2) (2014) 319–341, https://doi.org/10.1002/rob.21498.

[7] N. Miskovic, M. Bibuli, A. Birk, M. Caccia, S. Egi, K. Grammer, A. Marroni, J. Neasham, A. Pascoal, A. Vasilijevic, Z. Vukić, Caddy–cognitive autonomous diving buddy: two years of underwater human-robot interaction, Marine Technology Society Journal 50 (2016) 54–66, https://doi.org/10.4031/MTSJ.50.4.11.

[8] D.O. Jones, A.R. Gates, V.A. Huvenne, A.B. Phillips, B.J. Bett, Autonomous marine environmental monitoring: application in decommissioned oil fields, Science of the Total Environment 668 (2019) 835–853, https://doi.org/10.1016/j.scitotenv.2019.02.310.

[9] H.H. Wang, S.M. Rock, M.J. Lees, Experiments in automatic retrieval of underwater objects with an AUV, in: Challenges of Our Changing Global Environment. Conference Proceedings. OCEANS '95 MTS/IEEE, IEEE, 1995, pp. 366–373.

[10] S. Choi, G. Takashige, J. Yuh, Experimental study on an underwater robotic vehicle: ODIN, in: Proceedings of IEEE Symposium on Autonomous Underwater Vehicle Technology (AUV'94), IEEE, 1995, pp. 79–84.

[11] J. Evans, P. Redmond, C. Plakas, K. Hamilton, D. Lane, Autonomous docking for intervention-AUVs using sonar and video-based real-time 3D pose estimation, in: Oceans 2003. Celebrating the Past... Teaming Toward the Future (IEEE Cat. No. 03CH37492), IEEE, 2003, pp. 2201–2210.

[12] G. Marani, S.K. Choi, J. Yuh, Underwater autonomous manipulation for intervention missions AUVs, Ocean Engineering 36 (1) (2009) 15–23, https://doi.org/10.1016/j.oceaneng.2008.08.007.

[13] E. Simetti, G. Casalino, S. Torelli, A. Sperinde, A. Turetta, Floating underwater manipulation: developed control methodology and experimental validation within

the TRIDENT project, Journal of Field Robotics 31 (3) (2014) 364–385, https://doi.org/10.1002/rob.21497.
[14] F. Maurelli, M. Carreras, J. Salvi, D. Lane, K. Kyriakopoulos, G. Karras, M. Fox, D. Long, P. Kormushev, D. Caldwell, The PANDORA project: a success story in AUV autonomy, in: OCEANS 2016-Shanghai, IEEE, 2016, pp. 1–8.
[15] P. Cieslak, P. Ridao, M. Giergiel, Autonomous underwater panel operation by GIRONA500 UVMs: a practical approach to autonomous underwater manipulation, in: Proc. IEEE Int. Conf. Robotics and Automation (ICRA), 2015, pp. 529–536.
[16] D. Youakim, P. Ridao, N. Palomeras, F. Spadafora, D. Ribas, M. Muzzupappa, MoveIt!: autonomous underwater free-floating manipulation, IEEE Robotics & Automation Magazine 24 (3) (2017) 41–51, https://doi.org/10.1109/MRA.2016.2636369.
[17] E. Zereik, M. Bibuli, N. Mišković, P. Ridao, A. Pascoal, Challenges and future trends in marine robotics, Annual Reviews in Control 46 (2018) 350–368, https://doi.org/10.1016/j.arcontrol.2018.10.002.
[18] Wiley, INCOSE Systems Engineering Handbook: A Guide for System Life Cycle Processes and Activities, 4th edition, John Wiley & Sons, Incorporated, 2015. (Accessed 15 August 2019). ProQuest Ebook Central.
[19] DVN GL, Standard DNVGL-ST-0373. Hardware in the Loop Testing (HIL), http://rules.dnvgl.com/docs/pdf/DNVGL/ST/2016-05/DNVGL-ST-0373.pdf, 2016. (Accessed 15 August 2019).
[20] G. Booch, Object Oriented Design: With Applications, Benjamin/Cummings, 1991.
[21] M. Reckhaus, N. Hochgeschwender, J. Paulus, A. Shakhimardanov, G.K. Kraetzschmar, An overview about simulation and emulation in robotics, in: Proceedings of SIMPAR, 2010, pp. 365–374.
[22] C. Heckman, N. Keivan, G. Sibley, Simulation-in-the-loop for planning and model-predictive control, in: Robotics: Science and Systems, Realistic, Rapid, and Repeatable Robot Simulation Workshop, 2015.
[23] A. Iivari, J. Ronkainen, Building a simulation-in-the-loop sensor data testbed for cloud-enabled pervasive applications, Procedia Computer Science 56 (2015) 357–362.
[24] T. Cichon, C. Loconsole, D. Buongiorno, M. Solazzi, C. Schlette, A. Frisoli, Combining an exoskeleton with 3D simulation in-the-loop, in: 9th International Workshop on Human-Friendly Robotics, 2016.
[25] M. Quigley, K. Conley, B.P. Gerkey, J. Faust, T. Foote, J. Leibs, R. Wheeler, A.Y. Ng, ROS: an open-source robot operating system, in: ICRA Workshop on Open Source Software, 2009.
[26] M. Prats, J. Pérez, J.J. Fernández, P.J. Sanz, An open source tool for simulation and supervision of underwater intervention missions, in: Proc. IEEE/RSJ Int. Conf. Intelligent Robots and Systems, 2012, pp. 2577–2582.
[27] N. Koenig, A. Howard, Design and use paradigms for gazebo, an open-source multi-robot simulator, in: 2004 IEEE/RSJ International Conference on Intelligent Robots and Systems (IROS) (IEEE Cat. No.04CH37566), 2004, pp. 2149–2154.
[28] Epic Games, Inc., Unreal engine, http://unrealengine.com. (Accessed 7 January 2019).
[29] R. Diankov, Automated Construction of Robotic Manipulation Programs, PhD thesis, Carnegie Mellon University, Aug 2010, http://openrave.org. (Accessed 7 January 2019).
[30] S. Carpin, M. Lewis, J. Wang, S. Balakirsky, C. Scrapper, USARSim: a robot simulator for research and education, in: Proceedings 2007 IEEE International Conference on Robotics and Automation, 2007, pp. 1400–1405.

[31] O. Michel, Webots: professional mobile robot simulation, Journal of Advanced Robotics Systems 1 (1) (2004) 39–42, https://doi.org/10.5772/5618, http://cyberbotics.com. (Accessed 7 January 2019).

[32] E. Rohmer, S. Singh, M. Freese, V-REP: a versatile and scalable robot simulation framework, in: 2013 IEEE/RSJ International Conference on Intelligent Robots and Systems, 2013, pp. 1321–1326, http://coppeliarobotics.com. (Accessed 7 January 2019).

[33] R. Pereira, J. Rodrigues, A. Martins, A. Dias, J. Almeida, C. Almeida, E. Silva, Simulation environment for underground flooded mines robotic exploration, in: Proc. IEEE Int. Conf. Autonomous Robot Systems and Competitions (ICARSC), 2017, pp. 322–328.

[34] J.C. Garcia, B. Patrao, L. Almeida, J. Perez, P. Menezes, J. Dias, P.J. Sanz, A natural interface for remote operation of underwater robots, IEEE Computer Graphics and Applications 37 (1) (2017) 34–43, https://doi.org/10.1109/MCG.2015.118.

[35] E. Guerrero-Font, M. Massot-Campos, P.L. Negre, F. Bonin-Font, G.O. Codina, An USBL-aided multisensor navigation system for field AUVs, in: 2016 IEEE International Conference on Multisensor Fusion and Integration for Intelligent Systems (MFI), 2016, pp. 430–435.

[36] osgOcean – an ocean rendering nodekit for OpenSceneGraph, https://code.google.com/archive/p/osgocean/. (Accessed 29 August 2019).

[37] T. Watanabe, G. Neves, R. Cerqueira, T. Trocoli, M. Reis, S. Joyeux, J. Albiez, The Rock–Gazebo integration and a real-time AUV simulation, in: 2015 12th Latin American Robotics Symposium and 2015 3rd Brazilian Symposium on Robotics (LARS-SBR), 2015, pp. 132–138.

[38] K.J. DeMarco, M.E. West, A.M. Howard, A computationally-efficient 2D imaging sonar model for underwater robotics simulations in Gazebo, in: OCEANS 2015 – MTS/IEEE Washington, 2015, pp. 1–7.

[39] N. Palomeras, N. Hurtós, M. Carreras, P. Ridao, Autonomous mapping of underwater 3D structures: from view planning to execution, IEEE Robotics and Automation Letters 3 (3) (2018) 1965–1971, https://doi.org/10.1109/LRA.2018.2808364.

[40] O. Kermorgant, A dynamic simulator for underwater vehicle-manipulators, in: Simulation, Modeling, and Programming for Autonomous Robots, Springer International Publishing, 2014, pp. 25–36.

[41] M.M.M. Manhães, S.A. Scherer, M. Voss, L.R. Douat, T. Rauschenbach, UUV simulator: a Gazebo-based package for underwater intervention and multi-robot simulation, in: OCEANS 2016 MTS/IEEE Monterey, 2016, pp. 1–8.

[42] T. Fromm, C. Mueller, A. Birk, Unsupervised Watertight Mesh Generation for Physics Simulation Applications Using Growing Neural Gas on Noisy Free-Form Object Models, Tech. Rep., Jacobs University Bremen, 2016, https://arxiv.org/abs/1603.00663.

[43] D. Merkel, Docker: lightweight Linux containers for consistent development and deployment, Linux Journal 2014 (239) (Mar. 2014).

[44] M. Pfingsthorn, Docker networking simulation, https://github.com/maxpfingsthorn/mini-network-simulator, 2016.

[45] S. Hemminger, et al., Network emulation with NetEm, in: Linux Conf Au, 2005, pp. 18–23.

[46] T. Doernbach, A. Gomez Chavez, C.A. Mueller, A. Birk, High-fidelity deep-sea perception using simulation in the loop, in: IFAC-PapersOnLine – 11th IFAC Conference

on Control Applications in Marine Systems, Robotics, and Vehicles CAMS, 2018, pp. 32–37.
[47] S. Garrido-Jurado, R. Munoz-Salinas, F. Madrid-Cuevas, M. Marin-Jimenez, Automatic generation and detection of highly reliable fiducial markers under occlusion, Pattern Recognition 47 (6) (2014) 2280–2292, https://doi.org/10.1016/j.patcog.2014.01.005.
[48] A. Gomez Chavez, C.A. Mueller, T. Doernbach, A. Birk, Underwater navigation using visual markers in the context of intervention missions, International Journal of Advanced Robotic Systems 16 (2) (2019), https://doi.org/10.1177/1729881419838967.
[49] P. Letier, E. Motard, J. Verschueren, Exostation: haptic exoskeleton based control station, in: 2010 IEEE International Conference on Robotics and Automation, 2010, pp. 1106–1107.
[50] G. De Cubber, D. Doroftei, Y. Baudoin, D. Serrano, K. Chintamani, R. Sabino, S. Ourevitch, ICARUS: providing unmanned search and rescue tools, in: IROS Workshop on Robots and Sensors Integration in Future Rescue Information Systems (ROSIN'12), 2012.
[51] T. Luczynski, M. Pfingsthorn, A. Birk, The Pinax-model for accurate and efficient refraction correction of underwater cameras in flat-pane housings, Ocean Engineering 133 (2017) 9–22, https://doi.org/10.1016/j.oceaneng.2017.01.029.
[52] T. Luczynski, A. Birk, Underwater image haze removal with an underwater-ready dark channel prior, in: Oceans (Anchorage), 2017, pp. 1–6.
[53] K. Shoemake, Animating rotation with quaternion curves, in: Conference on Computer Graphics and Interactive Techniques, SIGGRAPH '85, 1985, pp. 245–254.
[54] D. Huynh, Metrics for 3D rotations: comparison and analysis, Journal of Mathematical Imaging and Vision 35 (2) (2009) 155–164, https://doi.org/10.1007/s10851-009-0161-2.
[55] D. Koehntopp, B. Lehmann, D. Kraus, A. Birk, Segmentation and classification using active contours based superellipse fitting on side scan sonar images for marine demining, in: 2015 IEEE International Conference on Robotics and Automation (ICRA), 2015, pp. 3380–3387.
[56] J. Kennedy, R. Eberhart, Particle swarm optimization, in: International Conference on Neural Networks, 1995, pp. 1942–1948.
[57] L. Zhang, L. Zhang, X. Mou, D. Zhang, FSIM: a feature similarity index for image quality assessment, Transactions on Image Processing 20 (8) (2011) 2378–2386.
[58] L. Zeng, Y. Dai, Single image dehazing based on combining dark channel prior and scene radiance constraint, Chinese Journal of Electronics 25 (6) (2016) 1114–1120, https://doi.org/10.1049/cje.2016.08.006.
[59] Y. Schechner, N. Karpel, Recovery of underwater visibility and structure by polarization analysis, Journal of Oceanic Engineering 30 (3) (2005) 570–587, https://doi.org/10.1109/JOE.2005.850871.
[60] T. Treibitz, Y. Schechner, Active polarization descattering, Transactions on Pattern Analysis and Machine Intelligence 31 (3) (2009) 385–399, https://doi.org/10.1109/TPAMI.2008.85.
[61] S. Lee, S. Yun, J. Nam, C. Won, S. Jung, A review on dark channel prior based image dehazing algorithms, EURASIP Journal on Image and Video Processing 2016 (1) (2016) 4, https://doi.org/10.1186/s13640-016-0104-y.
[62] L. Chen, S. Wang, K. McDonald-Maier, H. Hu, Towards autonomous localization and mapping of AUVs: a survey, International Journal of Intelligent Unmanned Systems 1 (2) (2013) 97–120, https://doi.org/10.1108/20496421311330047.

[63] D. Li, D. Ji, J. Liu, Y. Lin, A multi-model EKF integrated navigation algorithm for deep water AUV, International Journal of Advanced Robotic Systems 13 (1) (2016) 3, https://doi.org/10.5772/62076.
[64] T. Moore, D. Stouch, A generalized extended Kalman filter implementation for the robot operating system, in: Intelligent Autonomous Systems 13. Advances in Intelligent Systems and Computing, 2016, pp. 335–348.
[65] N. Palomeras, N. Hurtos, M. Carreras, P. Ridao, Autonomous mapping of underwater 3D structures: from view planning to execution, IEEE Robotics and Automation Letters 3 (3) (2018) 1965–1971, https://doi.org/10.1109/LRA.2018.2808364.
[66] A. Gomez Chavez, Q. Xu, C.A. Mueller, S. Schwertfeger, A. Birk, Adaptive navigation scheme for optimal deep-sea localization using multimodal perception cues, in: 2019 IEEE/RSJ International Conference on Intelligent Robots and Systems (IROS), 2019, pp. 7211–7218.
[67] N. Mayer, E. Ilg, P. Häusser, P. Fischer, D. Cremers, A. Dosovitskiy, T. Brox, A large dataset to train convolutional networks for disparity, optical flow, and scene flow estimation, in: IEEE Conference on Computer Vision and Pattern Recognition (CVPR), 2016, pp. 4040–4048.
[68] R. Mur-Artal, J.D. Tardós, ORB-SLAM2: an open-source slam system for monocular, stereo, and RGB-D cameras, IEEE Transactions on Robotics 33 (2017) 1255–1262, https://doi.org/10.1109/TRO.2017.2705103.
[69] H.Z. Nafchi, M. Cheriet, Efficient no-reference quality assessment and classification model for contrast distorted images, IEEE Transactions on Broadcasting 64 (2018) 518–523, https://doi.org/10.1109/TBC.2018.2818402.

CHAPTER THREE

Azimuth thruster single lever type remote control system

Guichen Zhang
Merchant Marine College, Shanghai Maritime University, Shanghai, PR China

Contents

Abstract

In this paper, azimuth thruster single lever type remote control system is designed to freely control the sailing direction and speed of a ship by controlling azimuth thrusters by using a lever mounted to the control stand. The control mechanism signals from the built-in single lever of the control stand are converted into port and starboard setting signals by the control unit with built-in microcomputer, and compared with the position signals of the follow-up unit. If there is a difference between these signals, a differential signal is fed to the power unit to rotate the azimuth thrusters until the error is eliminated. The control unit extracts the set direction and speed of the ship from the tilted position of the single lever as vector signals, and controls the composite vector of two propulsion directions. Thus, the ship's sailing direction is determined by the tilted direction of the single lever, while the ship's speed is determined by the tilted angle of the single lever without changing the revolution of the main engine.

https://doi.org/10.1016/B978-0-12-820271-5.00008-0

Keywords

Marine engineering equipment, Azimuthing ship, Azimuth thruster control, Marine remote control system

3.1. Introduction

Marine azimuth thruster is one of the most popular thruster devices, which is installed widely on tugboats, engineering ships, and special ships [1]. The full rotation propulsion device can deliver all-round thrust, so that the ship has sensitive maneuverability, which is widely used on a multi-purpose ship [2]. Scholte Company is one of the first manufacturers to produce a fully rotating propeller, its main products are Z-type full-rotary propeller and Z-type full-rotary double propeller [3]. The Z-type twin propeller has successfully optimized the performance of the ship's rudder propeller system, which can improve the propulsion efficiency by 20% [4]. ABB Azipod thruster has become an optimal propulsion solution in the field of luxury cruise [5,6]. The Azipull fully rotating propeller developed by the Rolls-Royce has streamlined appendages, it is very important to improve the propulsion efficiency of the marine supply ship by using the single low speed gear drive and the tractor propeller [7,8]. Because of the azimuth thruster, it has high flexibility, good efficiency, and low maintenance cost. However, the control of azimuth thruster cannot be performed with suitable precision since it is a nonlinear system [9]. Wang [10] mainly focuses on the nonlinear vibration characteristics of the spiral bevel gear transmission system in propulsion shafting of azimuth thrusters, to provide the optimization design for the vibration and noise reduction of the propulsion shafting of azimuth thrusters. Lin et al. [11] research on vibration characteristics of rudder–propeller propulsion shafting, Chen [12] provide a reference for the vibration design of a Z-type retractable full rotating propeller. With the application of the fully rotating propeller, the geometry of the stern of the ship is simplified, the hull structures such as the stern post and the stern tube are omitted, and the resistance of the ship is reduced. The assembly of the fully rotating propeller also reduces the maintenance work in case of failure and prolongs the service life of the marine diesel engine [2]. In this paper, azimuth thruster single lever type remote control system is designed and answers the open questions in this field – the difficulties and problems mentioned above are considered. The ship's sailing direction is determined by the tilted direction of the single lever, without changing the revolution of the main engine, and the navigation performance is improved.

3.2. Composition of control system

This control unit consists of the following components, as shown in Fig. 3.1. The single lever is mounted on the control stand. The ship's speed (stop – full speed) and direction (0°~360°) are controlled by the single lever positions (setting accuracy ±1.0°). This single lever is provided with two built-in potentiometers which convert the mechanical position of the single lever into an electrical signal and adjust trimmers for this electrical signal.

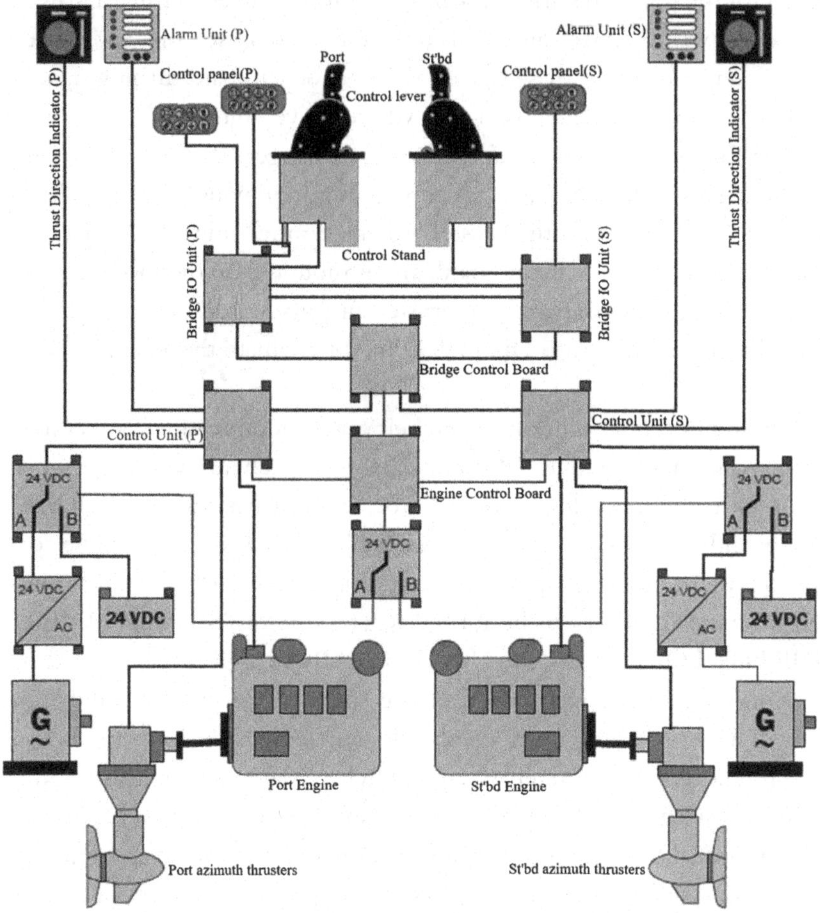

Figure 3.1 The control system for azimuth thrusters with FP propellers.

The direct lever is mounted below the front handrail of the control stand. It directly operates the solenoid valve of the power unit, and it can be used as the emergency device if the single lever control is defective. When turning this direct lever rightward, the azimuth thruster turns counterclockwise, and when turning this direct lever leftward, the azimuth thruster turns clockwise. When releasing this lever, it is reset to the original position by means of spring, and thus, it stops turning.

The control unit, which is built in the control stand, consists of the CPU circuit board with built-in microcomputer and the I/O circuit board. It calculates the signals electrically converted by the single lever as setting signals for the port and starboard azimuth thrusters. These setting signals are compared with the position signals from the position sensors, and if there is a difference between these signals, the port or starboard turning signal is fed to the solenoid of the power unit via the drive unit.

The follow-up unit is mounted to the worm gear shaft of the azimuth thruster. This unit contains a synchro transmitter which feeds back the position signals of azimuth thrusters to the control unit. The drive unit consists of one printed circuit board, and is mounted on the monitor panel backside. It amplifies the azimuth thruster turning. Signals which are output from the control unit into DC (0–999 mA) to actuate the solenoid of the power unit.

The annunciator unit is mounted below the monitor panel, and consists of one basic pack and up to 18 annunciator packs. There are two annunciator packs, namely, one for Contact A in case the alarm input is NORMAL OPEN and one for Contact B in case it is NORMAL CLOSED. The control mode changeover switch is mounted below the left handrail on the control stand front. It can be turned to any one of "SINGLE LEVER," "DIRECT LOW-SPEED," and "DIRECT HIGH-SPEED."

A monitor panel is arranged at the back of the control stand front door. The monitor panel is equipped with bar graph array for output current check, data channel selector switch, 4-digit numeral display unit for data display, 7 system trouble pilot lamps, single lever position pilot lamp, terminal for system power source check, direct low-speed adjusting volume, output current regulating volume, output current operation coefficient regulating volume, and digital switch for synchro transmitter position correction. These facilitate adjustment, maintenance, and inspection. Also there is the CPU rack, which contains the control unit, on the left, and the fuse panel and the annunciator unit below.

3.3. Control functions

This control equipment controls the composite vector of the propulsion forces and directions of two azimuth thrusters by operating the control lever. Speed control assumes straight ahead movement as a simple example, as shown in Figs. 3.2A–D.

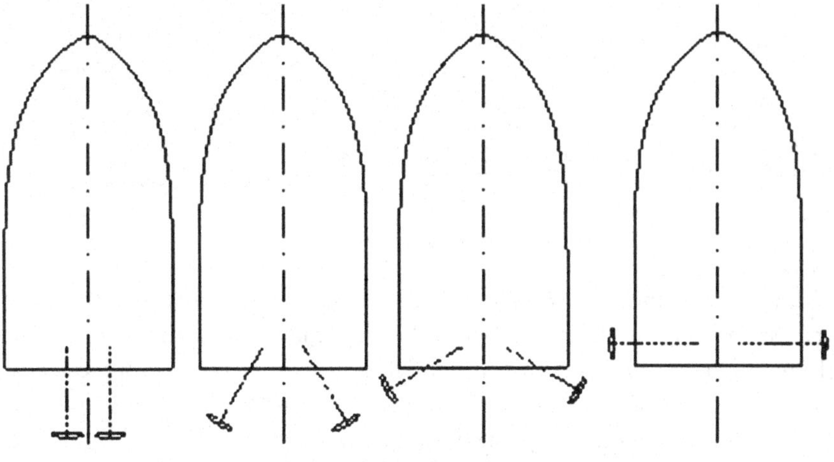

Figure 3.2 Thrust directions of the azimuth thrusters.

Fig. 3.2A illustrates a case where the thrust direction of the azimuth thrusters is parallel to the center line of the ship's hull. Since there is no composite vector in the traverse direction, the ship is cruising ahead at full thrust.

Fig. 3.2B illustrates a case where the thrust directions of the azimuth thrusters are tilted 30° with respect to the ship's hull center, and the ship is cruising ahead at about 7/8 full speed.

Fig. 3.2C illustrates a case where the thrust directions of the azimuth thrusters are tilted by 60° with respect to the ship's hull center, and the ship is cruising ahead at 1/2 full speed.

Fig. 3.2D illustrates a case where the thrust directions of the azimuth thrusters are perpendicular to the ship's hull center line. Since there is a composite vector in the traverse direction only, the directions of the propulsion forces of two azimuth thrusters are opposite to and cancel each other. Thus, the composite thrust is zero. The ship is stopping. As described above, this equipment can freely control the ship's speed over a range from the maximum speed to the stop without changing the engine revolutions.

The above operation is accomplished simply by tilting the lever forward. Notches (1), (2), (3), and (4) in Fig. 3.3 indicate the tilting positions of the control lever, and these positions correspond to (1), (2), (3), and (4) in Fig. 3.2, respectively. Position (1) in Fig. 3.3 indicates the maximum speed position and 30° from the vertical line.

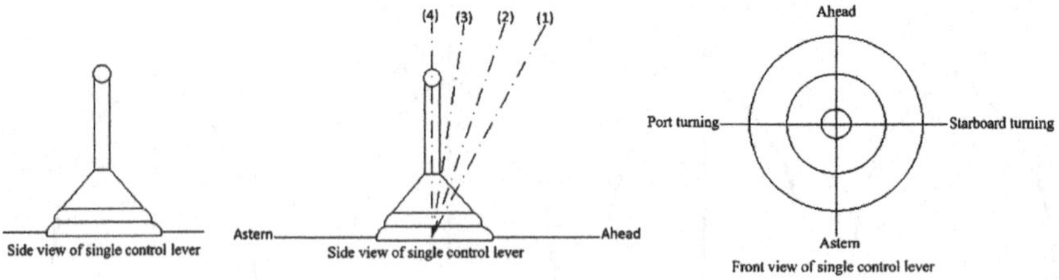

Figure 3.3 Tilting positions of the control lever.

Direction control is shown in Fig. 3.4, the directions other than the straight ahead are as described below. For turning the ship ahead starboard, tilt the single lever left forward, and the ship turns leftward within the arrow range in Fig. 3.4(1). This ship turns further leftward as this lever is tilted to the left. For turning the ship astern starboard, the lever is turned rightward within the arrow range shown in Fig. 3.4(2). The ahead starboard turning and astern port turning can be operated in the same manner as described above. For right ahead movement, tilt the lever straight forward. For right astern movement, tilt the lever straight backward (see Fig. 3.4(3)). The moving direction of the ship is determined simply by tilting the single lever in a desired direction. The tilting angle of the single lever determines the ship's speed in the chosen direction.

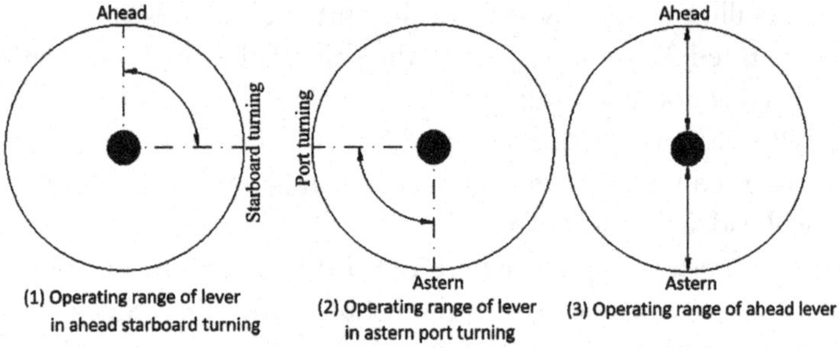

Figure 3.4 Tilting effect of the single lever.

As described above, the ship is manipulated by a single lever. In other words, the ship's moving direction is determined by the tilted direction of the single lever, while its speed is determined in the chosen direction by the magnitude of the tilting angle of the lever.

3.4. Going-sideways control

The going-sideways position is located on the X-axis of the single lever. When the single lever is tilted within the range of the going-sideways stroke, the going-sideways control is actuated with the going-sideways indicator lamp lit. When the lever is tilted within the going-sideways range, two propellers move by the same angle at a maximum of approx. 60° from the neutral position. This angle is controlled by the tilting angle of the lever as illustrated in Fig. 3.5.

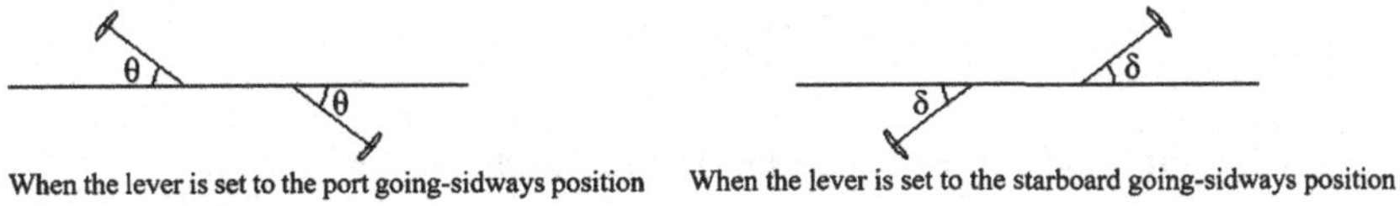

Figure 3.5 Tilting angle of the lever.

In making traversing motions, the vessel may possibly be influenced by the inborn characteristics of maneuvering and wind and waves. In order to make the vessel run free of such combined effects, fine adjustments of the direction of individual propellers can be executed on the going sideways adjustment dial combined with a slight control of the main engine revolution. The starboard propeller is controllable by this dial within a range of ±30° in case of the port going-sideways, or the port propeller is controllable by the same dial within a range of ±30° in case of the starboard going-sideways.

3.5. Relationship between single lever and azimuth thruster position

Fig. 3.6 indicates the directions of respective azimuth thrusters at respective lever settings in the setting directions of A, B, and C with the lever tilting angles 0, 1, 2, and 3. If the lever is moved along circumference 2 in the C direction, the ship turns ahead starboard at a speed of about 7/8 times the full speed.

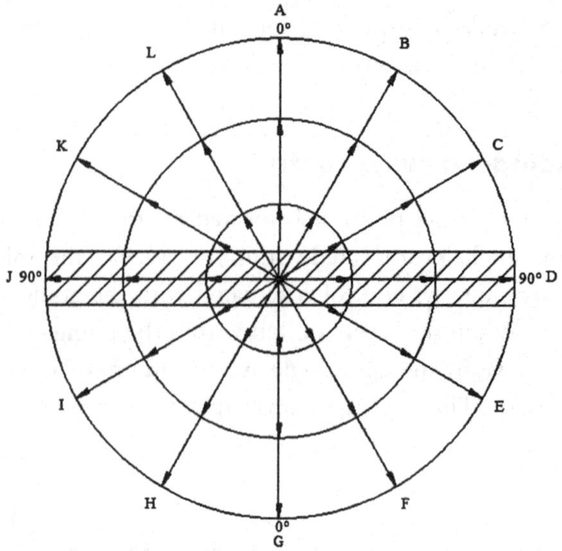

Figure 3.6 Interrelation between single lever and azimuth thruster.

Though D and J directions correspond to the boundary lines between the ahead and astern sides, there is no fear of erroneous operation since this system is designed for ahead direction as far as the lever is put on the line and a notch is provided on the line.

Notes:

① A-L: Tilting direction of lever

② 0, 1, 2, 3: Tilting angle of lever

③ ▨: Going –sideways position

④ This figure shows an example of operation. This control equipment can be operated steplessly in all directions, with a range of 360°.

3.6. Navigation

When steering ship port or starboard while cruising at full ahead, it is hard to attain a small angle by ordinary single lever operation; hunting of the ship may even result. This equipment is arranged so that it is easy to steer the ship when the single lever is within the shaded range shown in Fig. 3.7 with the illuminated push-button switch "NAVIGATION" being turned to ON. At this time, the "NAVIGATION" pilot lamp is lit.

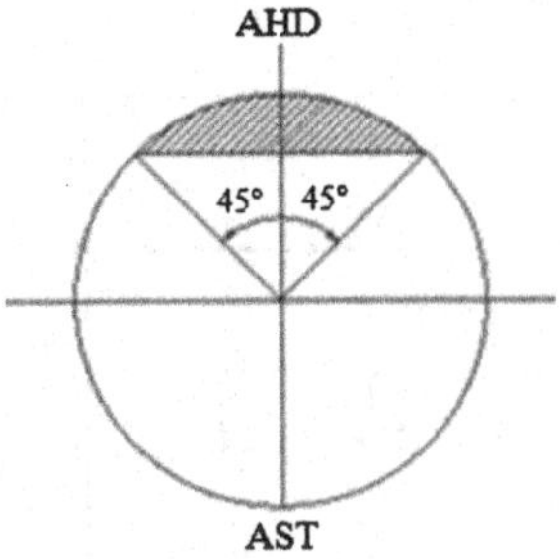

Figure 3.7 Single lever within the shaded range.

3.7. Turning direction control & direct lever control

Normally, the azimuth thrusters follow up the setting position by turning the outside instead of passing the inside (the position of 270°) in view of controllability and safety. This rule does not apply, however, when the difference between the present and setting positions is within 120°. Relations between single lever position and azimuth thruster direction are as shown in Fig. 3.8.

For using the direct lever, one can select the lever selector switch to either "direct low speed" or "direct high speed". Two direct levers are mounted for operating the port azimuth thruster and starboard azimuth thruster, respectively. The left direct lever is used to operate the port azimuth thruster, while the right direct lever is used to operate the starboard azimuth thruster. When tilting these levers leftward, the azimuth thrusters turn clockwise, and when tilting them rightward, the azimuth thrusters turn counterclockwise. One monitors the azimuth thruster direction receiver pointer without failure when operating these direct levers. Direct lever high speed turns are faster than direct lever low speed due to operating the abrupt turning solenoid.

3.8. Main engine revolution control

If the main engine must be stopped abruptly due to a certain circumstance, one can depress the push-button switch mounted at the front of the main engine lever. The engine revolution is controlled by the engine control lever, and the clutch can also be engaged or disengaged by this lever. When this lever is fully pulled toward you, the clutch is disengaged, so that

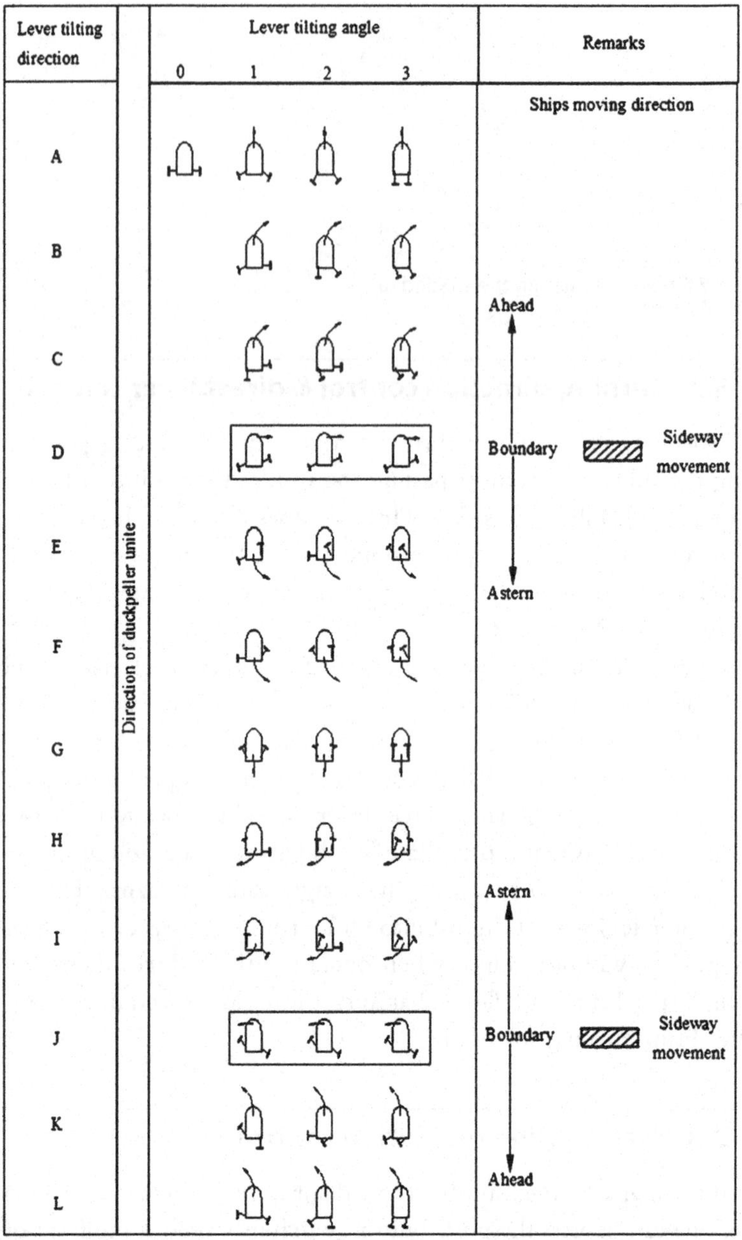

Figure 3.8 Relations between single lever position and azimuth thruster direction.

the engine runs idly. When this lever is tilted forward, the clutch is engaged to increase the engine revolution.

Two levers are mounted to be able to control the port and starboard engines separately. These two levers can operate both engines together by holding two levers by a single hand. Each lever is also provided with an engine tachometer at the front. One can adjust the main engine revolution, while monitoring this meter. The engaged and disengaged conditions of the clutch are indicated by the indicator lamp on the engine tachometer. For details, please refer to the instruction manual arranged by the main engine supplier.

3.9. Operating procedures

3.9.1 Single lever control

(1) Turn the AC power and DC power (battery) switches to ON.
(2) Start up the main engine. Previously set the clutch to the disengaged position, and keep the engine idling.
(3) Set the single control lever to the neutral position.
(4) Set the control mode changeover switch to "SINGLE LEVER".
(5) Adjust the main engine speed to the required revolution by operating the main engine revolution control lever.
(6) Tilt the single lever to the moving direction.
(7) By following the above steps, it becomes possible to operate the single lever.

3.9.2 Direct lever control

(1) Set the control mode changeover switch to "DIRECT LOW-SPEED" or "DIRECT HIGH-SPEED" when it becomes impossible to operate the single lever or when it is easier to operate the direct lever.
(2) Keep the port and starboard direct levers tilted until the required angle for azimuth thrusters is reached while watching the azimuth thruster direction indicators. The azimuth thrusters stop when the levers are released.

3.9.3 Cautions on operation

The adaptive control system is adopted in this control equipment. Therefore, the AC power source should not be turned off as long as possible.

Also, please, be careful not to depress these push-buttons unless the "system fault" or "CPU fault" state is assumed. If the above operation is made, all the data collected so far vanishes, and the system resumes initial values.

3.10. Conclusions

The fully rotating propeller is an all-round propulsion device which can provide thrust in any direction. The design of azimuth thruster single lever type remote control system includes the following: composition of control system, tilting positions and effect of the control lever, going-sideways control, relationship between single lever and azimuth thruster position, turning direction control and direct lever control, main engine revolution control, operating procedures and cautions. It is the best propulsion and control system which can meet the special requirements of marine equipment. With its considerable propulsion efficiency, excellent maneuverability and convenience of installation and maintenance, it has attracted more and more attention in the control field of marine engineering movement.

Acknowledgment

This paper was completed with the support of NSFC NO. 51779136 (The Adaptive Cooperative Optimum and Reconfigurable Control on the Multi-electrical Redundant Drive Hybrid Systems in Ship & Ocean Engineering).

References

[1] Dan-dan Li, Design features of the 4000 hp full rotation multi-purpose ship, Ship (2) (2007) 39–41.
[2] Dong Gui, Qi Gong, The design of the control system of the azimuth propulsion, Journal of Shanghai Ship and Shipping Research Institute 38 (3) (2015) 29–33.
[3] Yong-ping Ma, Da-kui Li, Xiao-yu Liu, The application of Z-type rotating device in the ferry, Journal of Jiamusi University (Natural Science Edition) 25 (5) (2007) 648–649.
[4] Zhi-xi Zhang, Schottle SRP1515FP full rotation rudder propeller flatulence clutch control system failure, Marine Technology (1) (2018) 51–52.
[5] Tong-zhou Lv, ABB launches the Azipot gearless POD propeller at the Expo, Maritime China (12) (2018) 44.
[6] ABB (China) Co., Ltd., ABB establishes the global production base of Azipod ship propulsion system in Shanghai, Motor and Control Application 18 (9) (2011) 66.
[7] Rolls-Royce, Azimuth thrusters/propulsion service manual aquapilot control system, Serial Nos. 61535, 61536.
[8] M. Fonte, L. Reis, M. Freitas, Failure analysis of a gear wheel of a marine azimuth thruster, Engineering Failure Analysis 18 (7) (2011) 1884–1888.

[9] Jing Dai, Research of Intelligent Control Technology for Marine Azimuth Thruster, Wuhan NO. 2 Ship Design and Research Institute, 2013, pp. 9–26.
[10] Shuai Wang, Research on the Nonlinear Vibration Characteristics of Spiral Bevel Gears in Propulsion Shaft of Azimuth Thrusters, Harbin Institute of Technology, 2017, pp. 49–53.
[11] T.R. Lin, J. Pan, Peter J. O'Shea, et al., A study of vibration and vibration control of ship structures, Marine Structures 22 (4) (2009) 730–743.
[12] Gang Chen, Yu-bo Wang, Dong-liang Shen, et al., Shafting torsional vibration calculation of Z-type retractable full rotating propeller, Ship Engineering 34 (3) (2012) 36–38.

CHAPTER FOUR

Autonomous environment and target perception of underwater offshore vehicles

Hongde Qin
College of Shipbuilding Engineering, Harbin Engineering University, Harbin, China

Contents

Abstract

An autonomous underwater vehicle (AUV) relies on an underwater camera and sonar for perception the surrounding environment including underwater image processing, object detecting and tracking. For underwater optical environment perception, the limited visual range of the camera severely limits the acceptance of detection information; in addition, underwater light is assimilated and scattered, which seriously deteriorate the underwater imaging, especially decreases the image contrast. To address these

Fundamental Design and Automation Technologies in Offshore Robotics
https://doi.org/10.1016/B978-0-12-820271-5.00009-2

issues, a joint framework based on CNN is proposed to improve underwater image quality and extract more matching feature points. For accurate registration between images, an underwater mosaicking technology is also involved, and a fusion algorithm is implemented to mitigate artificial mosaicking traces. Experimental results show that our presented framework can not only keep underwater image detail information, but also exact more matching feature points for registration and mosaicking. For underwater acoustic image environment perception, mechanically scanned imaging sonars (MSIS) are usually equipped on AUVs for avoiding obstacles and multiple-targets tracking. In this chapter, two multiple underwater object tracking methods are presented. The proposed methods are using the cloud-like model data association. Some sea trials are implemented to validate the effectiveness of the presented algorithms. Experiment results demonstrate that the presented cloud-like model data association method has the characteristics of more accurate clustering. Multiple targets were finally clustered for AUV stable tracking.

Keywords

AUV, Underwater environment perception, Underwater image enhancement, Underwater image mosaicking, Multiple underwater targets tracking

4.1. Introduction

The earth's land resources will be exhausted, and the ocean is rich in resources, so the marine resources exploration using underwater offshore vehicles is a necessary development strategy all over the world. However, the complex underwater environment limits the application of autonomous underwater sensing technology [1]. Underwater autonomous environment and target perception technology has been applied for long-term underwater autonomous detection operation or avoiding obstacles [2–5]. Whilst underwater exploration is implemented with AUVs, environmental perception is the prerequisite for underwater operation [6,7]. AUVs have been applied in many underwater fields, such as seabed topography drawing [8], identification and tracking of subsea pipelines [9], visual navigation [10], and exploration of seafloor mineral resources [11].

A lot of research on land or air moving targets tracking is in progress. However, there are few researches on underwater target tracking. Ocean observation has experienced a rapid development in the past decade [12–14]. Underwater image processing is the key for understanding underwater environmental information. Poor underwater image quality will undoubtedly affect the perception and detection of underwater environment [15]. The absorption and scattering of underwater light can make the color of underwater image appear blue-green [16]. The scattering of underwater transmitted light decreases the underwater image quality [17]. In addition, the uneven auxiliary lighting on the AUV often results in organic matter and suspended particles in the underwater image [18]. In an underwater image captured by one underwater image, due to the limited view angle, there is insufficient information obtained [19]. Therefore, we are increasingly interested in underwater image filtering, enhancement, registration and mosaicking [20].

The research on underwater image enhancement and mosaicking has developed rapidly. However, the complex underwater detection conditions pose a huge challenge to the autonomous environmental perception. Besides, serious degradation of underwater image lacks effective feature information [21]. Convolutional neural network (CNN) has developed with a high speed, and it has made great progress in the field of computer vision [22]. An underwater image CNN based enhancement is proposed and implemented to filter the noise and generate a clear underwater image. VGG net is involved to extract more exact matching points. Underwater target features are difficult to be extracted, so it is hard to track multiple targets using acoustic image features [23].

Nowadays, great progress has been made in the field of image enhancement and panoramic image mosaicking technology, such as classical fuzzy image enhancement, which have achieved good results in terrestrial or air images. However, these algorithms cannot compensate for the decrease of sonar image quality. In the same way, the adaptive thresholding segmentation will result in false segmentation of noise into targets, which makes it not suitable for underwater complex environment. Accurate image segmentation is the key factor of multiple target identification and tracking. High threshold region growing algorithm will lead to "oversegmentation" of an image, which results in some target regions with low gray level segmented as background regions.

Traditional data association methods include NNDA [24], PDA [25], JPDA [26], multiple hypothesis tracking method [27], and clustering method [28]. In addition, the minimum cost flow method is used for determining the optimal data association algorithm. J. Neira of the University of Zaragoza [29] proposed a new measurement system with joint compatibility, which adopts restrictive standards to effectively find the best solution for data association. The traditional feature selection and generation methods include the analysis of Fisher linear discriminant, clustering method, artificial neural network, and so on. Carin has presented a relevance vector machine feature selection and classification method [30]. Pezeshki used the spatiotemporal correlation to demonstrate sonar data orientation characteristics for extracting underwater target features [31]. However, these methods are hard when generating the stability features of sonar images.

4.2. Methodology

4.2.1 Underwater panoramic image enhancement and mosaicking method

4.2.1.1 Underwater image dataset and preprocessing method

(1) Diversifying an underwater image dataset

Due to the small dataset that can be applied to underwater CNN, it is too hard to collect degradation images or clear images without water at a fixed location based on existing technologies. Therefore, the establishment of underwater dataset is relatively difficult. In the field of CNN, a big dataset is a necessary condition in network weight optimization, and also a key difficulty that CNN cannot be widely used for underwater images. To address this bottleneck, we introduce the image blur method and generate some simulated underwater degradation images by using the clear image without water. Firstly, the transmission s in the image is estimated. Secondly, the original clear waterless image W is calculated. The background light coefficient is denoted by parameter C, and it is a random number within the range of [0.85,1]. Finally, the underwater fuzzy image I is generated as

$$I(x) = s(x)W(x) + C(1 - s(x)). \tag{4.1}$$

Then we attenuate the color of I^c_{water} as follows:

$$I^c_{\text{water}} = \alpha^c * I^c_{\text{blur}} \tag{4.2}$$

where α^c is the attenuation coefficient of each channel in the blurred image, "$*$" is the convolution symbol, I^c_{blur} is the blurred image, c means the channel number of the blurred image. The fuzzy image obtained has three channels: red, blue, and green. By changing the attenuation coefficient of each channel according to this formula, the simulated underwater degraded images which have color attenuation can be generated.

We expand underwater image dataset, the CycleGAN model is used to synthesize the underwater image, and the unsupervised training is carried out for the image in two corresponding domains, which increases the diversity of the underwater image. Our underwater image dataset is shown in Fig. 4.1. Some original waterless clear images are shown in the first line, and generated simulated underwater degradation images are shown in the following line. The two images on the left of the second line are the simulated images generated by combining the fuzzy processing method and the dark channel color attenuation. And the three images on the right are the images generated after training using the CycleGAN model.

Figure 4.1 Diverse underwater image dataset.

(2) Underwater low quality image enhancement method based on improved CNN

We can improve the learning ability of CNN by adjusting the convolution kernel and convolution layer feature map, so as to get the nonlinear mapping relationship between underwater degraded image and clear image accurately. The input is an underwater degraded image of any size. There are three different convolution kernels and symmetric convolution layer and deconvolution layer. Convolution layer extracts different features from underwater degraded images and generates different feature images. However, continuous convolution cannot restore the details of the degraded underwater image, while symmetrical deconvolution layer can optimize texture features and reconstruct the convolution layer of low-quality underwater image, so as to improve the visual effect of underwater image.

To evaluate the effect of the improved underwater low-quality image enhancement method, we improve the method of Mean Square Error (MSE), which can get the relationship between the pixel values of two images. The formula for calculating Loss is as follows:

$$\text{Loss} = \frac{1}{n}\sum_{i=1}^{n}\left(\frac{1}{wh}\sum_{j=1}^{w}\sum_{k=1}^{h}\left\| D_i(j,k) - X_i(j,k)\right\|^2\right). \tag{4.3}$$

Here, D represents output image, X represents original input image, n represents the sample size in the dataset, W represent the width, H represent the height. In our proposed method, W and H are fixed.

When the network model is trained, the network will calculate the relationship between the image in the sample dataset and the original clear image, to form the loss function. Then, the backpropagation network and random gradient descent method are combined to generate more appropriate weights in the structure. The weight optimization in the structure according to the following formulas:

$$W_{t+1}^{l} = W_{t}^{l} + \Delta_{t+1}^{l}, \tag{4.4}$$

$$\Delta_{t+1}^{l} = -\eta \frac{\partial L}{\partial W_{t}^{l}}. \tag{4.5}$$

Here, t means training times in the process of network training, l indicates that the network has l layer, Δ_{t}^{l} means the optimal parameters of the lth layer network after tth training, η represents learning rate set by the network in this layer, here it is 10^{-4}.

According to the principle of segmental learning, in order to obtain an effective network model, the degraded underwater image and the corresponding original image are randomly collected. During the training of CNN model, it is necessary to input the degraded underwater image and the original image without operation.

4.2.1.2 Improved underwater image registration method of CNN

At present, the accuracy of underwater image matching is not high, so an improved CNN-RANSAC registration algorithm based on VGGNet-16 model is proposed. CNN is used to generate multiscale feature points with high robustness. In the case of low accuracy requirements, the feature points are roughly matched first, and the method based on RANSAC is introduced to delete the point pairs that do not meet the conditions.

(1) VGGNet-16 model pretraining

VGGNet-16 is a network model for image classification, which can classify 1000 categories. In order to better extract the underwater image features, the network parameters are optimized and adjusted. The underwater dataset consists of 1,000 images and is divided into five goals: sea fish, urchins, octopus, coral reefs, and jellyfish. Each image corresponds to different color attenuation and blur characteristics. In order to get a better robustness and generalization ability with VGGNet, the dataset is extended by means of scaling, panning, flipping, and color dithering. After 200000 iterations, the accuracy of underwater target classification is 90.75%.

(2) The improved VGGNet-16 model generates feature points

In the last part of the network, all layers after the last convolution layer are deleted, then a pooling layer is added. When CNN convolution layers were stacked, the ability of the features extracted from the network to express the location information was weakened. Therefore, in order to cut down the calculation cost, and we fixed the input to 224 × 224, considering the input size and receiving field comprehensively. In order to cover the multisize receiving field, pool3, pool4, and pool5-conv1 layers are used for more comprehensive feature extraction from an image.

As far as the pool3 is concerned, the output is 28 × 28 × 256, and it uses a grid with 28 × 28 to segment the image blocks. Each block corresponds to the output 256-dimensional vector. Feature descriptors are generated by 8 × 8 blocks. Their centers represent a feature point. The output is the feature mapping M_1 with the size of 28 × 28 × 256. The 16 × 16 region of the image is further generated into a pool4 descriptor. The formula for M_2 calculation is

$$M_2 = O_{pool4} \otimes I_1. \tag{4.6}$$

Here, ⊗ denotes Kronecker product. The output of pool5_1 is a 7×7×512 feature map. The formula for pool5_1 feature map M_3 is shown as follows:

$$M_3 = O_{pool5\_1} \otimes I_2. \tag{4.7}$$

Then M_1, M_2, and M_3 are normalized according to formula

$$M_{\mathrm{n\_i}} \leftarrow \frac{M_i}{\sigma(M_i)}, \quad i = 1, 2, 3. \tag{4.8}$$

Here, $\sigma(\bullet)$ denotes the standard deviation of an element in the matrix, $M_{\mathrm{n\_i}}$ means feature graph after normalization.

(3) Rough registration feature point pair

Suppose there are feature points m and n, and calculate the feature distance between them as

$$d(m, n) = \sqrt{2}d_1(m, n) + d_2(m, n) + d_3(m, n), \tag{4.9}$$

then calculate d of each component:

$$d_i(m,n) = Euclidean - d(W_i(m), W_i(n)). \tag{4.10}$$

The matching of feature points A and B is based on the following:

(a) $d(m, n)$ is the minimum data among all $d(, n)$.

(b) There is no $d(q, n)$ that makes $d(q, n) < \theta \cdot d(m, n)$ tenable, and the threshold θ is greater than 1.

When the image is deformed, there are some identical image blocks and feature points between an untransformed image and a transformed image. And the overlap rate of feature points indicates the degree of matching. In order to achieve a higher matching degree, the matching degree of two images is calculated by the selection algorithm of dynamic interior points. We reselect internal points after every k steps. In the process of rough matching, the lower value θ_0 is used to filter out irrelevant points and obtain more feature points. Then a larger θ' is set to screen out the characteristic points with overlapping blocks, that is, the correct interior points. In order to obtain more influence of feature points for the transformation, when the number of iterations reaches K, subtract the corresponding threshold value θ from step size δ and repeat this step. Strong matching feature points can be used to estimate image transformation, so that the effect of rough matching is better.

(4) An improved method based on RANSAC algorithm

There are two defects in RANSAC algorithm: one is time-consumption as the number of iterations increases rapidly with the increase of registration pairs, while the other is the low precision since the algorithm randomly samples the sample data and estimates the initial model parameters. At the same time, the least subset is selected according to the efficiency, and finally the suboptimal model parameters are obtained. To avoid these two defects, this paper improves RANSAC algorithm in the following three aspects:

(a) We delete matched point pairs. The registration rate of the matched point pair in the front bit of the database is high, because the matched point pair is stored in the database regularly according to the distance. We extract the matched point pairs arranged in the top 80%, use these data as elements of a new database, and calculate all parameters with matrix H, so as to reduce the training times of the whole process and save time.

(b) We delete the matching points for feature duplication. If the corresponding feature points of two images are connected by a straight line, there will be no intersection lines between the images. If two lines intersect, there is a mismatch. When improving the use effect of the method, a fixed threshold value is used to evaluate the intersection between the line of the matching point pair and other matching point pairs, and a matching point pair whose intersection point number is above the threshold value is deleted.

(c) We filter unnecessary feature points. When feature points of the region of interest are matched with the image background, the matching speed and accuracy will be reduced. So we do not consider the image background feature points, only the region of interest is calculated. We set a threshold value here. If the logarithm of matching points in the range of a feature point is higher than a threshold value, we then store the point as the point of interest and join the database.

Using the proposed algorithm, we obtain the matching point pair database, and the matrix H can be calculated by formula:

$$P = HP' = \begin{bmatrix} h_0 & h_1 & h_2 \\ h_3 & h_4 & h_5 \\ h_6 & h_7 & 1 \end{bmatrix} P' \tag{4.11}$$

where P means a one-dimensional vector $(x, y, 1)^{\mathrm{T}}$, P' means a one-dimensional vector $(x', y', 1)^{\mathrm{T}}$; the vector of x, y and feature matching point pairs between two images.

4.2.1.3 Underwater image fusion method based on Laplace algorithm

Based on the Laplacian pyramid algorithm, the image fusion algorithm is further studied and improved. Firstly, the image is segmented according to the frequency, and then the segments are fused. The pyramid is divided into two layers, with the top layer representing the image as a whole and the bottom layer storing the image details.

The frame-by-frame mosaicking method of a panoramic image is to match the feature points of two adjacent images, obtain the position coordinate relationship between a single image and the panoramic image through a series of matrix operations, process image by image through the relationship between the coordinates, and finally obtain the panoramic image. However, the cumulative errors in this method will affect the number of mosaic images and reduce the visual effect of panoramic images.

4.2.2 Underwater multitarget image processing

4.2.2.1 C-means clustering and its improved algorithm

Cloud-like model is a model which can deal with the meaning of languages and realize the conversion between them. During the transformation, cloud-like droplets appear randomly in the model.

First, we introduce the traditional C-means clustering method, assuming that there is a sample $\{x_1, x_2, \ldots, x_n\}$, which has c categories, P dimensions, each category has a cluster center v_i $(i = 1, 2, \ldots d)$. The classification matrix $U = [u_{il}]_{dn}$ $(0 \leq u_{il} \leq 1)$ is used to save the probability that the samples belong to each category. In fact, the process of C-means clustering is to make the function $F(U, V)$ get the minimum value and search for the optimal U and $V = [v_i]$. The formula for calculating probability by Fuzzy C-means clustering (FCM) is shown as follows:

$$\mu_i(x_l) = u_{il} = \frac{\|x_l - v_i\|^{\frac{2}{m-1}}}{\sum_{c=1}^{d} \|x_l - v_c\|^{\frac{2}{m-1}}} \tag{4.12}$$

where $\|.\|$ means a norm, m is a parameter indicating the degree of fuzziness.

C-means clustering uses a cloud-like model to represent all kinds of clusters. Expectations are at the center of each category. The cloud-like model obtains a classification matrix according to expectation, which includes mathematical expectation (E_x), calculated entropy (E_n), and hyper-entropy (H_e). Here E_x is the expectation of various water droplets in the sample space, E_n is a water drop, which can represent the fuzzy random qualitative concept, and entropy represents the dispersion degree and range of the water drop, H_e indicates the condensation and dispersion of water droplets, which is related to the fuzzy randomness of entropy.

The calculation formula of each element in the classification matrix is

$$u_{ik} = \bigwedge_{i=1}^{p} \exp\left[-\frac{(x_{ik} - E_{x_{ik}})^2}{(E_{n_{ik}})^2}\right]. \tag{4.13}$$

The objective function of CCM clustering is shown as

$$F(U, V) = \sum_i \sum_l u_{il}^m \|x_l - v_i\|^2. \tag{4.14}$$

The FCM method is based on the Lagrange multiplier method in the case of constraints. CCM method is a direct use of cloud transformation in the case of unconstrained solution. The samples used in the two clustering methods are underwater sonar images and the fuzzy random uncertainty of sonar images which can be expressed by a cloud-like model. Therefore, an improved C-means clustering method combined with cloud-like model is proposed, which can also cluster under unconstrained conditions.

4.2.2.2 Improved C-means clustering model clustering association algorithm

In order to study the effective echo in underwater sonar image, this chapter first uses the improved cloud model clustering algorithm to complete the clustering. Then, considering the applicability of the nearest neighbor method, it is used to complete the correlation work for the effective echo in the image. Finally, the clustering centers of multiple effective echoes are calculated and the data are evaluated. The complete research process is as follows:

(1) The echo number is m and the target number is t. First, calculate the state vector $\hat{A}_c(l+1\mid l)$ and measurement vector $\hat{B}_c(l+1)$ in the data, as shown in formulas (4.15) and (4.16). Then calculate the innovation covariance $D_c(l+1)$ and gain G_c in the target c $(c=1,2,\ldots t)$, as shown in formulas (4.17) and (4.19):

$$\hat{A}_c(l+1\mid l)=F_c\hat{A}_c(l\mid l), \tag{4.15}$$

$$\hat{B}_c(l+1\mid l)=H_c\hat{A}_c(l+1\mid l), \tag{4.16}$$

$$P_c(l+1\mid l)=F_cP_c(l\mid l)F_c'+Q_c(l), \tag{4.17}$$

$$D_c(l+1)=H_cP_c(l+1\mid l)H_c'+R_c(l+1), \tag{4.18}$$

$$G_c=P_c(l+1)H_c'D_c^{-1}(l+1). \tag{4.19}$$

(2) This step computes all measurements to see if they enter the tracking gate. Set the threshold value of the tracking gate as γ; if $\nu_{ic}'(l+1)D_c^{-1}(l+1)\nu_{ic}(l+1)<\gamma$, the measurement should be saved. The formula of $\nu_{ic}(l+1)$ is as follows:

$$\nu_{ic}(l+1)=B_c(l+1)-\hat{B}_c(l+1\mid l). \tag{4.20}$$

(3) By using the improved CCM clustering algorithm, the effective targets saved in the previous step are clustered and their centers $V_i(l+1)$ are calculated. The calculation method is as follows:

① Use $U(0)=|u_{ic}^0|_{t\times m_l}$ and initialize it, using

$$u_{ic}^0=e^{-\nu_{ic}'(l+1)D_c^{-1}(l+1)\nu_{ic}(l+1)}. \tag{4.21}$$

② Using $U^{(b)}$, the targets $\{B_1(l+1), B_2(l+1),\ldots, B_{mq}(l+1)\}$ are divided into t kinds. Then calculate b-dimensional model of each type of target by the nondeterministic inverse cloud-like generator.

③ The classification matrix $U^{(b+1)} = |u_{ic}^{(b+1)_i}|$ is updated according to the calculation of various cloud models, as shown below:

$$u_{ic}^{b+1} = \bigwedge_{k=1}^{p}\left\{\exp\left[-\frac{(B_{ik}(l+1) - E_{x_{ck}}^{(b)})^2}{2(E_{n_{ck}}^{(b)})^2}\right]\right\}. \tag{4.22}$$

④ The improved clustering objective function formula is shown as

$$F(U^{(b+1)}, V^{(b+1)}) = \sum_{c=1}^{t}\sum_{i=1}^{ml}(u_{ci}^{b+1})^m \| B_i(l+1) - E_{x_c}^{(b)} \| \tag{4.23}$$

where $V^{(b+1)} = |E_{x_c}^{(b)}|$ is a mean vector matrix.

⑤ Set a predetermined ε. If $|F(U^{(b+1)}, V^{(b+1)}) - F(U^{(b)}, V^{(b)})| < \varepsilon$, the clustering is completed and $V_c^{(l+1)} = E_{x_c}^{(b)}$. Otherwise, update $b = b+1$ and return to step ①.

(4) A suitable method is needed for the final measurement of target trajectory after pairing. Here, the nearest neighbor search method is introduced.

(5) The state and covariance are calculated as follows:

$$N_c(l+1) = V_c(l+1) - \hat{B}(l+1 \mid l), \tag{4.24}$$

$$\hat{A}(l+1 \mid l+1) = \hat{A}(l+1 \mid l) + G_c N_c, \tag{4.25}$$

$$\hat{P}_c(l+1 \mid l+1) = \hat{P}_c(l+1 \mid l) + G_c H_c P_c(l+1 \mid l). \tag{4.26}$$

4.3. Experimental results and analysis

4.3.1 Results and analysis of underwater sequence images stitching experiment

4.3.1.1 Underwater image enhancement experiment

In this chapter, some classical methods are chosen for comparison, including CLAHE, DCP, HE, MSRCR, IFE, and WCID. We chose to use different real underwater images to verify the results of the enhancement method. The effect of this method to improve image clarity is different from MSRCR, HE, DCP methods. The HE method deviates from the normal color due to some color distortions. The test result of the DCP and CLAHE methods is attenuation, so these two methods can be used to enhance underwater blurred images. The fogging caused the image details using the MSRCR enhancement method is unclear. The enhancement results of IFE and WCID methods obviously improve the problems of sharpness and color attenuation, but some image details still have problems. It can be seen from the enhancement results of the method proposed in this chapter that, compared with other enhancement methods, this method has higher contrast, clearer details, and less noise.

We compare the enhancement results of 20 underwater images using 7 evaluation methods, including average, standard, information entropy, blur, average gradient, UCIQE, and UIQM. It can be seen that this method has the highest average value, information entropy, average gradient, UCIQE and UIQM values, but the standard result is not as good as that of HE, and the blur value is the smallest. Therefore, the proposed method has better, clearer, enhanced processing effects and can display more details.

4.3.1.2 Results and analysis of underwater image registration

Figs. 4.2–4.4 show the results of underwater image registration based on SIFT, SURF, and our proposed method, respectively.

The SURF method takes less time and has better real-time performance, but can only generate about half the number of feature points of the SIFT method. The method proposed in this chapter extracts fewer feature points and takes less time than the SIFT method. In addition, this method can extract more feature points than SURF method in less time. In order to improve the accuracy of coarse matching feature points and generate more coarse matching point pairs, this method increases the selection of dynamic internal points in the coarse matching stage. In Figs. 4.2–4.4, in the coarse matching stage, the proposed registration method is more effective than the other two registration methods. There are fewer mismatched point pairs, and more accurate matching point pairs can be retained during the exact matching phase.

In Fig. 4.4C, the conventional RANSAC method generates 54 pairs of exact matching point pairs with two pairs of mismatch points. Using the improved RANSAC method, 76 pairs of exact matching pairs can be accurately obtained without errors. The improved method has better performance than the old algorithm.

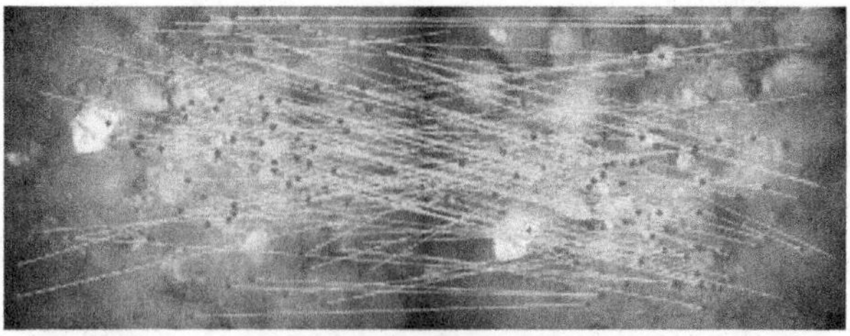

(A) Roughly matched pairs

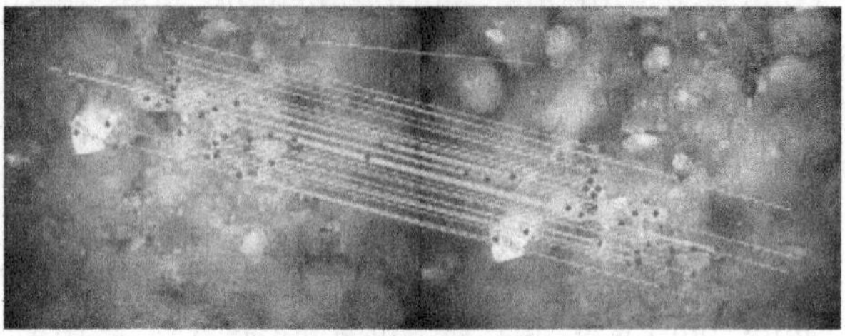

(B) Accurately matched pairs

Figure 4.2 Matching result of SIFT method.

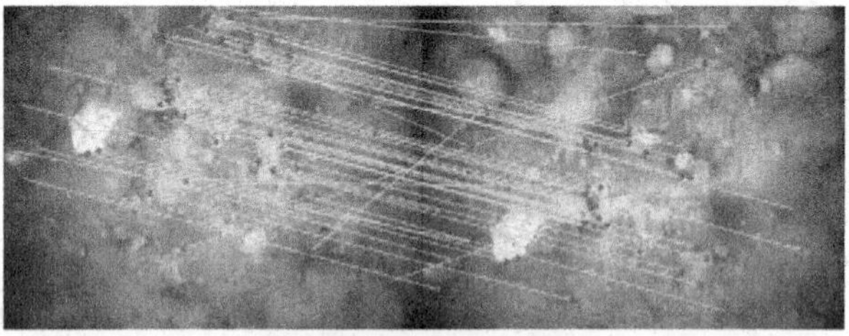

(A) Roughly matched pairs

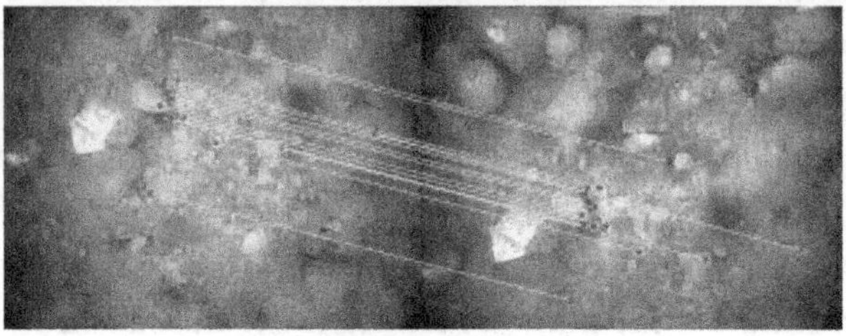

(B) Accurately matched pairs

Figure 4.3 Matching result of SURF method.

4.3.1.3 Results and analysis of underwater panoramic image stitching

In order to reflect the actual effect of the proposed method, some underwater sequence images taken by an AUV under typical underwater environ-

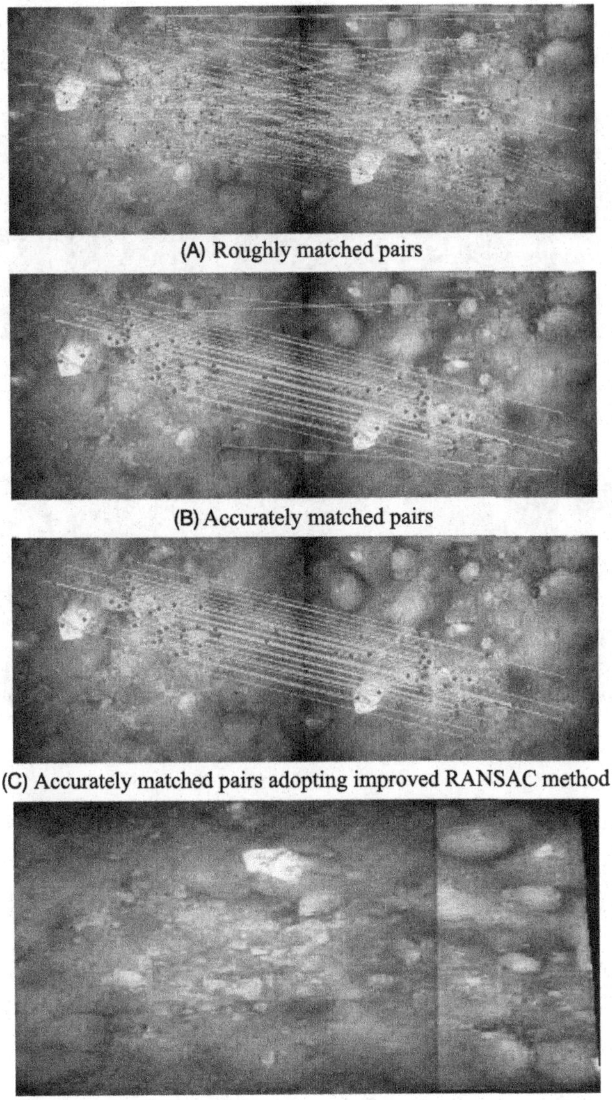

(A) Roughly matched pairs

(B) Accurately matched pairs

(C) Accurately matched pairs adopting improved RANSAC method

(D) Proposed method results

Figure 4.4 Matching result of the proposed method.

ment were chosen to generate a wide-field underwater panoramic image. Fig. 4.5 shows the five sequence images of a lake bottom. Figs. 4.6 shows the results of the CNN-based algorithm using the above 5 underwater images. It shows that our proposed image enhancement method can significantly improve the brightness and clarity, highlight detailed information, and can eliminate the effect of blue-green hue caused by absorbed light.

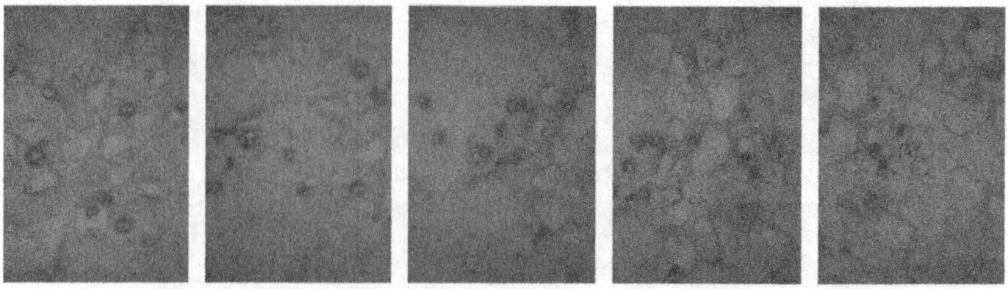

Figure 4.5 Underwater bottom experiment image.

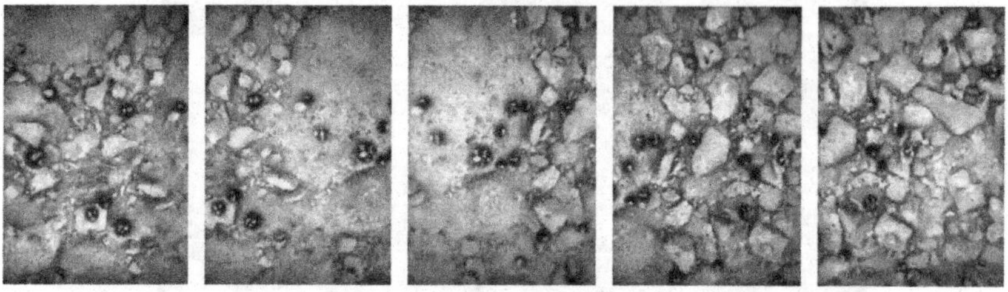

Figure 4.6 Underwater bottom image enhancement.

The underwater image stitching results using the CNN-RANSAC registration method and fusion method proposed in this paper are shown in Fig. 4.7, and the results of the image fusion method for the panoramic images of the underwater bottom are shown in Fig. 4.8. It shows that the Laplacian pyramid image fusion method effectively eliminates the staged changes in illumination and the gaps between the images, making the mosaic area smoother.

4.3.2 Result and analysis of the experiment of multiple target tracking

4.3.2.1 Experimental results and analysis of non-cross-movement

The above three data association algorithms track the trajectory of two non-crossing moving targets as shown in Figs. 4.9–4.11. When target 1 moves from left to right, and target 2 moves from top to bottom, we compare the actual position obtained by these three algorithms and the previously predicted position.

Figure 4.7 Image stitching results without image fusion.

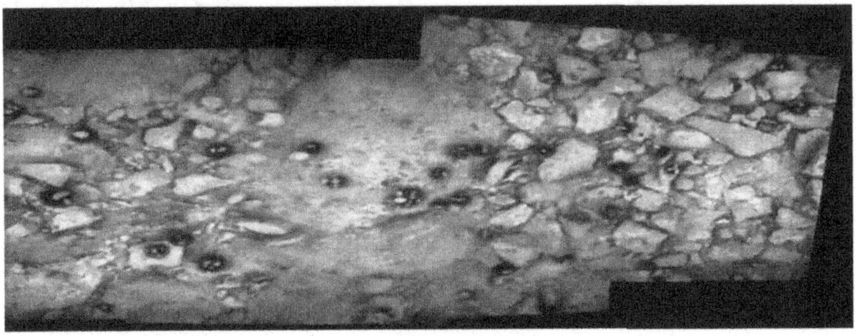

Figure 4.8 Panoramic image using image fusion method.

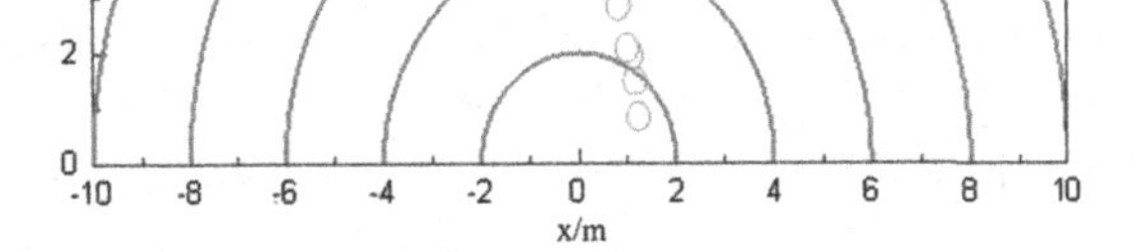

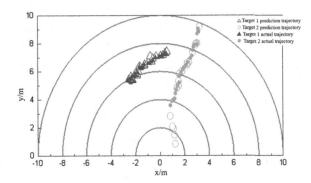

Figure 4.9 NNDA tracking graph of two non-crossing moving targets.

Comparing the actual and predicted graphs of multitarget tracking results obtained by three different algorithms in Figs. 4.12–4.13, it can be

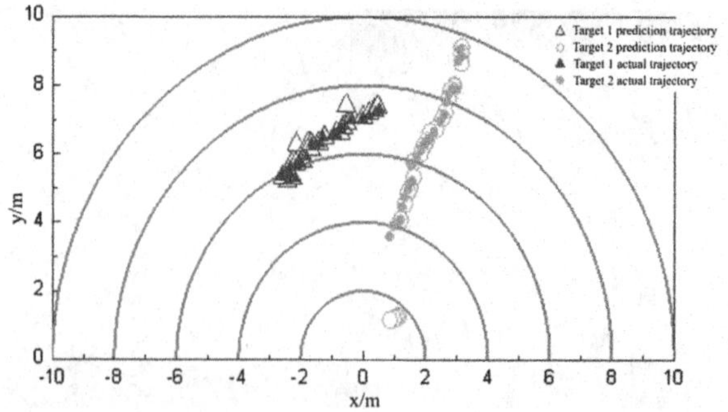

Figure 4.10 JPDA tracking graph of two non-crossing moving targets.

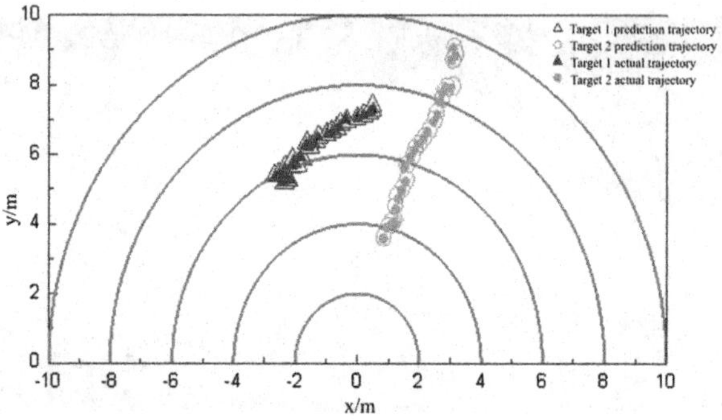

Figure 4.11 CCM tracking graph of two non-crossing moving targets.

found that when targets 1 and 2 do not cross, these three algorithms can roughly complete the tracking of the target. When tracking target 1, the CCM algorithm is the most stable among several algorithms, with the smallest tracking error. The JPDA algorithm has error with actual trajectory of the tracked target in some area, while the NNDA algorithm also has a slight problem in some area. CCM algorithm can accurately track the positions of two targets throughout the experiment.

A comparison in Figs. 4.12–4.13 shows that when the two targets are in non-crossover motion, the three algorithms can complete the tracking of target 1, and the CCM algorithm is more stable and the tracking error is smallest. For target 2 tracking, in the $x \in [0.8, 1.5]$ interval, NNDA and PDA tracking target 2 deviated from the real track of target 2 actual trajectory, but CCM algorithm could track the target more accurately in the whole tracking period, with good tracking performance.

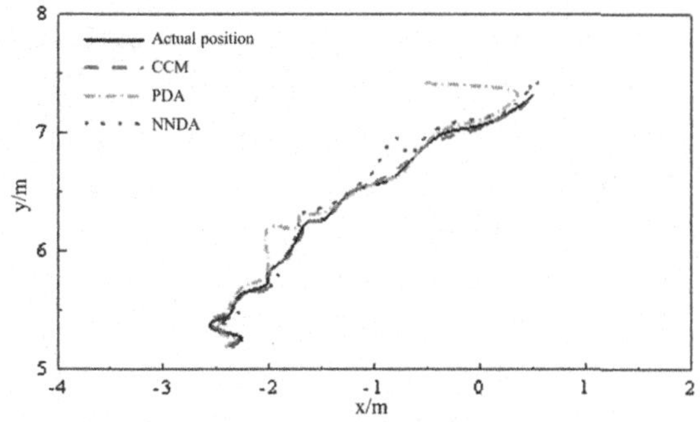

Figure 4.12 Target 1 actual and predicted trajectory.

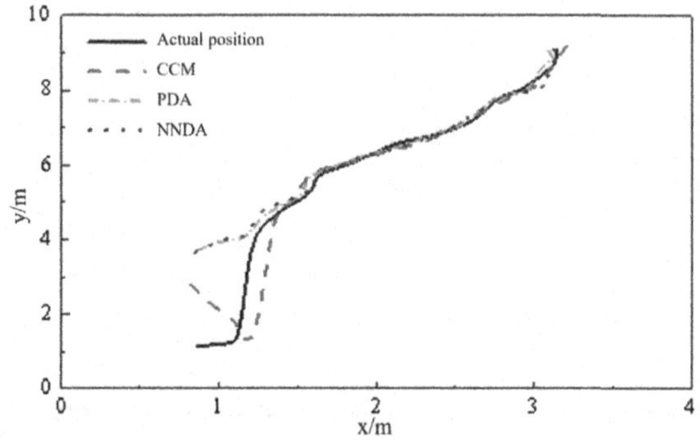

Figure 4.13 Target 2 actual and predicted trajectory.

4.3.2.2 Experimental results and analysis of cross-movement

The sonar data of crossover movement of two targets were processed by the above three data association algorithms, and the track tracking trace diagram is shown in Figs. 4.14–4.16. The comparison results of predicted position and actual position of target 1 and target 2 obtained by the three algorithms are shown in Figs. 4.17–4.18. The motion track of target 1 is from left to right, and that of target 2 is from top right to bottom left, approaching target 1 slowly. Compared with the three track tracking graphs in Figs. 4.14–4.16 and the actual position, when the two targets are moving without crossing, the three algorithms can basically complete the tracking of target 1, but the CCM algorithm is more stable and the tracking error is smaller. Target 2 enters from the top right and intersects with target 1 in the interior.

As shown in Figs. 4.17–4.18, NNDA algorithm loses track of target 2 due to error association near the intersection area; PDA algorithm mistak-

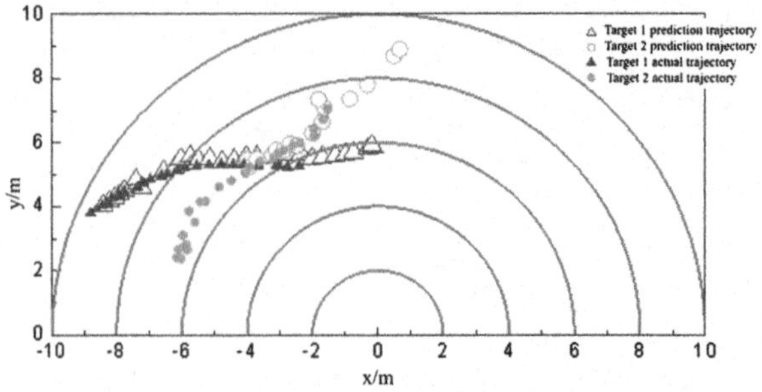

Figure 4.14 Cross-moving NNDA tracking diagram of multiple targets.

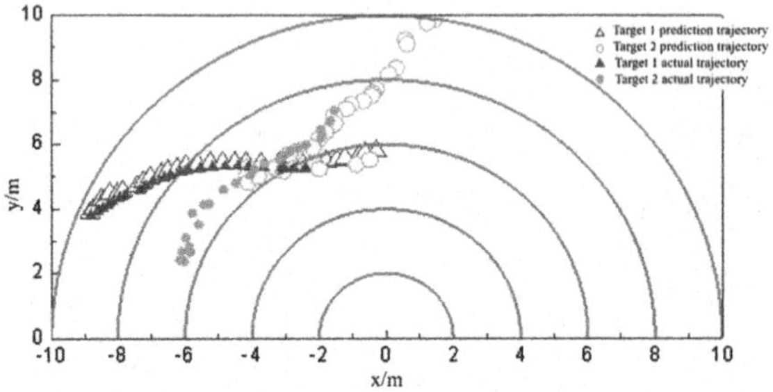

Figure 4.15 Cross-moving JPDA tracking diagram of multiple targets.

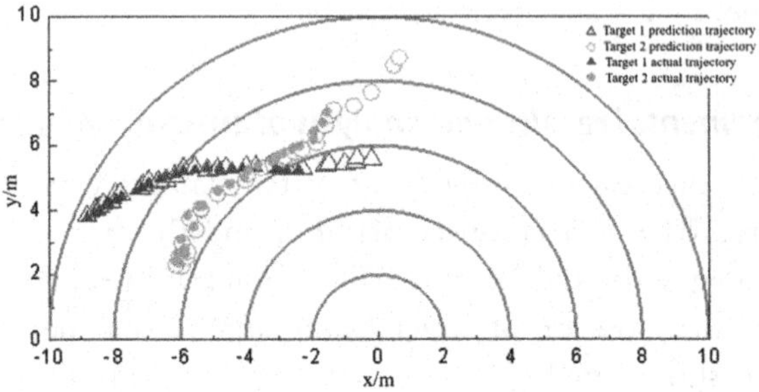

Figure 4.16 Cross-moving CCM tracking diagram of multiple targets.

enly tracks target 1 as target 2 due to error association near the intersection area; through comparison, CCM algorithm has good correlation performance in this region, it can correctly distinguish the target to track, and its tracking effect is good. For target 2 tracking, in the $x \in [0.8, 1.5]$ interval, NNDA and PDA tracking target 2 deviated from the real track of the tar-

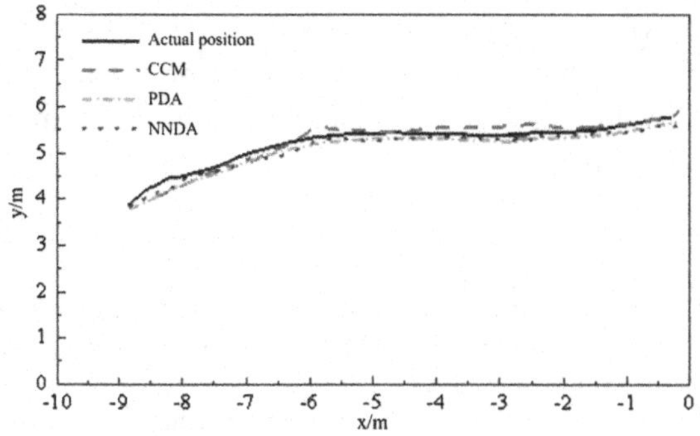

Figure 4.17 Target 1 trajectory comparison.

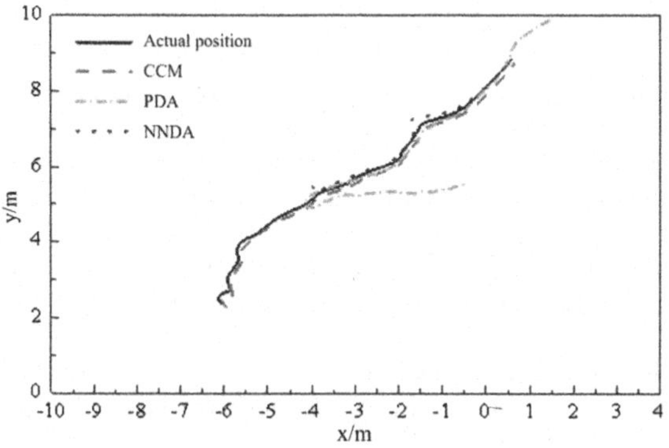

Figure 4.18 Target 2 trajectory comparison.

get. However, CCM algorithm could track the target more accurately in the whole interval, with good tracking performance.

4.4. Conclusions

In this chapter, we proposed solutions for underwater perception of environment and object. A joint framework for underwater imagery mosaicking was proposed for AUV perception of underwater wider visual range. Different underwater scenarios experiments have demonstrated the effective image enhancement, feature points generation, and mosaicking performance. We presented a method for underwater multiple target tracking for environment perception using underwater acoustic image. This method is based on the clustering cloud-like model (CCM). The experimental results demonstrated that the association failure of NNDA and JPDA is caused by the wrong association with other targets near the intersection area in the crossover case. On the contrary, the proposed CCM algorithm presented herein could produce correct association results in both crossover and non-crossover situations.

References

[1] D.X. Ji, H.C. Li, C.W. Chen, et al., Visual detection and feature recognition of underwater target using a novel model-based method, International Journal of Advanced Robotic Systems 15 (6) (2018) 1–11.

[2] X. Liang, X.R. Qu, N. Wang, et al., Swarm control with collision avoidance for multiple underactuated surface vehicles, Ocean Engineering 191 (106516) (2019) 1–10.

[3] H.D. Qin, X. Yu, Z.B. Zhu, et al., An expectation-maximization based single-beacon underwater navigation method with unknown ESV, Neurocomputing (2019), https://doi.org/10.1016/j.neucom.2019.10.066.

[4] X. Liang, X.R. Qu, N. Wang, et al., A novel distributed and self-organized swarm control framework for underactuated unmanned marine vehicles, IEEE Access 7 (2019) 112703–112712.

[5] H.D. Qin, H. Chen, Y.C. Sun, et al., Distributed finite-time fault-tolerant containment control for multiple ocean Bottom Flying node systems with error constraints, Ocean Engineering (2019), https://doi.org/10.1016/j.oceaneng.2019.106341.

[6] X. Yang, Z.Y. Liu, H. Qiao, et al., Underwater image matching by incorporating structural constraints, International Journal of Advanced Robotic Systems 14 (6) (2017) 1–10.

[7] J. Jung, J. Choi, Y. Lee, et al., AUV localization using depth perception of underwater structures from a monocular camera, in: OCEANS 2016 MTS/IEEE Monterey, Monterey, USA, 19–23 September, 2016.

[8] J. Li, M. Lee, J. Kim, et al., Development of P-SURO II hybrid AUV and its experimental study, in: OCEANS 2013 MTS/IEEE Bergen, Bergen, Norway, 10–13 June, 2013.

[9] A. Khan, S. Ali, F. Meriaudeau, et al., Visual feedback-based heading control of autonomous underwater vehicle for pipeline corrosion inspection, International Journal of Advanced Robotic Systems 14 (3) (2017) 1–5.

[10] F. Hover, R. Eustice, A. Kim, et al., Advanced perception navigation and planning for autonomous in-water ship hull inspection, International Journal of Robotics Research 31 (12) (2012) 1445–1464.

[11] S. Yokota, K. Kim, M. Imasato, Development and sea trial of an Autonomous Underwater Vehicle equipped with a sub-bottom profiler for surveying mineral resources, in: IEEE/OES Autonomous Underwater Vehicles (AUV), Tokyo, Japan, 6–9 November, 2016.

[12] X. Xiang, G. Yu, L. Lapierre, Survey on fuzzy-logic-based guidance and control of marine surface vehicles and underwater vehicles, International Journal of Fuzzy Systems 20 (2018) 572–586.

[13] A. Guillaume, M. Hua, K. Szymon, Pipeline following by visual servoing for Autonomous Underwater Vehicles, Control Engineering Practice 82 (2019) 151–160.

[14] X. Xiang, L. Lapierre, Robust magnetic tracking of subsea cable by AUV in the presence of sensor noise and ocean currents, IEEE Journal of Oceanic Engineering 43 (2018) 311–322.

[15] R.S. Liu, X. Fan, M. Zhu, et al., Real-world underwater enhancement: challenges, benchmarks, and solutions, IEEE Transactions on Image Processing (2019) 1–12.
[16] A. Galdran, Alvarez-G, Automatic red channel underwater image restoration, Journal of Visual Communication and Image Representation 26 (2015) 132–145.
[17] Y. Peng, P. Cosman, Underwater image restoration based on image blurriness and light absorption, IEEE Transactions on Image Processing 26 (4) (2017) 1579–1594.
[18] M. Sheng, S. Tang, Z. Cui, et al., A joint framework for underwater sequence images stitching based on deep neural network convolutional neural network, International Journal of Advanced Robotic Systems 17 (2) (2020) 1729881420915062.
[19] H.Y. Li, J. Luo, C.J. Huang, An adaptive image-stitching algorithm for an underwater monitoring system, International Journal of Advanced Robotic Systems 11 (2014) 1–8.
[20] M. Sheng, S. Tang, H. Qin, et al., Clustering cloud-like model-based targets underwater tracking for AUVs, Sensors 19 (2) (2019) 370.
[21] J.X. Ruan, L.Y. Xie, Y.Y. Ruan, Image stitching algorithm based on SURF and wavelet transform, in: 7th International Conference on Digital Home (ICDH), 30 November – 1 December, 2018.
[22] S. Liu, X. Liang, L. Liu, et al., Matching-CNN meets KNN: quasi-parametric human parsing, in: IEEE Conference on Computer Vision and Pattern Recognition, Boston, USA, 7–12 June, 2015.
[23] C. Chang, Q. Du, Interference and noise-adjusted principal components analysis, IEEE Transactions on Geoscience and Remote Sensing 37 (5) (1999) 2387–2396, https://doi.org/10.1109/36.789637.
[24] K. Zhang, H. Lin, G. Liu, Improved interacting multiple model-new nearest neighbor data association algorithm, in: 2018 IEEE 4th International Conference on Control Science and Systems Engineering, Wuhan, China, 21–23 August, 2018, pp. 499–504.
[25] Q. Chen, Y. Yan, Y. Dai, Joint nearest neighbor data association based on interacting multiple model Kalman filtering, in: Proceedings of 2016 2nd IEEE International Conference on Computer and Communications, Chengdu, China, 14–17 October, 2016, IEEE, New York, NY, USA, 2016, pp. 75–79.
[26] D. Clark, Multiple target tracking with the probability hypothesis density filter, PhD Thesis, Heriot-Watt University, 2006.
[27] J. Luom, Data association for AUV localization and map building, in: 2010 International Conference on Measuring Technology and Mechatronics Automation, Changsha, China, 13–14 March, 2010, pp. 2231–2240.
[28] Y. Haanju, K. Kikyung, B. Moonsub, Online scheme for multiple camera multiple target tracking based on multiple hypothesis tracking, IEEE Transactions on Circuits and Systems for Video Technology 27 (2017) 454–469.
[29] L. Zhang, Y. Li, R. Nevatia, Global data association for multi-object tracking using network flows, in: Global Data Association for Multi-Object Tracking Using Network Flows, 2008, pp. 31–38.
[30] A. Pezeshki, M. Azimi-Sadjadi, L. Scharf, et al., A canonical correlation-based feature extraction method for underwater target classification, Oceans 1 (2002) 29–37, https://doi.org/10.1109/OCEANS.2002.1193244.
[31] Q. Huynh, L. Cooper, N. Intrator, et al., Classification of underwater mammals using feature extraction based on time-frequency analysis and BCM theory, IEEE Transactions on Signal Processing 46 (5) (1997) 1202–1207, https://doi.org/10.1109/TFSA.1996.547456.

CHAPTER FIVE

Autonomous control of underwater offshore vehicles

Hongde Qin, Yanchao Sun
College of Shipbuilding Engineering, Harbin Engineering University, Harbin, China

Contents

Fundamental Design and Automation Technologies in Offshore Robotics
https://doi.org/10.1016/B978-0-12-820271-5.00010-9

Abstract

Motion control for underwater offshore vehicles have been an active research field since there have been many multiple autonomous underwater vehicles (AUVs) and remotely operated vehicles (ROVs) working in the offshore regions. Modern underwater vehicle control systems are based on a variety of design methods such as PID control, prescribed performance control, nonlinear neural networks control theory and so on. In Section 5.1 of the chapter, dynamic models for underwater vehicle are presented. Many presented design methods have been successfully implemented and tested on AUVs and ROVs. Therefore, Section 5.2 presents the PID control, observer method, and prescribed performance control methods for trajectory-tracking and of underwater vehicle, which includes brief introduction and algorithm design. Moreover, algorithm extension and case study are discussed in Section 5.3.

Keywords

Autonomous underwater vehicle, Trajectory tracking control, Prescribed performance, Disturbance observer, Neural network

5.1. Introduction

5.1.1 Preface

In recent years, autonomous underwater vehicle (AUV) has been widely applied in the fields of marine environmental monitoring and military intelligence gathering [1–3]. It is intended to provide scientists and researchers with simple, low-cost, medium, and long-range, appropriate time response capability to implement different functions. With the expansion of the ocean development scale, the applications of the AUVs gradually extend from observation to operation, such as underwater infrastructure inspection, survey on underwater animals and plants, operations in dangerous waters, photometric survey, pipeline route survey, seabed mapping, environmental monitoring, chemical plume tracing, and gas exploration [4–7]. These underwater vehicles are required to execute different types of missions without the interaction of human operators while performing well under a variety of load conditions and with unknown sea currents. Motion control systems for AUV have been an active field of research. Modern control systems are based on a variety of design techniques such as PID control, sliding mode control, fuzzy systems, neural networks control, and prescribed performance control, to mention only a few.

An AUV motion control system is usually constructed as two independent blocks denoted as the navigation and control systems. These systems interact with each other through data and signal transmission. The navigation and control systems makes use of the estimated alternatively measured positions and velocities. This is referred to as a closed-loop guidance system. To the contrary, a guidance system that only uses reference feedforward is an open-loop guidance system. The motion control for AUV is the action of determining the necessary control forces and moments to be provided by the vehicle in order to satisfy a certain control objective. The desired control objective is usually seen in conjunction with the guidance system. Examples of control objectives are minimum energy, path following, and trajectory tracking. Constructing the control algorithm involves the design of feedback and feedforward control laws. The outputs from the navigation system, position, velocity, and acceleration are used for feedback control while feedforward control is implemented using signals available in the guidance system and other external sensors [8].

The position and velocity of the AUV should track desired time-varying position and velocity reference signals. The corresponding feedback controller is a trajectory tracking controller. Tracking control can be used for course-changing maneuvers, speed-changing, and attitude control. An advanced guidance system computes optimal time-varying trajectories from a dynamic model for a predefined control objective. If a constant set-point is used as input to a reference model in an open-loop guidance system, the outputs of the filter will be smooth time-varying reference trajectories for position, velocity, and acceleration.

Motion control of AUVs may have different control objectives or strategies such as trajectory tracking, path following, and way-point tracking. The trajectory tracking refers to the design of control laws so that the vehicle tracks a desired and time-parametrized trajectory while the path following requires the path to be independent of time and is expressed in terms of its geometric description. For the way-point tracking control problem, it is required to guide the AUV through a series of way points between the vehicle's starting position and a desired final position.

The subsequent contents in this chapter are as follows: Section 5.1.2 describes the control difficulties for underwater vehicles. Section 5.1.3 constructs the dynamic model of fully actuated AUV. Several common AUV control methods and some simulation cases are described in Section 5.2.

5.1.2 Description of control difficulties for underwater vehicles

Trajectory tracking is a fundamental element of the AUV control system. However, highly nonlinear and cross-coupled characteristics of system dynamics, model uncertainties introduced by unpredictable underwater environment, and external disturbances bring challenges for the AUV control algorithms design. In addition, according to the complicated missions, the requirement for the control precision will be further increased.

Ocean currents are horizontal and vertical circulation systems of ocean waters produced by gravity, wind friction, and water density variation in different parts of the ocean. Besides wind-generated currents, the heat exchange at the sea surface, together with salinity changes, develop an additional sea current component, usually referred to as thermohaline currents. Since the Earth is rotating, the Coriolis force will try to turn the major currents to the east in the northern hemisphere and west in the southern hemisphere. Finally, the major ocean circulations will also have a tidal component arising from planetary interactions like gravity. In coastal regions and bays, tidal components can reach very high speeds. In fact, speeds of 3 m/s or more have been measured [8]. Therefore, ocean current which is a common disturbance in marine environment affects the control precision. In order to simulate ocean currents and their effect on AUV motion, the model can be expressed in Section 5.1.3.

Additionally, the parameters which depend on the experimental data or the computational fluid dynamics (CFD) commercial software contain modeling errors. Therefore, it is more suitable to consider model uncertainties for the control laws design, which has actual engineering value. Because of the complexity of the maritime environment, the AUV may break down in operation process while the thruster fault is representative. Since this study investigates the trajectory tracking control for the AUVs, the thruster fault case is worth considering.

5.1.3 Models for underwater vehicles

This chapter presents a hydrodynamic model for AUV. The foundations, including kinematic equations, rigid-body kinetics, hydrostatics, seakeeping theory, and maneuvering theory, are not repeated. A 6-degree-of-freedom (DOF) model is usually implemented in a computer to describe all dynamics effects as accurately as possible. This is referred to as the simulation model. The simulation model should be able to reconstruct the time responses of the physical system. Model-based controllers and observers, however, can be designed using reduced-order or simplified models.

When designing motion control systems for AUV, it is important to distinguish between underactuated AUV and fully actuated AUV. For an underwater vehicle, DOF is the set of independent displacements and rotations that completely specify the displaced position and orientation of the vehicle. A vehicle that can move freely in the 3D space has a maximum of 6 DOFs, that is, three translational and three rotational components. For underwater vehicle with actuation in all DOFs, such as fully actuated AUV and redundantly actuated AUV, a model-based controller and observer design requires a 6 DOF model, while underactuated AUV control systems can be designed using a 3 or 4 DOF model.

5.1.3.1 Basic model

We introduce the 6 DOF AUV model with disturbances to represent dynamic model of the AUV, as shown below [8]:

$$M_\eta \ddot{\eta} + C_{RB\eta}\dot{\eta} + C_{A\eta}\dot{\eta}_r + g_\eta = \tau \tag{5.1}$$

where $M_\eta = MJ^{-1}$, $C_{RB\eta} = \left[C_{RB}(\nu) - MJ^{-1}\dot{J}\right]J^{-1}$, $C_{A\eta} = C_A(\nu_r)J^{-1}$, $D_\eta = D(\nu_r)J^{-1}$, $g_\eta = g(\eta)$, $\dot{\eta}_r = J(\eta)\nu_r$, and $\nu_r = \nu - \nu_c$; M is the inertial matrix including the added mass, $\eta = [x, y, z, \phi, \theta, \psi]^\mathrm{T}$ is the position and orientation vector in the earth-fixed frame, $\nu = [u, v, w, p, q, r]^\mathrm{T}$ is the velocity vector in the body-fixed frame, J is the transformation matrix between the earth-fixed frame and the body-fixed frame, C_{RB} is the rigid-body Coriolis and centripetal matrix, C_A is the added Coriolis and centripetal matrix, D is the drag coefficient matrix, g_η is the vector of gravitational/buoyancy forces and moments, and τ is the control forces and moments acting on the AUV; ν_r is the vehicle velocity relative to ocean current, and ν_c is the ocean current velocity in the body-fixed frame.

5.1.3.2 Dynamic model including thruster faults

The thruster, an important part of the AUV, is the main resource of faults [9]. The variation of thrust allocation matrix is adopted to represent the effect of thruster fault, denoted as ΔB. Therefore, the real control force/moment is changed to $\tau + \Delta\tau$ [10],

$$\tau + \Delta\tau = (B_0 - K_1 B)u = (B_0 + \Delta B)u, \tag{5.2}$$

where B_0 is the nominal value of the thrust allocation matrix, u is the control input of thrusters, and K_1 is a diagonal matrix while the element $k_{ii} \in [0, 1]$, which represents the level of the corresponding thruster fault.

5.2. Advanced motion control methods

5.2.1 A review of AUV control methods

From a general view, the adaptive control is a type of nonlinear control using a system with uncertainty or time-varying parameters. It is implemented on plants with a definite structure with unspecified fixed or slowly varying parameters. Adaptive method is useful for AUVs because of variation of real model parameters. The controller can adapt itself according to the level or characteristics of waves and currents or to the changing weight of AUV. Also, neural network has some weak points which bind its improvement. It converges to a precise model with long training time and slow rate, which is not acceptable by many systems. Also, classical neural network does not qualify the main requirements such as fast response. SMC is an earlier method that is a good solution for nonlinear system but it can cause chattering on actuators, waste energy, and make fault on fins. However, there are some methods like combination with fuzzy or changing the sign function by saturation function to reduce chattering.

Modeling uncertainties and ocean current disturbances are common influence factors in trajectory tracking control for AUVs. Observers and neural networks are often used to deal with external disturbances [11,12]. Chen et al. [13] proposed an under-actuated AUV robust control scheme with modeling uncertainties and environmental disturbances. This method adopted an adaptive fuzzy control algorithm and a sliding-mode control approach to compensate for modeling uncertainties and disturbances, respectively. Elmokadem et al. [14] developed a robust control scheme based on terminal sliding-mode control methods to overcome the influence of uncertainties and disturbances. Xu et al. [15] used radial basis function neural networks to approximate the nonlinear uncertainties and enhance the robustness of the AUV against the uncertainties and disturbances. Joe et al. [16] designed a second-order sliding-mode control strategy which comprised an equivalent controller and a switching controller. The control algorithm could suppress the parameter uncertainties and eliminate the unpredictable disturbance effects caused by ocean currents. Zhang et al. [17] proposed an adaptive output feedback control approach and introduced an observer to reconstruct the full states. This method could make AUV track desired target in external disturbances. Lakhekar and Waghmare [18] developed an adaptive fuzzy PI sliding-mode control strategy based on approximately known inverse dynamic model output while the continuous adaptive PI term overcame the influence of disturbances and uncertainties.

The above references contained many trajectory tracking control strategies to deal with the uncertainties and disturbances and obtained some good effects. However, these control strategies were designed under fault-free assumptions. Owing to the complexity of underwater environment, faults may occur in several components of the AUV, especially in thruster. Since the AUVs are applied to large-scale deployment, thruster faults are worth considering. Liu et al. [19] proposed a fault isolation issue for the redundant thrusters. This approach got rid of some fault-free terms from the given control input equations, and adopted consistency check to achieve control task. Wang et al. [20] developed a novel fault detection observer with a nonsingular structure. Ismail et al. [21] proposed a control technique based on fault-tolerant decomposition for thruster force allocation to deal with thruster fault for redundant-thruster AUV system. Sun et al. [22] developed an improved Elman neural network which had stronger identification ability when applied to the AUV. This strategy calculated and analyzed the residual by comparing the model output with the actual measured values based on fault judging criteria to obtain fault diagnosis results. Chen and Shan [23] proposed a distributed fault-tolerant controller with the feedback of the information of rigid bodies only based on the sign function. Hai et al. [24] combined infinity-norm optimization with 2-norm optimization for the optimal allocation of thrust to construct a bi-criteria primal–dual neural network fault-tolerant control method, and enhanced the robustness with respect to nonlinear characteristics for ROV. It is obvious that design ideas of above references were to design the fault diagnosis schemes separately. Zhang et al. [25] proposed an adaptive terminal sliding-mode fault-tolerant control technique, and introduced the adaptive strategy to estimate the upper bounds of the system general uncertainties which included uncertainties, disturbances, and thruster faults. Therefore, this strategy could handle the thrust faults more flexibly so that it is more suitable to be utilized when designing the fault-tolerant control schemes for the AUVs.

The references mentioned gave the corresponding methods to deal with disturbances, uncertainties, as well as thruster faults, and made the systems have certain stability and robustness. However, according to the AUV special work requests which include large-scale deployment, high accuracy tracking, and landing on the seabed, we should not only consider the influence caused by above factors but also make trajectory tracking system have desired performance. Additionally, the overshoots also need to be limited to avoid hitting each other or other objects. In 2008, Bechlioulis proposed a prescribed performance control method, which introduced performance function and corresponding error transformation to make convergence rate, overshoot, and tracking error to obtain preestablished performances [26]. This algorithm was initially used to general nonlinear systems research. The prescribed performance method gradually extends to many other fields in recent years, such as chaotic system [27,28], spacecraft system [29,30], electro hydraulic system [31], and marine system [32]. According to the high performance of prescribed performance approach, we can introduce it into trajectory tracking control system of the AUV.

The thruster saturation is a common problem which influences control effects in actual AUV systems. If the thruster saturation is ignored, it would bring adverse effects, such as reducing the system tracking precision or making the system instability. Wu et al. [33] designed an L^1 adaptive control architecture with antiwindup to guarantee the robustness of the AUV with input saturation. Sarhadi et al. [34] proposed an adaptive control approach which combined a model reference adaptive algorithm, an integral state feedback, and a modern antiwindup compensator to accomplish adaptive autopilot in the presence of input saturations. The same authors [35] developed an adaptive PID control method with a dynamic antiwindup compensator for AUVs to improve the quality of the adaptive controller when the saturation occurred. Rezazadegan et al. [36] proposed an adaptive control strategy based on Lyapunov theory and the backstepping technique. This algorithm used saturation functions to bound control signals, and designed another adaptive strategy to deal with actuator saturation.

5.2.2 Common method

5.2.2.1 PID control

This section discusses Proportional Integral Derivative (PID) control design for AUV motion control systems. The PID controller, which has been used for half a century, is widely used in metallurgical, mechanical, chemical, thermal, light industrial, electrochemical, and other industrial process control. PID control is by far the most general control method since its control structure is simple, easy to operate and adjust, and has certain robustness characteristics. In recent years, great efforts have been made to develop and disseminate it as a major and reliable technical project and tool for industrial process control. In PID control, a crucial problem is the tuning of PID parameters (proportional coefficient, integral time, and differential time). The typical tuning method of PID parameters is to determine the PID parameters according to a whole set principle on the basis of acquiring the mathematical model of the object.

The quality of parameter setting affects not only the control quality, but also the stability and robustness of the control system. Due to the complexity, diversity, variability, and uncertainty of modern industrial process, it is possible to cause the change of model parameters and model structure. Therefore, the system cannot work under the original set condition, deviating from the control performance index.

Proportional-derivative (PD) and PID controllers are the most commonly used techniques to control the position and orientation of commercial underwater vehicles, this is due to their design simplicity and their good performance, especially when some system parameters are unknown.

However, it is well-known that when the plant's dynamics is highly nonlinear, time-varying, or with significant time delays, the PID controls performance is often degraded. The impact of these drawbacks can be reduced by using adaptive, saturated, or nonlinear PD/PID strategies. Inspired by this problem, several advanced PD/PID control schemes for underwater vehicles have been proposed in previous literature and some of them are summarized below.

The basic PID controller can be expressed as

$$u(t) = K_{\mathrm{p}}e(t) + K_{\mathrm{i}}\int_0^t e(t)dt + K_{\mathrm{d}}\frac{de(t)}{dt} \tag{5.3}$$

where the proportional term $e(t)$ represents the current error of the control system. Increasing the proportional coefficient K_{p} can make the system more responsive, speed up the adjustment, and reduce the steady-state error. However, a too large proportional coefficient will lead to many problems, such as the overshoot increases, the number of vibrations increases, the adjustment time increases, and the dynamic performance is bad. If the proportional coefficient K_{p} is too large, even the closed-loop system will be unstable. Proportional control is difficult to ensure proper adjustment and complete elimination of errors. The integral term K_{i} represents the accumulated error of the control system. When the control error exists, the integral adjustment continues until the system error reaches zero. When the system is in a stable state, the error is always zero, the proportional and differential parts are both zero, and the integral part is no longer changing, which is exactly equal to the output value of the controller required in the steady state. Therefore, the integral term K_{i} can be used to eliminate the steady-state error and improve the control accuracy. The differential term

$\frac{de(t)}{dt}$ represents the rate of change of system error $e(t) - e(t-1)$, which is predictive and can predict the trend of deviation change. Before the error is formed, it is eliminated by differential adjustment. Therefore, the differential coefficient K_d can improve the dynamic performance of the system. However, the differential coefficient K_d amplifies the noise interference. Because the differential of the error is the change rate of the error, the faster the error changes, the bigger the absolute value of the differential, so strengthening the differential is adverse to the system antiinterference.

5.2.2.2 Observer technique

The research on observer theory started in the 1960s, and it can be divided into linear and nonlinear systems. In the past few decades, the observer-theoretical research on these two kinds of system has achieved abundant results. The observer can be regarded as a closed-loop information reformer based on the system model and measurement information, which can be the input or output variable of the system. The observer design problem is to reconstruct a new system by taking the directly measurable information in the original system as the input signal and to make the output signal of the new system converge to the state of the original system under certain conditions. The new system is called the observer of the original system.

Observers can be divided into state and functional observers. The state output of the state observer is asymptotically equivalent to the state of the original system, while the state output of the functional observer is asymptotically equivalent to a function of the state of the original system. The dimension of state observer is generally higher than that of function observer. The state observer can also be divided into full- and reduced-dimensional observer. The former has the dimensionality of the original system, while the latter is of dimensionality less than that of the original system. The reduced-dimension observer is relatively simple in structure, while the full-dimension observer has advantages in noise suppression.

In 1964, Luenberger proposed the design method of a linear system observer based on the traditional control principle, and obtained the so-called Luenberger observer, which is not only the main method of linear system observer design, but also has important guiding significance for nonlinear system observer design. Therefore, this section will introduce the design idea and method of an observer selectively, which will pave the way for the design of AUV advanced controller.

Consider an nth-order continuous-time linear time-invariant system

$$\begin{cases} \dot{x} = Ax(t) + Bu(t), & x(0) = x_0, \ t \geq 0 \\ y(t) = Cx(t) \end{cases} \tag{5.4}$$

where $A \in R^{n\times n}$, $B \in R^{n\times p}$, $C \in R^{q\times n}$, state $x(t)$ cannot be measured directly, while input $u(t)$ and output $y(t)$ can be measured and utilized.

The general idea of state observer design can be divided into two steps. First, A, B, C are used to duplicate a system with the same structure as the original:

$$\begin{cases} \dot{\hat{x}} = A\hat{x}(t) + Bu(t), & \hat{x}(0) = \hat{x}_0, \ t \geq 0 \\ \hat{y}(t) = C\hat{x}(t) \end{cases} \tag{5.5}$$

where $\hat{x}(t)$ is the reconstructed state of $x(t)$. Second, the correction term $y(t) - \hat{y}(t)$ is fed back to the input end of the integrator set in system (5.4) through the gain matrix L of the observer. Through these two steps, the full-order state observer can be obtained. The general structure of the traditional Luenberger full-order state observer is

$$\dot{\hat{x}}(t) = A\hat{x}(t) + Bu(t) + L\left(y(t) - C\hat{x}(t)\right), \quad \hat{x}(0) = \hat{x}_0. \tag{5.6}$$

The difference between the Luenberger observer (5.6) and the replication system (5.5) is the introduction of feedback terms for correction. In general, for some ideal cases, the open loop replication system (5.5) can also realize the state reconstruction. In particular, if the original state of the replication system and the original state can be equal, it can theoretically realize the reconstruction of all states. The goal of the design of the full-order state observer is to select the gain matrix L appropriately and make the reconstructed state $\hat{x}(t)$ asymptotically converge to the original state $x(t)$.

5.2.2.3 Radial basis function neural network

Radial basis function neural network (RBFNN) has been widely applied in the fields of function approximation, pattern recognition, signal processing, and system identification due to its deep physiological basis, simple network structure, fast learning ability, and excellent approximation performance. At present, it is still an important part of neural network research.

Due to the strong nonlinear nature of the AUV model and the complex working environment, it is often necessary to deal with various uncertainties in the design of the controller, such as ocean current disturbances, modeling uncertainties, and so on. In order to deal with the uncertainty easily, we can use the RBFNN to approximate the upper bound of the system uncertainty, and then eliminate the interference of the uncertainty in the controller design.

RBFNN is mainly a three-layer forward network composed of input, hidden, and output layers. In the hidden layer, a radial basis function is adopted as the excitation function. The functional relation between output and input of RBFNN is

$$y_i(X) = \sum_{j=1}^{M} \omega_{ij}\phi_j(X) + b_j, \quad i = 1, 2, \cdots, n \tag{5.7}$$

where $X = (x_1, x_2, \cdots, x_n)^{\mathrm{T}}$ is the input vector and y_i is the output value of the ith output unit; M is the number of centers, and ω_{ij} is the weight of the jth hidden neuron to the ith output unit; b_j is the bias value and $\phi\,(.)$ is the nonlinear transfer function of the RBF layer. The commonly used RBF layer transport functions are as follows:

Thin plate spline

$$\varphi(X) = X^2 \lg(X); \tag{5.8}$$

Multiple quadratic function

$$\varphi(X) = \left(X^2 + c\right)^{\frac{1}{2}}, \quad c > 0; \tag{5.9}$$

Gaussian function

$$\varphi_j(X) = \exp\left(-\frac{\|X - c_j\|}{2\sigma_j^2}\right) \tag{5.10}$$

where $c_j \in R^n$ $(1 \le j \le M)$ is the center of RBF; σ is the width parameter, which can adjust the sensitivity of RBF neuron. The RBF layer transfer function $\varphi(\cdot)$ is generally a Gaussian function. In this case, the output function is

$$y_i(X) = \sum_{j=1}^{M} \omega_{ij} \exp\left(-\frac{\|X - c_j\|^2}{2\sigma_j^2} + b_j\right), \quad i = 1, 2, \cdots, n. \tag{5.11}$$

In order to train an RBFNN well, it can be known from the output function of the network that the learning of RBFNN parameters consists of two parts: one is the determination of vector c and normalized parameter vector σ in the hidden layer neuron center, and the other is the determination of weights matrix w in the output layer. The influence of the form of nonlinear function used in RBFNN on the network performance is not crucial. The key factor is the selection of the center of the basis function, and the performance of RBFNN constructed by improper center selection is generally unsatisfactory. For example, some centers are too close to each other to produce approximate linear correlations, which lead to numerical pathological conditions. For the design of how RBFNN is applied to AUV control algorithm, please refer to case study 1 in Section 5.3.

5.2.3 Prescribed performance control

5.2.3.1 Introduction

In this chapter, we consider the problem of robustly adaptively controlling MIMO feedback linearizable nonlinear systems, with unknown nonlinearities, capable of allowing attributes such as maximum overshoot, a lower bound on the convergence rate and maximum permissible steady state error to be formally specified, via the introduction of performance functions. Visualizing the prescribed performance characteristics as tracking error constraints, the key idea was to transform the constrained system into an equivalent unconstrained one, via an appropriately defined output error transformation. We have proven that stabilizing the unconstrained system is sufficient to solve the stated problem. This is critical property as it leads to the design of less complex control algorithms. Further, the designed controller, which is based on the transformed error system, is smooth, with easily selected parameter values, that successfully bypasses the loss of controllability issue.

5.2.3.2 Basic theory

In order to achieve the desired control objective, we introduce the performance function as prescribed performance bound. The definition of the performance function is given as follows.

Definition 5.1. A smooth function $\rho(t) : \mathrm{R}_+ \rightarrow \mathrm{R}$ will be called a performance function if

(1) $\rho(t)$ is decreasing and positive;
(2) $\lim_{t\rightarrow\infty} \rho(t) = \rho_\infty > 0$.

The performance function is usually designed as follows:

$$\rho(t) = (\rho_0 - \rho_\infty)e^{-kt} + \rho_\infty \tag{5.12}$$

where ρ_0, ρ_∞, and k are preset positive constants.

Based on performance function, the tracking error can be expressed as follows:

$$\begin{aligned} -\delta_i\rho_i(t) < e_i(t) < \rho_i(t), \quad e_i(0) \geq 0; \\ -\rho_i(t) < e_i(t) < \delta_i\rho_i(t), \quad e_i(0) < 0, \end{aligned} \tag{5.13}$$

where $e_i(t)$, $i = 1$–6 denotes the ith trajectory tracking error of the AUV in the earth-fixed frame, $0 \leq \delta_i \leq 1$. If the initial value of the tracking error $e_i(t)$ satisfies $0 \leq |e_i(0)| < \rho_i(0)$, parameter k_i regulates the minimum convergence speed of tracking error $e_i(t)$, and the terminal value of the performance function ρ_∞ restrains the maximum bound of steady state tracking error. Therefore, the expected tracking error can be obtained by designing appropriate $\rho_i(t)$.

In order to solve prescribed performance control problem represented by Eq. (5.13), this research adopts an error transformation to transform the constrained trajectory tracking control problem into an equivalent unconstrained one. We define a function $S_i(\varepsilon_i)$ as follows:

(1) $S_i(\varepsilon_i)$ is smooth and strictly increasing;

(2) $\begin{array}{l} -\delta_i < S_i(\varepsilon_i) < 1, \ e_i(0) \geq 0 \\ -1 < S_i(\varepsilon_i) < \delta_i, \ e_i(0) < 0 \end{array}$;

(3) $\left.\begin{array}{l} \lim\limits_{\varepsilon_i \to -\infty} S_i(\varepsilon_i) = -\delta_i \\ \lim\limits_{\varepsilon_i \to +\infty} S_i(\varepsilon_i) = 1 \end{array}\right\}, \ e_i(0) \geq 0, \quad \left.\begin{array}{l} \lim\limits_{\varepsilon_i \to -\infty} S_i(\varepsilon_i) = -1 \\ \lim\limits_{\varepsilon_i \to +\infty} S_i(\varepsilon_i) = \delta_i \end{array}\right\}, \ e_i(0) < 0$

where $\varepsilon_i \in (-\infty, +\infty)$ denotes the transformed error. A function $S_i(\varepsilon_i)$ which satisfies the above conditions can be expressed as

$$S_i(\varepsilon_i) = \begin{cases} \dfrac{e^{\varepsilon_i} - \delta_i e^{-\varepsilon_i}}{e^{\varepsilon_i} + e^{-\varepsilon_i}}, & e_i(0) \geq 0; \\ \dfrac{\delta_i e^{\varepsilon_i} - e^{-\varepsilon_i}}{e^{\varepsilon_i} + e^{-\varepsilon_i}}, & e_i(0) < 0. \end{cases} \tag{5.14}$$

From the characteristic of $S_i(\varepsilon_i)$, Eq. (5.13) can be equivalently expressed as

$$e_i(t) = \rho_i(t) S_i(\varepsilon_i). \tag{5.15}$$

Owing to the strict increasingness property of $S_i(\varepsilon_i)$, there exists an inverse function expressed as follows:

$$\varepsilon_i = S_i^{-1}\left(\frac{e_i(t)}{\rho_i(t)}\right). \tag{5.16}$$

If we are able to keep ε_i bounded, then we can guarantee that Eq. (5.13) is satisfied. Owing to the constraint of prescribed performance $\rho_i(t)$, the tracking error could obtain expectation objective. The trajectory tracking control problem can be transformed into the stabilization control problem of closed-loop system with respect to variable ε_i.

If we define $S_i(\varepsilon_i)$ as in Eq. (5.14), then

$$\varepsilon_i = S_i^{-1}\left(\frac{e_i(t)}{\rho_i(t)}\right) = \begin{cases} \frac{1}{2}\ln\frac{z_i+\delta_i}{1-z_i}, & e_i(0)\geq 0; \\ \frac{1}{2}\ln\frac{1+z_i}{\delta_i-z_i}, & e_i(0)<0; \end{cases} \tag{5.17}$$

where $z_i = e_i(t)/\rho_i(t)$.

Remark 5.1. From Eq. (5.17), δ_i cannot be chosen equal to zero with $e_i(0)=0$ otherwise $e_i(0)$ will tend to infinity.

Differentiating (5.17) with respect to time, we have

$$\dot{\varepsilon}_i = \frac{\partial S_i^{-1}}{\partial z_i}\cdot\dot{z}_i = \frac{\partial S_i^{-1}}{\partial z_i}\cdot\frac{\dot{e}_i\rho_i - e_i\dot{\rho}_i}{\rho_i\cdot\rho_i} = r_i\left(\dot{e}_i - \frac{e_i\dot{\rho}_i}{\rho_i}\right) \tag{5.18}$$

where $r_i = \left(\partial S_i^{-1}/\partial z_i\right)\cdot(1/\rho_i)$ by calculation using Eq. (5.17); r_i is positive because of $\left(\partial S_i^{-1}/\partial z_i\right) > 0$ and $\rho_i(t) > 0$. In addition, r_i is bounded, $\underline{r} < r_i < \overline{r}$, when trajectories of the error $e_i(t)$ are limited to the bounds of Eq. (5.13), where $\underline{r}$ and $\overline{r}$ are positive constants.

Differentiating (5.18) with respect to time, we have

$$\begin{aligned}\ddot{\varepsilon}_i &= \dot{r}_i\left(\dot{e}_i - \frac{e_i\dot{\rho}_i}{\rho_i}\right) + r_i\left(\ddot{e}_i - \frac{\dot{e}_i\dot{\rho}_i\rho_i + e_i\ddot{\rho}_i\rho_i + e_i\dot{\rho}_i^2}{\rho_i^2}\right) \\ &= \dot{r}_i\left(\dot{e}_i - \frac{e_i\dot{\rho}_i}{\rho_i}\right) - r_i\cdot\frac{\dot{e}_i\dot{\rho}_i\rho_i + e_i\ddot{\rho}_i\rho_i + e_i\dot{\rho}_i^2}{\rho_i^2} + r_i(\ddot{\eta}_i - \ddot{\eta}_{di})\end{aligned} \tag{5.19}$$

where $\ddot{\eta}_i$, $\ddot{\eta}_{di}$, $i = 1, 2, 3, 4, 5, 6$ denote actual and desired trajectories for the AUV, respectively.

The error variance $s \in \mathrm{R}^6$ can be denoted as follows:

$$s = \lambda\varepsilon + \dot{\varepsilon} \tag{5.20}$$

where $\varepsilon = [\varepsilon_1, \varepsilon_2, \varepsilon_3, \varepsilon_4, \varepsilon_5, \varepsilon_6]^{\mathrm{T}}$, and $\lambda = \mathrm{diag}[\lambda_1, \lambda_2, \lambda_3, \lambda_4, \lambda_5, \lambda_6] > 0$ are predefined design parameters.

5.2.3.3 The design of performance function

To overcome the drawbacks of the traditional prescribed performance technique using an exponential performance function, a novel performance function is introduced such that the convergence speed of the tracking error can be set and the convergence time is known based on the design parameters.

It is obvious that the convergence rate of the traditional performance function depends on the exponential term. However, it is difficult to establish an explicit mathematical relationship between the parameter and the convergence rate. In addition, the selection rule of the parameter is unclear. Motivated by the above discussions, a novel performance function is proposed as follows:

$$\rho(t) = \begin{cases} a_1 + a_2 \sin\left(\dfrac{\pi t}{2t_f}\right) + a_3 \cos\left(\dfrac{\pi t}{2t_f}\right) + a_4 e^{-kt}, & 0 \le t \le t_f \\ \rho_{tf}, & t > t_f \end{cases} \tag{5.21}$$

where a_1, a_2, a_3, and a_4 are design parameters. The definition of parameters k and $\rho_{tf} = \rho_\infty$ are similar to that of a traditional performance function (5.12). The preset parameter t_f denotes the assignable terminal time at which the function (5.21) reaches ρ_∞. In this paper, we can design (5.21) according to Definition 5.1 in two steps.

Step 1. We need calculate parameters a_1, a_2, a_3, and a_4 by performing the following restricted conditions.

The initial and terminal conditions of the new performance function (5.21) are the same as those of the traditional function (5.12), and they can be expressed as $\rho(0) = \rho_0$ and $\rho(t_f) = \rho_{tf}$. Additionally, the first and second derivatives of $\rho(t)$ are continuous functions. This means that $\lim_{t \to t_f^-} \dot{\rho}(t) = \lim_{t \to t_f^+} \dot{\rho}(t) = 0$ and $\lim_{t \to t_f^-} \ddot{\rho}(t) = \lim_{t \to t_f^+} \ddot{\rho}(t) = 0$.

Substituting the above initial conditions into the new performance function $\rho(t) = a_1 + a_2 \sin\left(\frac{\pi t}{2t_f}\right) + a_3 \cos\left(\frac{\pi t}{2t_f}\right) + a_4 e^{-kt}, 0 \le t \le t_f$, we get

$$\begin{cases} a_1 + a_3 + a_4 = \rho_0 \\ a_1 + a_2 + a_4 e^{-kt_f} = \rho_{t_f} \\ -a_3 \dfrac{\pi}{2t_f} - a_4 k e^{-kt_f} = 0 \\ -a_2 \dfrac{\pi^2}{4t_f^2} + a_4 k^2 e^{-kt_f} = 0 \end{cases} \tag{5.22}$$

Thus we can solve for the four unknown coefficients a_0, a_1, a_2, a_3, and a_4 based on the above conditions. Letting $a_0 = 2t_f k/\pi$, we have

$$\begin{aligned} a_1 &= \rho_0 + (a_0 e^{-kt_f} - 1)a_4 \\ a_2 &= a_0^2 a_4 e^{-kt_f} \\ a_3 &= -a_0 a_4 e^{-kt_f} \\ a_4 &= \frac{\rho_0 - \rho_{t_f}}{1 - (a_0^2 + a_0 + 1)e^{-kt_f}} \end{aligned} \tag{5.23}$$

Noting that $a_0 = 2t_f k/\pi$, the value of a_0 is fixed by the maximum convergence time t_f and parameter k which controls the convergence rate at the initial stage.

Step 2. This step aims to prove that $\rho(t)$ is a positive and monotonically decreasing function.

Note that $\rho(0) = \rho_0 > 0$ and $\rho(t_f) = \rho_{t_f} > 0$. If $\dot{\rho}(t) < 0$ is proved to be true for all $t \in [0, t_f)$, the monotonic decreasingness and positivity properties of $\rho(t)$ are satisfied.

Proof. Take the time derivative of (5.21) on $t \in [0, t_f)$ and substitute a_0, a_1, a_2, a_3, and a_4 into it to get

$$\begin{aligned} \dot{\rho}(t) &= a_2 \frac{\pi}{2t_f} \cos\left(\frac{\pi}{2t_f}\right) - a_3 \frac{\pi}{2t_f} \sin\left(\frac{\pi t}{2t_f}\right) - k a_4 e^{-kt} \\ &= \frac{2t_f k^2}{\pi} a_4 e^{-kt_f} \cos\left(\frac{\pi t}{2t_f}\right) + k a_4 e^{-kt_f} \sin\left(\frac{\pi t}{2t_f}\right) - k a_4 e^{-kt}, \quad 0 \le t \le t_f \end{aligned} \tag{5.24}$$

From the calculation, one has $ka_4 > 0$. Therefore, the problem is transformed into proving that the function $y < 0$ on $[0, t_f)$ and

$$\begin{aligned} y = {} & \frac{2t_f k}{\pi} e^{-kt_f} \cos\left(\frac{\pi t}{2t_f}\right) + e^{-kt_f} \frac{\pi}{2t_f} \sin\left(\frac{\pi t}{2t_f}\right) \\ & - e^{-kt}, \quad 0 \le t \le t_f \end{aligned} \tag{5.25}$$

Letting $c = t_f k$ and $x = t/t_f$, (5.25) can be rewritten as

$$y = \frac{2}{\pi} c \cdot \cos\left(\frac{\pi}{2}x\right) + \sin\left(\frac{\pi}{2}x\right) - e^{c(1-x)}, \quad 0 \le x \le 1. \tag{5.26}$$

Noting that the initial value $y(0) = 2c/\pi - e^c < 0$ and $y(1) = 0$, the derivative of $y(x)$ with respect to x is given by

$$\dot{y}(x) = -c \sin\left(\frac{\pi}{2}x\right) + \frac{\pi}{2} \cos\left(\frac{\pi}{2}x\right) + ce^{c(1-x)}, \quad 0 \le x \le 1. \tag{5.27}$$

Taking the second order derivative of $y(x)$ based on $\dot{y}(0) = \pi/2 + ce^c > 0$ and $\dot{y}(1) = 0$, we get

$$\ddot{y}(x) = -\frac{\pi}{2} c \cdot \cos\left(\frac{\pi}{2}x\right) - \frac{\pi^2}{4} \sin\left(\frac{\pi}{2}x\right) - c^2 e^{c(1-x)}, \quad 0 \le x \le 1. \tag{5.28}$$

It is obvious that $\ddot{y}(x) < 0$ so that $\dot{y}(x)$ is a monotonically decreasing function. Since $\dot{y}(0) > 0$ and $y(1) = 0$, we can conclude that $\dot{y}(x) \ge 0$ within the domain of definition so that $y(x)$ is also a monotonically decreasing function. Similarly, $y(x) \le 0$ since $y(0) < 0$ and $y(1) = 0$. In conclusion, $\dot{\rho}(t) < 0$ for all $0 \le t \le t_f$ ($t = t_f$ if and only if $\dot{\rho}(t) = 0$). In other words, $\rho(t)$ is a positive and monotonically decreasing function.

Therefore, (5.21) can be used as a performance function whose parameter is set according to (5.23). From the above analysis, Step 2 demonstrates that both parameters, t_f and k, affect the convergence rate of the performance function (5.21), and can be chosen without constraints between them. The performance function (5.21) possesses some important properties: First, we can preset the maximum convergence time t_f. Second, we can change parameter k to adjust the convergence rate of performance function (5.21) given the steady-state convergence time. □

5.3. Case study

5.3.1 Prescribed performance neural network adaptive trajectory tracking control for AUV

5.3.1.1 Algorithm design

In case study 1, we consider using traditional performance functions to design control strategy for a fully actuated AUV. We use the AUV dynamic model provided in Section 5.1 for algorithm design. Moreover, the modeling uncertainties and the thruster faults are considered. Therefore, the basic dynamic model Eq. (5.1) can be transformed as

$$\ddot{\eta} = M_{\eta 0}^{-1}\left(B_0 u - C_{RB\eta 0}\dot{\eta} - C_{A\eta 0}\dot{\eta} - D_{\eta 0}\dot{\eta} - g_{\eta 0}\right) - F \tag{5.29}$$

where subscript "0" denotes the nominal value; F is the system general uncertainty expressed as

$$\begin{aligned} F = M_{\eta 0}^{-1}(&\Delta M_\eta \ddot{\eta} - \Delta Bu + \Delta C_{RB\eta}\dot{\eta} + \Delta C_{A\eta}\dot{\eta} + \Delta D_\eta \dot{\eta} + \Delta g_\eta \\ &+ \overline{C_{A\eta}\eta_r + D_\eta \eta_r}) \end{aligned} \tag{5.30}$$

where $\overline{C_{A\eta}\eta_r + D_\eta \eta_r}$ denotes the effect of ocean current disturbance, and Δ denotes the uncertainty.

The objective of this study is to design a control strategy for the AUV to track the desired trajectory and to ensure a desired transient and steady response of the tracking error when considering system uncertainty and thruster fault.

We make the following assumptions based on practical project background:

Assumption 5.1. The trajectory vector η and its first derivative are available for measurement.

Assumption 5.2. The desired trajectory η_d and its first and second derivatives are known bounded functions.

Remark 5.2. Since $\dot{\eta} = J(\eta)v$, the first derivative of trajectory vector η can be calculated by the velocity vector v in the body-fixed frame. Then Assumption 5.1 is equivalent to that η and v are available for measurement.

Remark 5.3. The AUV needs track predefined trajectory to arrive at a target point. Therefore, the trajectory is known, and then its first and second derivatives are known and bounded.

The error variance $s \in \mathrm{R}^6$ can be expressed as follows:

$$s = \lambda\varepsilon + \dot{\varepsilon} \tag{5.31}$$

where $\varepsilon = [\varepsilon_1, \varepsilon_2, \varepsilon_3, \varepsilon_4, \varepsilon_5, \varepsilon_6]^{\mathrm{T}}$, and $\lambda = \mathrm{diag}[\lambda_1, \lambda_2, \lambda_3, \lambda_4, \lambda_5, \lambda_6] > 0$ are predefined design parameters. According to the dynamic model (5.29) for the AUV,

$$\ddot{\eta} = M_{\eta 0}^{-1}\left(B_0 u - C_{RB\eta 0}\dot{\eta} - C_{A\eta 0}\dot{\eta} - D_{\eta 0}\dot{\eta} - g_{\eta 0}\right) - F \tag{5.32}$$

where $A = -M_{\eta 0}^{-1}\left[\left(C_{RB\eta 0}\dot{\eta} + C_{A\eta 0}\dot{\eta} + D_{\eta 0}\dot{\eta}\right)\dot{\eta} + g_{\eta 0}\right]$, $H = M_{\eta 0}^{-1}B_0$, and $D = -F$. Eq. (5.29) can be expressed as

$$\ddot{\eta} = A + Hu + D. \tag{5.33}$$

In addition,

$$\dot{s} = \lambda\dot{\varepsilon} + \ddot{\varepsilon} = L + R(A + Hu + D) \tag{5.34}$$

where $L = [l_1, l_2, l_3, l_4, l_5, l_6]^{\mathrm{T}}$, $l_i = (\lambda_i r_i + \dot{r}_i)\left(\dot{e}_i - \frac{e_i\dot{\rho}_i}{\rho_i}\right) - r_i \cdot \frac{\dot{e}_i\dot{\rho}_i\rho_i + e_i\ddot{\rho}_i\rho_i + e_i\dot{\rho}_i^2}{\rho_i^2} - r_i\ddot{\eta}_{di}$, $i = 1$–6, and $R = \mathrm{diag}[r_1, r_2, r_3, r_4, r_5, r_6]$. If we design a controller u to guarantee s to be bounded, then ε_i and $\dot{\varepsilon}$ are also bounded.

We utilize the capabilities of the RBFNN to approximate uncertain nonlinearity D in the system (5.33),

$$D = W^{*\mathrm{T}}h(x) + \mu, \tag{5.35}$$

where $x \in \Omega_x \subset \mathrm{R}^q$ is RBFNN input vector, $h(x) = [h_1(x), h_2(x), \ldots, h_j(x), \ldots, h_m(x)]^{\mathrm{T}} \in \mathrm{R}^m$, m is the number of RBFNN's hidden layer nodes. By using Gaussian basis functions, $h_j(x)$ can be expressed as follows:

$$h_j(x) = \exp\left(-\frac{x - c_j^2}{2b_j^2}\right), \quad j = 1, 2, \ldots, m \tag{5.36}$$

where c_j is the jth center parameter of RBFNN, $c_j = [c_{j1}, c_{j2}, \ldots, c_{jq}]^{\mathrm{T}}$; $b_j > 0$ is the jth parameter of RBFNN; $W^* = [W_1^*, W_2^*, W_3^*, W_4^*, W_5^*, W_6^*] \in \mathrm{R}^{m\times 6}$ is the desired hidden-to-output layer interconnection weighted matrix; $\mu \in \mathrm{R}^6$ is the RBFNN estimation error, and $\|\mu\| \leq \mu^*$, where μ^* is an unknown positive constant. The desired weighted matrix $W \in \mathrm{R}^{m\times 6}$ can be defined as

$$W^* = \arg \min_{W \in \mathrm{R}^{m\times 6}} \left\{ \sup_{x\in\Omega_x} \|D - W^{\mathrm{T}}h(x)\| \right\}. \tag{5.37}$$

The input vector of RBFNN can be defined as $x = \left[e^{\mathrm{T}}, \dot{e}^{\mathrm{T}}\right]^{\mathrm{T}}$, then the estimated value of the uncertain nonlinearity D can be written as follows:

$$\hat{D} = \hat{W}^{\mathrm{T}}h(x) \tag{5.38}$$

where $\hat{W} = \left[\hat{W}_1, \hat{W}_2, \hat{W}_3, \hat{W}_4, \hat{W}_5, \hat{W}_6\right]$ is the estimated value of the weights matrix W^*.

To consider unknown upper bounds μ^* of an approximation error, we propose the adaptive control law as follows:

$$u = H^{-1}\left(-A - R^{-1}L - \hat{D} - \hat{\mu}^2 \frac{\|s\|}{\hat{\mu}\|s\| + \sigma} - K_2 s\right), \tag{5.39}$$

$$\dot{\hat{W}}_i = \tau_{wi}\left[s_i h(x) - \beta \hat{W}_i\right], \quad i = 1, 2, 3, 4, 5, 6, \tag{5.40}$$

$$\dot{\hat{\mu}} = \tau_\mu(\|s\| - \gamma\hat{\mu}) \tag{5.41}$$

where $\hat{\mu}$ denotes the estimated value of the upper bounds μ^*, $K_2 > 0$, $\sigma > 0$, $\tau_{wi} > 0$, $\beta > 0$, $\tau_\mu > 0$, and $\gamma > 0$ are designed control parameters and adaptive gains, respectively.

Theorem 5.1 ([39]). *For the error system (5.34) from AUV dynamic model (5.29) by the error transformation (5.16), according to the control law (5.39) and adaptive law (5.40) and (5.41), the transforming error ε_i and tracking error e_i can be guaranteed to be uniformly ultimately bounded and satisfy prescribed performance (5.13), respectively.*

5.3.1.2 Algorithm proof

Since matrix R is symmetric positive definite and bounded, we consider the Lyapunov function candidate

$$V = \frac{1}{2}s^{\mathrm{T}}R^{-1}s + \frac{1}{2}\mathrm{tr}\left(\tilde{W}^{\mathrm{T}}\Gamma_w^{-1}\tilde{W}\right) + \frac{1}{2}\cdot\frac{1}{\tau_\mu}\tilde{\mu}^2 \tag{5.42}$$

where $\tilde{W} = \hat{W} - W^*$, $\tilde{\mu} = \hat{\mu} - \mu^*$ are estimation errors, and $\Gamma_w = \mathrm{diag}[\tau_{w1}, \tau_{w2}, \tau_{w3}, \tau_{w4}, \tau_{w5}, \tau_{w6}]$. Differentiating V with respect to time and substituting (5.34) and (5.39)–(5.41) into it, we obtain

$$\begin{aligned}
\dot{V} &= s^{\mathrm{T}}R^{-1}\dot{s} + \mathrm{tr}\left(\tilde{W}^{\mathrm{T}}\Gamma_w^{-1}\dot{\hat{W}}\right) + \frac{1}{\tau_\mu}\tilde{\mu}\dot{\hat{\mu}} \\
&= s^{\mathrm{T}}R^{-1}\left[Q + R(A + Hu + D)\right] + \sum_{i=1}^{6}\frac{1}{\tau_{wi}}\tilde{W}_i^{\mathrm{T}}\dot{\hat{W}}_i + \frac{1}{\tau_\mu}\tilde{\mu}\dot{\hat{\mu}} \\
&= s^{\mathrm{T}}\left[-\tilde{W}^{\mathrm{T}}h(x) + \mu - \frac{\hat{\mu}^2 s}{\hat{\mu}s + \sigma} - K_2 s\right] + \sum_{i=1}^{6}\tilde{W}_i^{\mathrm{T}}\left[s_i h(x) - \beta\hat{W}_i\right] \\
&\quad + \tilde{\mu}(s - \gamma\hat{\mu}) \\
&\le -s^{\mathrm{T}}K_2 s + s\hat{\mu} - \frac{\hat{\mu}^2 s^2}{\hat{\mu}s + \sigma} - \beta\sum_{i=1}^{6}\tilde{W}_i^{\mathrm{T}}\hat{W}_i - \gamma\tilde{\mu}\hat{\mu}
\end{aligned} \tag{5.43}$$

Applying Young's inequality, we obtain

$$-\beta\sum_{i=1}^{6}\tilde{W}_i^{\mathrm{T}}\hat{W}_i \le \frac{1}{2}\beta\sum_{i=1}^{6}\tilde{W}_i^{\mathrm{T}}\tilde{W}_i + \frac{1}{2}\beta\sum_{i=1}^{6}\tilde{W}_i^{*\mathrm{T}}W_i^{*}, \tag{5.44}$$

$$-\gamma\tilde{\mu}\hat{\mu} \le -\frac{1}{2}\gamma\tilde{\mu}^2 + \frac{1}{2}\gamma\mu^{*2}. \tag{5.45}$$

By adaptive law (5.41), we can have $\hat{\mu} > 0$, and then

$$\|s\|\hat{\mu} - \frac{\hat{\mu}^2\|s\|^2}{\hat{\mu}\|s\| + \sigma} = (\hat{\mu}\|s\|)\cdot\frac{\sigma}{\hat{\mu}\|s\| + \sigma} < \sigma. \tag{5.46}$$

To further simplify (5.43), one has that

$$\dot{V} \le -s^{\mathrm{T}}K_2 s - \frac{1}{2}\beta\sum_{i=1}^{6}\tilde{W}_i^{\mathrm{T}}\tilde{W}_i - \frac{1}{2}\gamma\tilde{\mu}^2 + \frac{1}{2}\beta\sum_{i=1}^{6}W_i^{*\mathrm{T}}W_i^{*} + \frac{1}{2}\gamma\mu^{*2} + \sigma. \tag{5.47}$$

Letting $\kappa = \frac{1}{2}\beta\sum_{i=1}^{6}W_i^{*\mathrm{T}}W_i^{*} + \frac{1}{2}\gamma\mu^{*2} + \sigma$, $\dot{V} \le 0$ when $\|s\| > \sqrt{\kappa/\lambda_{\min}(K_2)}$ or $\|\tilde{W}_i\| > \sqrt{2\kappa/\beta}$, or $|\tilde{\mu}| > \sqrt{3\kappa/\gamma}$. Thus the variable s, matrix $\tilde{W}_i$, and estimation error $\tilde{\mu}$ are uniformly ultimately bounded with respect to the sets:

$$\begin{aligned}
\mathrm{N}_1 &= \left\{s \in \mathrm{R}^6 : s \le \sqrt{\kappa/\lambda_{\min}(K_2)}\right\}\\
\mathrm{N}_2 &= \left\{\tilde{W}_i \in \mathrm{R}^m : \tilde{W}_i \le \sqrt{2\kappa/\beta}\right\}\\
\mathrm{N}_3 &= \left\{\tilde{\mu} \in \mathrm{R} : \tilde{\mu} \le \sqrt{2\kappa/\gamma}\right\}
\end{aligned} \tag{5.48}$$

where $\lambda_{\min}(K_2)$ denotes the minimal eigenvalue of the matrix K_2. The transforming error is uniformly ultimately bounded with respect to the set

$$\mathrm{N}_4 = \left\{\varepsilon_i \in \mathrm{R} : \varepsilon_i \le \sqrt{\kappa/\lambda_{\min}(K_2)}/\lambda_i\right\}. \tag{5.49}$$

Owing to the smoothness and strict increasingness properties of $S_i(\varepsilon_i)$, the performance constraint Eq. (5.13) is satisfied. It means that the tracking error e_i achieves prescribed dynamic performance and steady-state response.

Remark 5.4. According to the definition of prescribed performance method and the proof of Theorem 5.1, we give the guidelines on how to choose appropriate parameters and gains of performance function (5.12), RBFNN (5.36), and control strategy (5.39)–(5.41) in practical applications as follows:

(1) The parameters ρ_{i0}, $\rho_{i\infty}$, and k_i of the performance function should be selected to satisfy practical mission requirements. Especially, ρ_{i0} needs to meet the initial conditions of the trajectory tracking control system.

(2) The number of the RBFNN hidden nodes is approximately seven and the matrix of the RBFNN center needs to be symmetric.

(3) The parameter σ and matrix λ of the controller and adaptive gains β and γ should be small enough. Additionally, the choice of matrix K_2 and adaptive gains τ_{wi} and τ_μ is not specified.

5.3.1.3 Simulation results

In order to verify the effectiveness of the proposed control method, simulations are applied to the AUV system which considers ocean current disturbances and model uncertainties, as well as thruster faults. The hydrodynamic parameters, inertia coefficients, and the initial values of η are tabulated in Tables 5.1–5.3, respectively.

Table 5.1 The hydrodynamic parameters of the AUV.

Name	Coefficients	Value
	$X_{\dot{u}}$ / kg	40.6
Surge	X_u / kg · s^{-1}	21.9
	$X_{u\|u\|}$ / kg · m^{-1}	24.5
	$Y_{\dot{v}}$ / kg	82.3
Lateral	Y_v / kg · s^{-1}	47.4
	$Y_{v\|v\|}$ / kg · m^{-1}	34.7
	$Z_{\dot{w}}$ / kg	114.6
Heave	Z_w / kg · s^{-1}	53.5
	$Z_{w\|w\|}$ / kg · m^{-1}	40.2
	$N_{\dot{r}}$ / kg · m^2	59.1
Yaw	N_r / kg · m^2 · s^{-1}	86.9
	$N_{r\|r\|}$ / kg · m^2	39.3
	$K_{\dot{p}}$ / kg · m^2	55.7
Roll	K_p / kg · m^2 · s^{-1}	71.9
	$K_{p\|p\|}$ / kg · m^2	42.1
	$M_{\dot{q}}$ / kg · m^2	76.5
Pitch	M_q / kg · m^2 · s^{-1}	97.3
	$M_{q\|q\|}$ / kg · m^2	48.3

Table 5.2 The inertia coefficients of the OBFN.

Mass (kg)	I_x $(Nm \cdot s^2)$	I_y $(Nm \cdot s^2)$	I_z $(Nm \cdot s^2)$	I_{xy} $(Nm \cdot s^2)$	I_{yz} $(Nm \cdot s^2)$	I_{xz} $(Nm \cdot s^2)$
92	49	84	79	0	0	0

Table 5.3 The initial values of η.

x (m)	y (m)	z (m)	ϕ (rad)	θ (rad)	ψ (rad)
0.02	0	0.01	0	0	0

The model uncertainties are quantified for convenience in this paper. Also 20% nominal values serve as model error and consider it as a part of the disturbance in the simulation.

A first-order Gauss–Markov process is applied in simulation to represent ocean current disturbance, denoted as follows:

$$\dot{V}_c + \mu V_c = \omega \tag{5.50}$$

where V_c is the speed of ocean current in the earth coordinate system, ω is Gaussian white noise with mean 1 and variance 1, and $\mu = 3$. In this study, we assume that the direction of ocean current parallels the x-axis positive direction in the earth coordinate system.

Thruster configuration of the AUV is adopted in the fully-actuated form, as shown in Fig. 5.1.

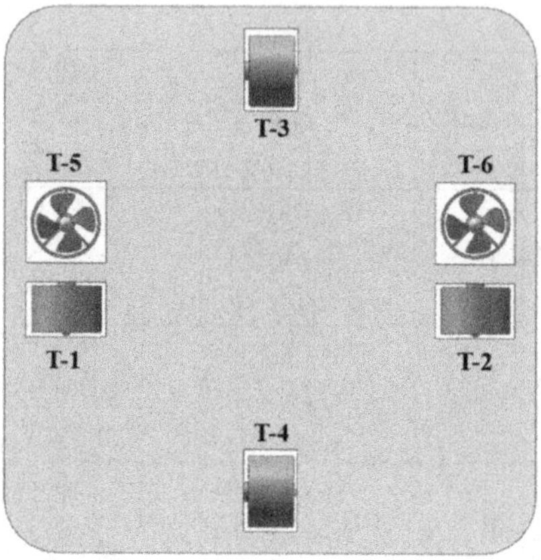

Figure 5.1 Thruster configuration of the AUV.

From Fig. 5.1, it can be found that the thruster configurations are the same in each direction. Therefore, we consider that the fault is only occurred in the thruster T-1, which can represent any thruster fault form.

Two kinds of thruster fault for the thruster T-1 can be expressed as

$$k_{11} = \begin{cases} 0 & t < 30 \\ -0.5 & t \geq 30 \end{cases} \tag{5.51}$$

$$k_{12} = \begin{cases} 0 & t < 15 \\ -0.5(1 - \exp(-(t-15)/5)) & t \geq 15 \end{cases} \tag{5.52}$$

where Eqs. (5.51) and (5.52) represent abrupt and incipient thruster fault cases, respectively.

The simulation requires that the AUV trajectory tracking precision should be limited to at most 0.0035. Therefore, the desired trajectory tracking control performance for the AUV can be designed as follows:

(1) Steady-state tracking error is limited to be at most 0.0035;
(2) Convergence speed is faster than $e^{-0.15t}$;
(3) System response has no overshoot.

The performance function $\rho_i(t)$ and parameter δ_i are given in Table 5.4.

Table 5.4 Parameters of the prescribed performance function.

Parameters	ρ_{i0}	$\rho_{i\infty}$	k_i	δ_i
Values	1.6	0.0035	0.15	0.2

The controller parameters can be denoted as $\lambda = \text{diag}[0.125, 0.125, 0.125, 0.125, 0.125, 0.125]$, $K_2 = \text{diag}[0.6, 0.6, 0.6, 0.6, 0.6, 0.6]$, $\sigma = 0.01$. Adaptive gains can be chosen as $\tau_{wi} = \tau_\mu = 0.5$, and $\beta = \gamma = 0.01$. The number of the RBFNN hidden nodes can be designed as $j = 7$, $b_j = 0.09$. The center parameter of RBFNN is $c = [c_1, \ldots, c_7]$, which can be expressed as follows:

$$c = \begin{bmatrix} -0.3 & -0.17 & -0.08 & 0 & 0.08 & 0.17 & 0.3 \\ -0.15 & -0.1 & -0.05 & 0 & 0.05 & 0.1 & 0.15 \\ -0.2 & -0.13 & -0.07 & 0 & 0.07 & 0.13 & 0.2 \\ -0.03 & -0.02 & -0.01 & 0 & 0.01 & 0.02 & 0.03 \\ -0.15 & -0.1 & -0.05 & 0 & 0.05 & 0.1 & 0.15 \\ -0.03 & -0.02 & -0.01 & 0 & 0.01 & 0.02 & 0.03 \\ -0.3 & -0.17 & -0.08 & 0 & 0.08 & 0.17 & 0.3 \\ -0.15 & -0.1 & -0.05 & 0 & 0.05 & 0.1 & 0.15 \\ -0.2 & -0.13 & -0.07 & 0 & 0.07 & 0.13 & 0.2 \\ -0.03 & -0.02 & -0.01 & 0 & 0.01 & 0.02 & 0.03 \\ -0.15 & -0.1 & -0.05 & 0 & 0.05 & 0.1 & 0.15 \\ -0.03 & -0.02 & -0.01 & 0 & 0.01 & 0.02 & 0.03 \end{bmatrix}. \tag{5.53}$$

After the supporting ship arrives at the target exploration area, AUVs are successively detached from the geophysical vessel. Then the AUV need to overcome the influence of ocean current and track predefined trajectory to arrive at target points. The downward spiral is beneficial to dive the AUV deep in a small area. Therefore, this study chooses a spiral trajectory as the desired trajectory, expressed as follows:

$$x_d = 2\sin(0.1t), \qquad y_d = 2\cos(0.1t) + 2, \qquad z_d = -0.5144t,$$
$$\phi_d = 0, \qquad \theta_d = 0, \qquad \psi_d = 0,$$
$$\eta_d = [x_d;\ y_d;\ z_d;\ \phi_d;\ \theta_d;\ \psi_d].$$

We simulate for two different thruster fault cases of the AUV. Besides, in order to make a comparative study, we compare the proposed control algorithm (5.39)–(5.41) with a unprescribed-performance controller with the same control structure and parameters. This controller can be called an adaptive tracking controller and is expressed as in Eqs. (5.54)–(5.56):

$$u = H^{-1}\left(-A + \ddot{\eta}_d - \lambda\dot{e} - \hat{D} - \hat{\mu}^2\frac{s_e}{\hat{\mu}\|s_e\| + \sigma} - K_2 s\right), \tag{5.54}$$

$$\dot{\hat{W}}_i = \tau_{wi}\left[s_{ei}h(x) - \beta\hat{W}_i\right], \quad i = 1, 2, 3, 4, 5, 6, \tag{5.55}$$

$$\dot{\hat{\mu}} = \tau_\mu\left(\|s_e\| - \gamma\hat{\mu}\right) \tag{5.56}$$

where $s_e = [s_{e1}, s_{e2}, s_{e3}, s_{e4}, s_{e5}, s_{e6}]^{\mathrm{T}} = \lambda e + \dot{e}$.

In this part, thruster fault is based on Eq. (5.51). Figs. 5.2–5.7 represent 6-DOF trajectory tracking error of the AUV. The prescribed performance adaptive tracking controller (5.39)–(5.41) is denoted as PPATC, the adaptive tracking controller (5.54)–(5.56) is denoted as ATC, and the prescribed performance bounds is denoted as ρ. The comparisons of actual and desired 3D trajectories is shown in Fig. 5.8.

From Figs. 5.2–5.7, it can be found that the proposed PPATC control method (5.39)–(5.41) could keep the trajectory tracking error inside the prescribed performance bounds and obtain desired transient and steady responses. However, the ATC control strategy with the same parameters could not meet predefined precision and had an obvious overshoot. The simulation results indicate that the prescribed performance method could make the tracking error achieve desired control requirements.

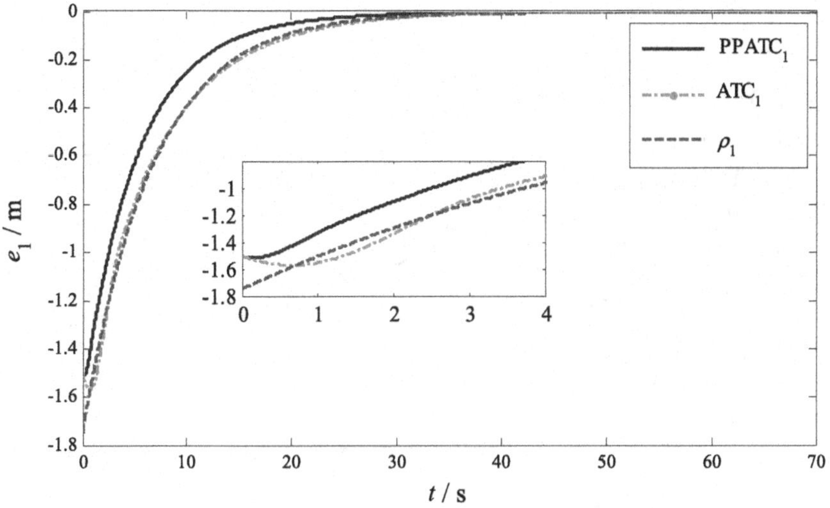

Figure 5.2 Tracking error in surge e_1 with thruster abrupt fault.

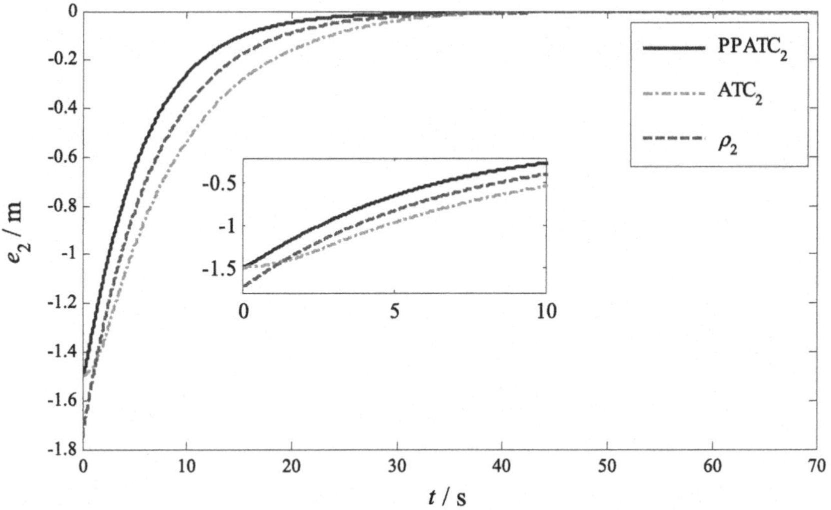

Figure 5.3 Tracking error in lateral e_2 with thruster abrupt fault.

5.3.2 Disturbance-observer-based prescribed performance trajectory tracking control for AUV

5.3.2.1 Algorithm design

In case study 2, we use a novel performance function to design a control strategy for a fully actuated AUV. Similar to case study 1, in addition to the

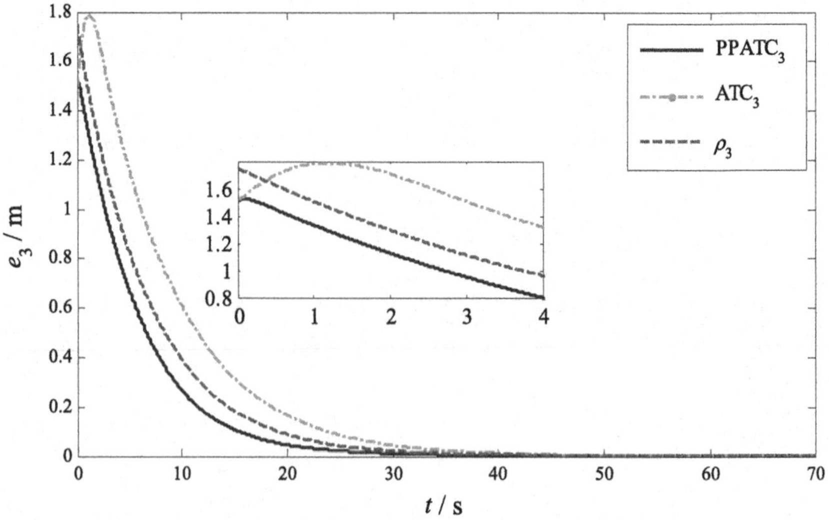

Figure 5.4 Tracking error in heave e_3 with thruster abrupt fault.

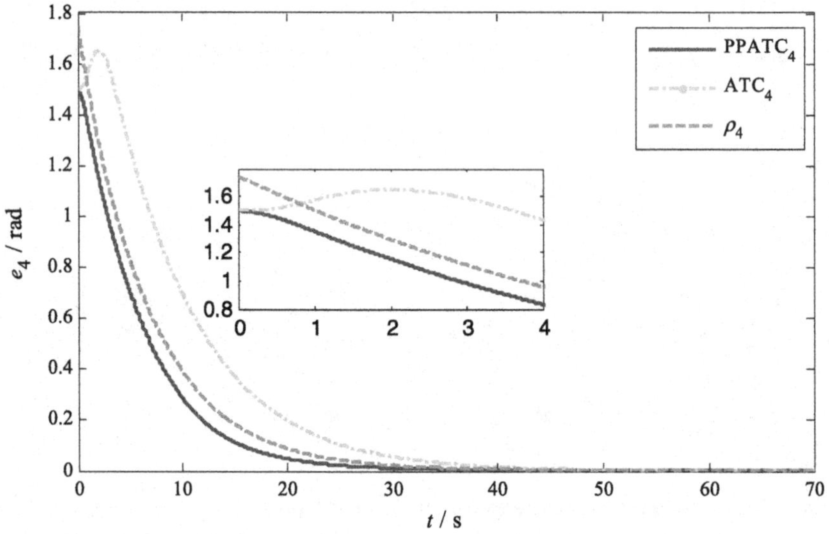

Figure 5.5 Tracking error in roll e_4 with thruster abrupt fault.

modeling uncertainties and thruster faults, the saturation of the thruster is also considered in this case.

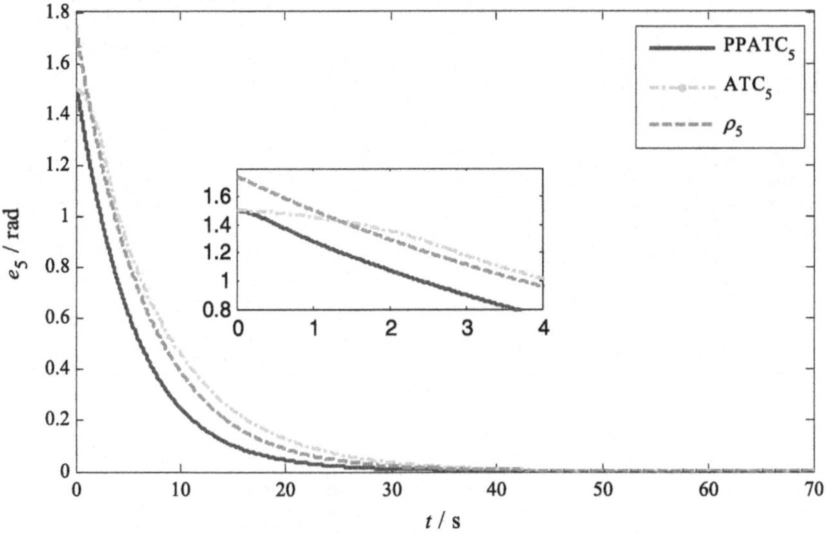

Figure 5.6 Tracking error in pitch e_5 with thruster abrupt fault.

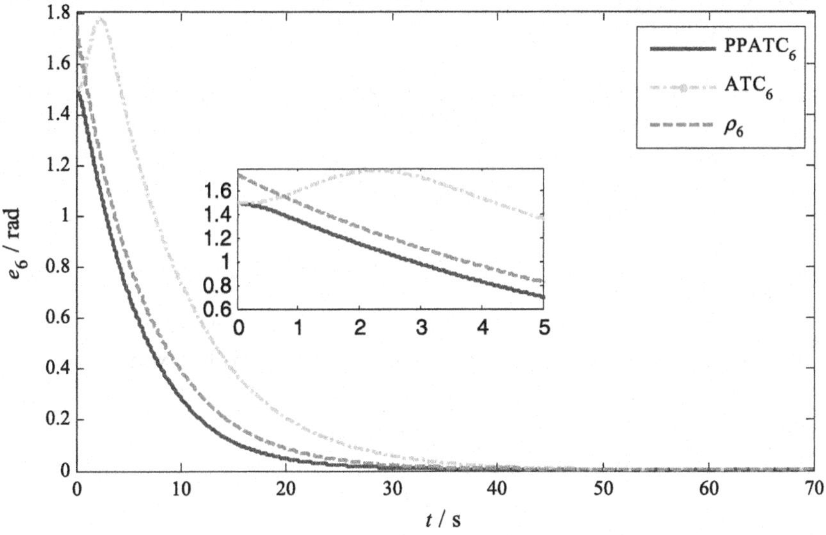

Figure 5.7 Tracking error in yaw e_6 with thruster abrupt fault.

Based on the dynamic model (5.1) and new performance function (5.21), the error variable $s \in R^6$ can be written as

$$s = \lambda\varepsilon + \dot{\varepsilon} \tag{5.57}$$

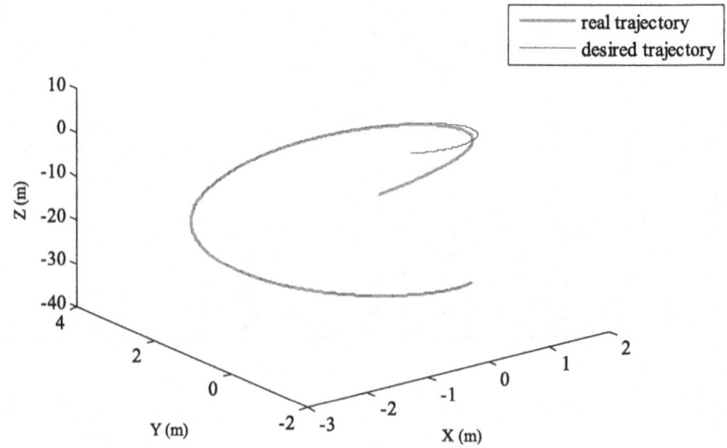

Figure 5.8 The real and desired trajectories in the simulation case 1.

where $\varepsilon = [\varepsilon_1, \varepsilon_2, \varepsilon_3, \varepsilon_4, \varepsilon_5, \varepsilon_6]^{\mathrm{T}}$ and $\lambda = \mathrm{diag}[\lambda_1, \lambda_2, \lambda_3, \lambda_4, \lambda_5, \lambda_6] > 0$ are design parameters. According to the dynamic model (5.1) of the AUV, Eq. (5.57) can be rewritten as

$$\dot{v} = M^{-1}\left[B_0 u - C_{v0} v - D_{v0} v - g_{\eta 0}\right] - F.$$

It can further be rewritten as

$$\begin{aligned}\ddot{\eta}_e &= \dot{J}(\eta)v_e + J(\eta)\dot{v}_e \\ &= \dot{J}(\eta)v_e + J(\eta)[M_0^{-1}(B_0 u - C_{v0} v - D_{v0} v - g_{\eta 0}) - F] - J(\eta)\dot{v}_d\end{aligned} \tag{5.58}$$

where $\ddot{\eta}_e = \ddot{\eta} - \ddot{\eta}_d$, $v_e = v - v_d$, and $\dot{v}_e = \dot{v} - \dot{v}_d$; J denotes the transformation matrix between the inertial frame and the body-fixed frame. Let $G = \dot{J}(\eta)v_e - J(\eta)M_0^{-1}(B_0 u + C_{v0} v + D_{v0} v + g_{\eta 0}) - J(\eta)\dot{v}_d$, $H = J(\eta)M_0^{-1}B_0$, and $D = -J(\eta)F$. Eq. (5.58) can be abbreviated as

$$\ddot{\eta}_e = G + Hu + D. \tag{5.59}$$

Then

$$\dot{s} = \lambda\dot{\varepsilon} + \ddot{\varepsilon} = V + R(G + Hu + D) \tag{5.60}$$

where $V = [v_1, v_2, v_3, v_4, v_5, v_6]^{\mathrm{T}}$, $v_i = (\lambda_i r_i + \dot{r}_i)\left(\dot{e}_i - \frac{e_i \dot{\rho}_i}{\rho_i}\right) - r_i \cdot \frac{\dot{e}_i \dot{\rho}_i \rho_i + e_i \ddot{\rho}_i \rho_i + e_i \dot{\rho}_i^2}{\rho_i^2}$, and $R = \mathrm{diag}[r_1, r_2, r_3, r_4, r_5, r_6]$. If we design a controller u to make s bounded, ε_i and $\dot{\varepsilon}$ will all be bounded based on Eq. (5.57).

Thruster saturation must exist in actual systems. It is obvious that thruster is easier to reach saturation when control system obtains better control results. Therefore, in order to achieve desired control performance, it is worth to investigate the design of a trajectory tracking controller when considering the thruster saturation.

We use variable u_c to replace the original control variable u when thruster saturation occurs. Then $u_c = sat(u) = [sat(u_1), sat(u_2), sat(u_3), sat(u_4), sat(u_5), sat(u_6)]^{\mathrm{T}}$, where u_c represents the actual output value of the thrusters, and $sat(u_i) = \min\{|u_i|, u_{i\max}\} \cdot \mathrm{sgn}(u_i)$, where $u_{i\max}$ represents the maximum output value of each axis. Then, the error system can be rewritten as

$$\dot{s} = \lambda\dot{\varepsilon} + \ddot{\varepsilon} = V + R(G + Hu_c + D). \tag{5.61}$$

Assumption 5.3. The change rate of the system general uncertainties is bounded, and then $\|\dot{D}\| \leq \chi$, where χ is an unknown positive constant.

Assumption 5.4. The actual control output can compensate the influence of the system general uncertainties D and control error variable s to be bounded.

We introduce an auxiliary system (5.61) to deal with thruster saturation as follows:

$$\dot{z}_a = \begin{cases} -K_3 z_a - \dfrac{\|H\|^2 \|\Delta u\|^2}{2\|z_a\|^2} z_a - H\Delta u, & \|z_a\| \geq \sigma \\ 0, & \|z_a\| < \sigma \end{cases} \tag{5.62}$$

where z_a, σ, and K_3 are an auxiliary variable, a small positive vector, and a gain matrix, respectively; $\Delta u = u - u_c$. When the auxiliary variable of Eq. (5.62) satisfies $\|z_a\| \geq \sigma$, the auxiliary system works, and vice versa.

Remark 5.5. Note that the auxiliary system based on the mathematical treatment method is to handle thruster saturation. The control input must be sufficient to achieve proposed control objective under ocean current disturbances, modeling uncertainties, and thruster faults which are reasonable in practical engineering. Therefore, the auxiliary system is invalid when the value increases above the saturation limit.

The system observer and controller are designed as follows:

$$\begin{cases} \dot{z}_D = -Lz_D - L(G + Hu_c + R^{-1}V + K_1 s + P\int_0^t s d\tau) \\ \hat{D} = z_D + K_1 s + P\int_0^t s d\tau \end{cases} \tag{5.63}$$

$$u_c = H^{-1}(-R^{-1}V - G - K_2 s - K_4 z_a - \hat{D}) \tag{5.64}$$

where P, K_1, and $L = K_1 R$ are the observer gain matrices; K_2 and K_4 are the control gain matrices.

Remark 5.6. In most underwater vehicle trajectory tracking control strategies, the ocean current and other disturbances are handled. The ocean current disturbances described in this chapter can also be estimated by establishing the disturbance observer. In this study, the disturbances and faults are treated as total uncertainties, and an observer is introduced to estimate them. This strategy could handle the different disturbances more flexibly.

Theorem 5.2 ([40]). *Considering the trajectory tracking error system (5.61) under thruster saturation, if the controller u, observer, and auxiliary system are designed as Eqs. (5.64), (5.63), and (5.62), respectively, and the gain matrices P, K_1, K_2, K_3, and K_4 are chosen to satisfy the following inequalities:*

$$\begin{aligned} \kappa_1 &= \lambda_{\min}(PK_2) - \frac{1}{2}\lambda_{\max}^2(P) - \frac{1}{2}\lambda_{\max}^{+}(P\dot{W}) > 0 \\ \kappa_2 &= \lambda_{\min}(L) - \frac{1}{2\mu_2} > 0 \\ \kappa_3 &= \lambda_{\min}(K_3) - \frac{1}{2}\lambda_{\max}^2(K_4) - \frac{1}{2} > 0 \end{aligned} \tag{5.65}$$

where μ_2 is a positive constant, then the transformed error ε_i is uniformly ultimately bounded, and tracking error e_i satisfies the prescribed performance constraint Eq. (5.13).

5.3.2.2 Algorithm proof

Since R is a symmetric positive definite matrix, and r_i is bounded, the corresponding Lyapunov function candidate can be designed as follows when the auxiliary system (5.63) is working:

$$V_1 = \frac{1}{2}s^{\mathrm{T}}PR^{-1}s + \frac{1}{2}D_e^{\mathrm{T}}D_e + \frac{1}{2}z_a^{\mathrm{T}}z_a. \tag{5.66}$$

Taking the derivative of V_1 with respect to time and substituting Eqs. (5.61)–(5.64) into it, we obtain

$$\begin{aligned}
\dot{V}_1 &= s^{\mathrm{T}}PR^{-1}\dot{s} + \frac{1}{2}s^{\mathrm{T}}P\dot{W}s + D_e^{\mathrm{T}}\dot{D}_e + z_a^{\mathrm{T}}\dot{z}_a \\
&= s^{\mathrm{T}}PR^{-1}\left[V + R(G + Hu_c + D)\right] + \frac{1}{2}s^{\mathrm{T}}P\dot{W}s \\
&\quad + D_e^{\mathrm{T}}(\dot{D} - LD_e - Ps) + z_a^{\mathrm{T}}\dot{z}_a \\
&= s^{\mathrm{T}}P(D_e - K_2 s - K_4 z_a) + \frac{1}{2}s^{\mathrm{T}}P\dot{W}s \\
&\quad + D_e^{\mathrm{T}}(\dot{D} - LD_e - Ps) + z_a^{\mathrm{T}}\dot{z}_a \\
&= -s^{\mathrm{T}}PK_2 s - s^{\mathrm{T}}PK_4 z_a + \frac{1}{2}s^{\mathrm{T}}P\dot{W}s + D_e^{\mathrm{T}}\dot{D} \\
&\quad - D_e^{\mathrm{T}}LD_e - \lambda_{\min}(K_3)z_a^{\mathrm{T}}z_a - z_a^{\mathrm{T}}H\Delta u - \frac{1}{2}\|H\|^2\|\Delta u\|^2
\end{aligned} \tag{5.67}$$

Applying Young's inequality to Eq. (5.67) gives

$$\begin{aligned}
-s^{\mathrm{T}}PK_4 z_a &\le \frac{1}{2}\lambda_{\max}^2(P)s^{\mathrm{T}}s + \frac{1}{2}\lambda_{\max}^2(K_4)z_a^{\mathrm{T}}z_a \\
-z_a^{\mathrm{T}}H\Delta u &\le \frac{1}{2}z_a^{\mathrm{T}}z_a + \frac{1}{2}\|H\|^2\|\Delta u\|^2
\end{aligned} \tag{5.68}$$

Substituting Eq. (5.68) into Eq. (5.67), we have

$$\begin{aligned}
\dot{V}_1 &\le -\left[\lambda_{\min}(PK_2) - \frac{1}{2}\lambda_{\max}^2(P) - \frac{1}{2}\lambda_{\max}^{+}(P\dot{W})\right]s^{\mathrm{T}}s \\
&\quad - \left[\lambda_{\min}(L) - \frac{1}{2\mu_2}\right]D_e^{\mathrm{T}}D_e + \frac{1}{2}\mu_2\chi^2 \\
&\quad - \left[\lambda_{\min}(K_3) - \frac{1}{2}\lambda_{\max}^2(K_4) - \frac{1}{2}\right]z_a^{\mathrm{T}}z_a \\
&= -\kappa_1 s^{\mathrm{T}}s - \kappa_2 D_e^{\mathrm{T}}D_e - \kappa_3 z_a^{\mathrm{T}}z_a + \gamma
\end{aligned} \tag{5.69}$$

where $\gamma = \frac{1}{2}\mu_2\chi^2$. When we choose appropriate gain matrices P, K_1, K_2, K_3, and K_4 to satisfy condition (5.65), the error s, observation error D_e, and auxiliary variable z_a are uniformly ultimately bounded, which respectively converge to the sets:

$$\begin{aligned} M_1 &= \left\{ s \in R^6 : \|s\| \le \sqrt{\gamma/\kappa_1} \right\} \\ M_2 &= \left\{ D_e \in R^6 : \|D_e\| \le \sqrt{\gamma/\kappa_2} \right\} \\ M_3 &= \left\{ z_a \in R^6 : \|z_a\| \le \sqrt{\gamma/\kappa_3} \right\} \end{aligned} \tag{5.70}$$

Additionally, the transformed error ε_i is uniformly ultimately bounded, which converges to the set

$$M_4 = \left\{ \varepsilon_i \in R : |\varepsilon_i| \le \sqrt{\gamma/\kappa_1} / \lambda_i \right\}. \tag{5.71}$$

Finally, the prescribed performance constraint Eq. (5.13) is obtained based on $S_i(\varepsilon_i)$, that is, the trajectory tracking error e_i achieves prescribed dynamic performance and steady state response represented by Eq. (5.21).

Consider the case that the thruster saturation never happens. Then, $\dot{z}_a = 0$ and $\Delta u = 0$. Similarly, to when $\|z_a\| \ge \sigma$, the new result is as follows:

$$\dot{V}_1 = -\kappa_1 s^T s - \kappa_2 D_e^T D_e + \frac{1}{2} \lambda_{\max}^2 (K_4) \sigma^2 + \gamma \tag{5.72}$$

The conclusion is similar with the case of $\|z_a\| \ge \sigma$, then all the signals of the trajectory tracking closed-loop system are uniformly ultimately bounded.

Remark 5.7. The proposed observer (5.63) is used to approximate the system general uncertainties D. Based on the auxiliary variable z_D, the observation error D_e in the Lyapunov function candidate can converge to the set $M_2 = \left\{ D_e \in R^6 : \|D_e\| \le \sqrt{\gamma/\kappa_2} \right\}$ when the gain matrices P, K_1, and L are chosen to satisfy Eq. (5.65).

Remark 5.8. From the results of references [37] and [38], it can be known that the disturbances observer (5.63) in this paper is much better at dealing with low frequency disturbances. The performance of the observer is limited by the choice of gain matrices, as shown in Eq. (5.65). Additionally, the change rate of system general uncertainties D must be bounded. Therefore, the proposed observer is difficult to be extended to deal with multiple disturbances mentioned in the references above. The future work might be extending this current study to multiple disturbance systems.

Remark 5.9. The parameter σ can avoid singularity based on the auxiliary system (5.62). In reality, parameter σ is usually designed as a small positive constant to make the initial value $z_a(0)$ of auxiliary variable satisfy $\|z_a(0)\| \geq \sigma$ so that it will ensure auxiliary system works at the initial time.

According to the definition of the prescribed performance method and the proof of Theorem 5.1, we propose the parameter selection guidelines for the performance function (5.21), observer (5.63), and control strategy (5.64):

1. The performance function parameters ρ_{i0}, t_{if}, ρ_{it_f}, and k_i should satisfy specific mission needs. Especially, parameter ρ_{i0} should satisfy the initial condition of the trajectory tracking control system, such as $0 \leq |e_i(0)| < \rho_{i0}$.
2. The observer gain matrix K_1 should be small enough so that $\kappa_2 > 0$ is satisfied in Eq. (5.65).
3. The observer and controller should choose appropriate gain matrices P and K_2, respectively, to satisfy $\kappa_1 > 0$ in Eq. (5.65).
4. The auxiliary system gain matrix K_3 and controller gain matrix K_4 should satisfy $\kappa_3 > 0$ in Eq. (5.65).

5.3.2.3 Simulation results

In this part, a redundantly actuated AUV is introduced to demonstrate the effectiveness of the proposed control method. The AUV nonlinear dynamic model is given in Section 5.1.3. All the thrusters work independently from each other, and provide double-direction thrust. The thruster maximum output is set as ± 60 N under the thruster saturation. The initial position and attitude vector is set as $\eta(0) = [0; 0; 0; 1.5; 1.5; 1.5]$ in the inertial frame. The initial velocity and angular velocity vector is set as $v(0) = [0.1; 0; -0.1; 0; 0; 0]$ in the body-fixed frame. Additionally, the hydrodynamic and inertia coefficients are given in Tables 5.1 and 5.2, respectively.

We assume the ocean current orientation parallels the x-axis positive direction in the earth coordinate system. The current velocity can be expressed as follows:

$$V_c = 2\sin(0.1t - \frac{\pi}{2}) + 2. \tag{5.73}$$

In this section, the modeling uncertainties are quantified. The 20% nominal values are used to represent modeling uncertainties.

In order to demonstrate that the proposed controller can deal with thruster faults, we introduce three common thruster faults, including incipient thruster fault, intermittent fault, and abrupt fault, as follows [25]:

$$k_{11}^1 = \begin{cases} 0, & t < 20 \\ \frac{0.5}{40}t - \frac{0.5}{2} + 0.1\sin\left(\frac{\pi}{5}t - 4\pi\right), & 20 \le t < 60 \\ 0.5 + 0.1\sin\left(\frac{\pi}{10}t - 6\pi\right), & t \ge 60 \end{cases} \tag{5.74}$$

$$k_{11}^2 = \begin{cases} 0, & t < 20 \\ 0.6, & 20 \le t < 50 \\ 0, & 50 \le t < 70 \\ 0.6, & 70 \le t \end{cases} \tag{5.75}$$

$$k_{11}^3 = \begin{cases} 0, & t < 50 \\ 0.6, & t \ge 50 \end{cases} \tag{5.76}$$

At the same time, we also introduce two curves such as the desired trajectories that include straight and spiral lines. The corresponding expressions are shown below:

$$\eta_{d1} = [1.5 + 0.1t; 1.5; 1.5 - 0.1t; 0; 0; 0], \tag{5.77}$$

$$\eta_{d2} = [2\sin(0.1t); 2\cos(0.1t) + 2; -0.5144t; 0; 0; 0]. \tag{5.78}$$

For each axis, the desired control performances are designed as follows: (1) the steady-state tracking error is less than 0.01; (2) the maximum convergence time is not more than 20 s; (3) the system response has no overshoot. According to the above conditions, $\rho_i(t)$ and δ_i are given in Table 5.5. Additionally, the parameters of the controller and the observer are given in Table 5.6.

Table 5.5 Parameters of the new prescribed performance function.

Parameters	ρ_{i0}	$\rho_{i\infty}$	k_i	t_{if}, s	δ_i
Value	1.8	0.011	0.08	80	0

Table 5.6 Parameters of the controller and observer.

Parameters	λ	P	K_1	K_2	K_3	K_4
Value	$0.125I_6$	$0.1I_6$	$0.5I_6$	$0.8I_6$	$0.6I_6$	$0.1I_6$

Case 1. Straight line trajectory tracking

In this part, the desired trajectory is based on Eq. (5.77). Considering the influence of uncertainties, disturbances, and thruster saturation, we give three kinds of tracking error and observation error curves based on the three fault cases represented by Eqs. (5.74)–(5.76), respectively.

Remark 5.10. In order to avoid occupying a great deal of space, the 6-DOF trajectory tracking error curves are drawn together. Additionally, the prescribed performance constraint curve of each axis is the same to make the simulation result more visual.

The prescribed performance constraint curve is denoted as ρ, and the description of other curves is illustrated in corresponding legends.

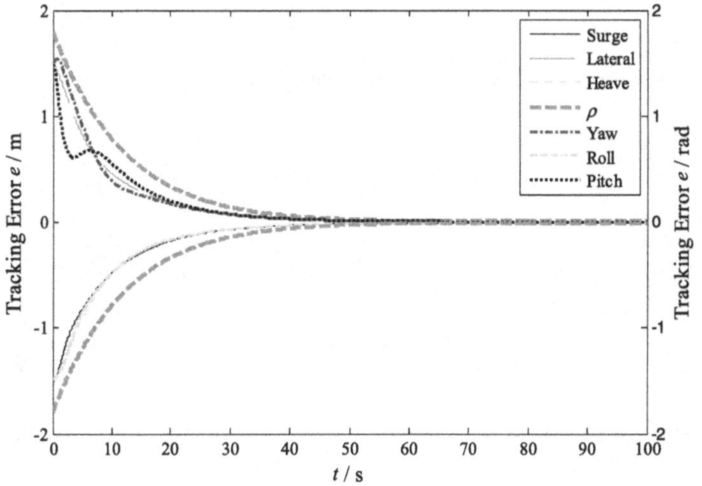

Figure 5.9 Tracking error with thruster incipient fault.

From Figs. 5.9–5.14, we can conclude that proposed general uncertainties observer can effectively observe the influence caused by disturbances, uncertainties, and thruster faults. The proposed prescribed performance control method can limit tracking error within the boundary created by the performance function. Additionally, the tracking errors converge to prescribed steady-state precision within the preestablished time.

Case 2. Spiral line trajectory tracking

In this part, the desired trajectory adopts the spiral line based on Eq. (5.78). The rest of the simulation is similar to the straight line case.

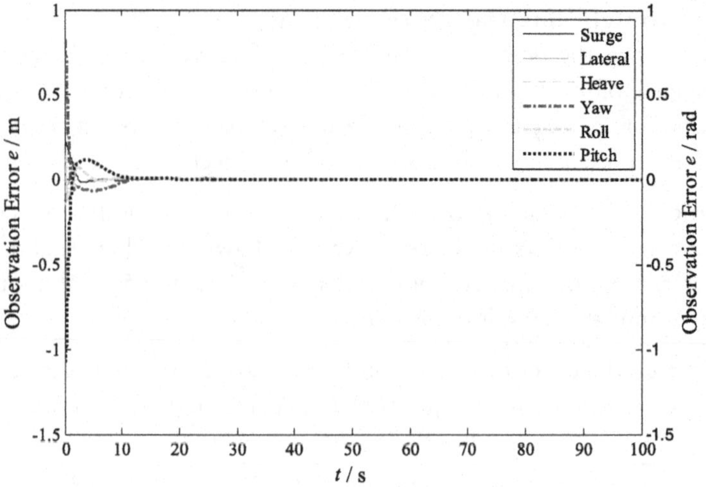

Figure 5.10 Observation error with thruster incipient fault.

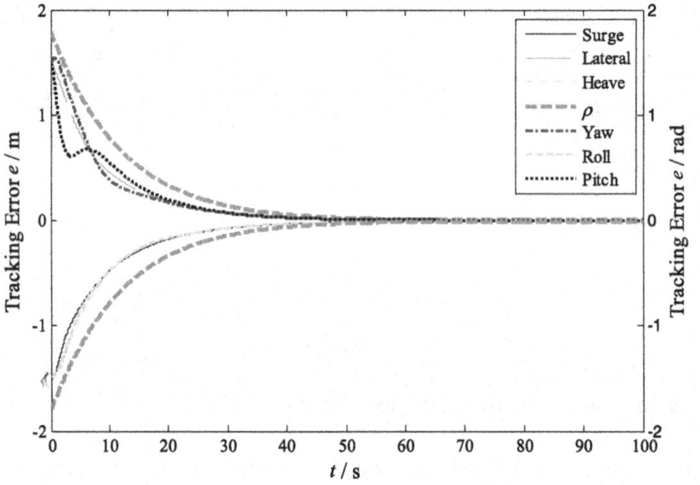

Figure 5.11 Tracking error with thruster intermittent fault.

As shown in Figs. 5.15–5.20, we can obtain similar conclusions that the proposed general uncertainties observer and prescribed performance control method are still valid.

Additionally, in order to make comparative research, we compare the proposed control algorithm (5.62)–(5.64) with a traditional prescribed performance controller with the same parameters of the performance function.

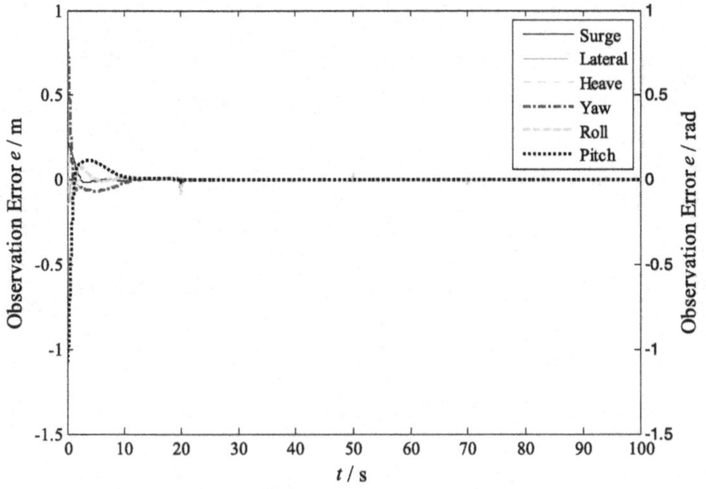

Figure 5.12 Tracking error with thruster intermittent fault.

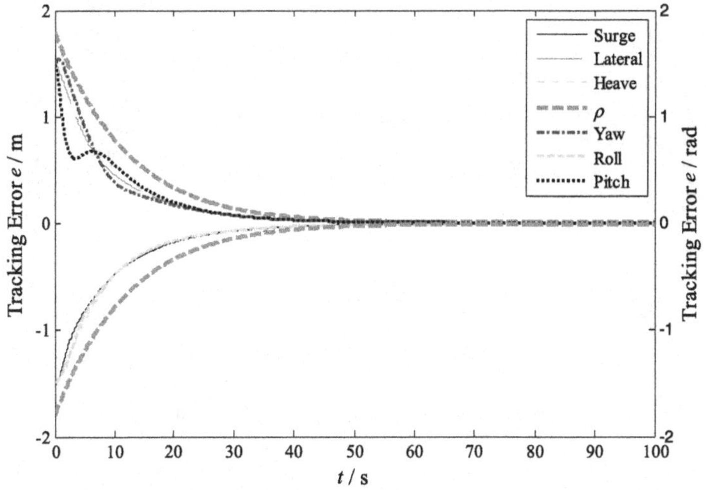

Figure 5.13 Tracking error with thruster intermittent fault.

In this part, thruster fault is based on Eq. (5.76). The thruster maximum output is set as ±85 N under the thruster saturation. Since the simulation results are similar, this section only provides surge and yaw trajectory tracking errors.

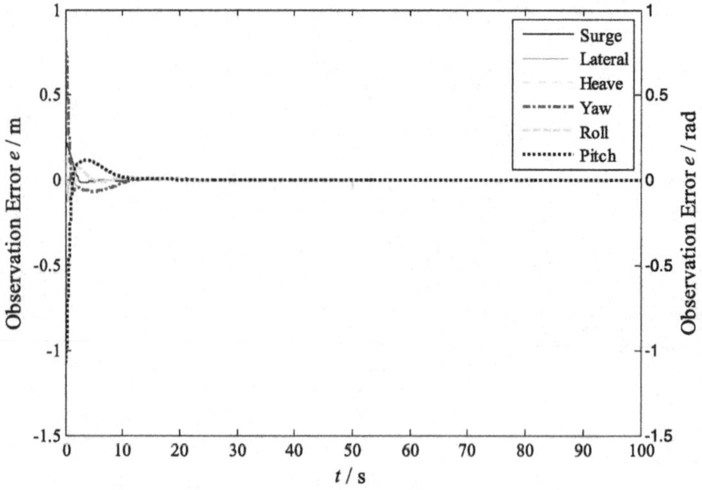

Figure 5.14 Tracking error with thruster intermittent fault.

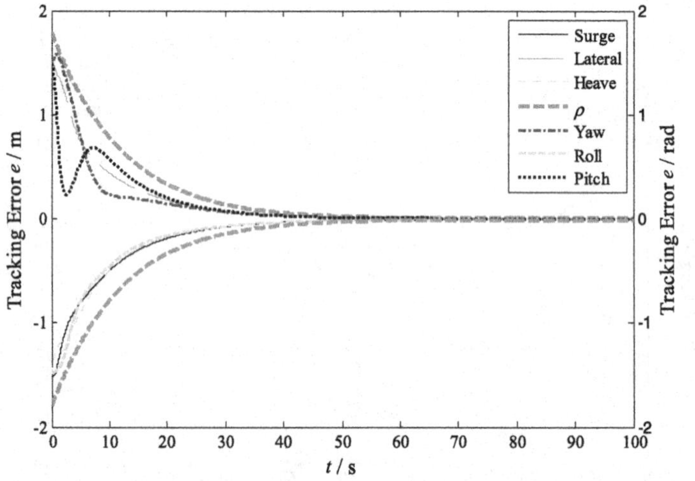

Figure 5.15 Tracking error with thruster incipient fault.

Figs. 5.21–5.22 represent surge and yaw trajectory tracking errors of the AUV, respectively. The proposed control algorithm (5.62)–(5.64) is denoted as TSPPC, the traditional prescribed-performance controller is denoted as PPC.

From Figs. 5.21–5.22, it can be found that the proposed TSPPC method (5.62)–(5.64) has better transient and steady responses with the help

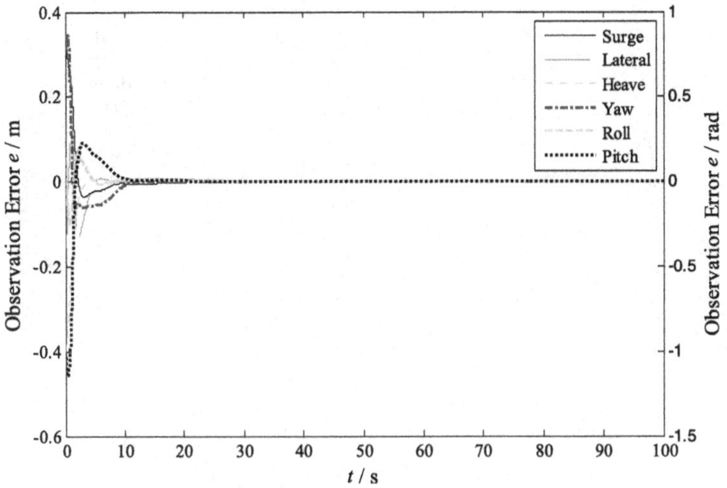

Figure 5.16 Observation error with thruster incipient fault.

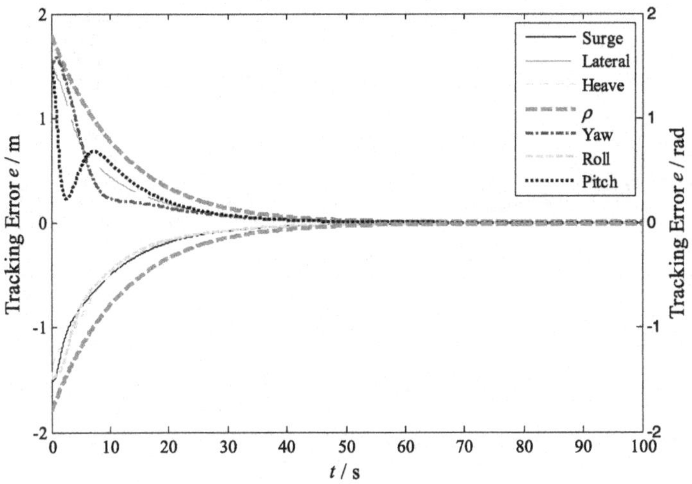

Figure 5.17 Tracking error with thruster intermittent fault.

of the saturation auxiliary system and new performance function. However, the traditional prescribed performance controller with the same parameters can only satisfy the thruster saturation condition by reducing the global convergence rate of the tracking error. The simulation results indicate that the proposed method is better than the traditional prescribed performance method under the actual situation.

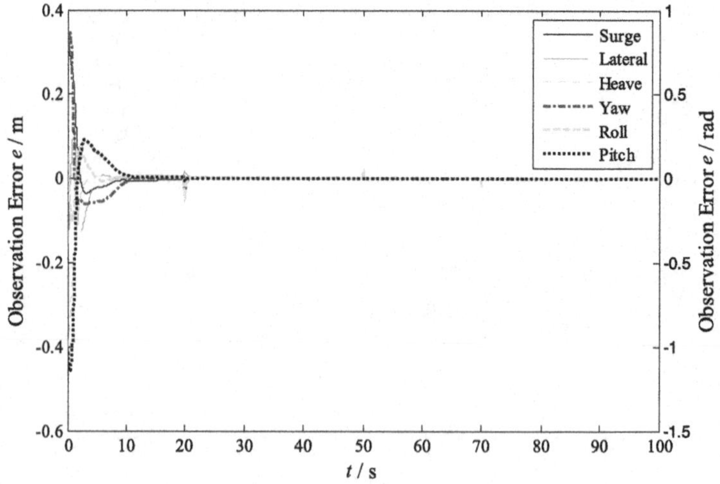

Figure 5.18 Observation error with thruster intermittent fault.

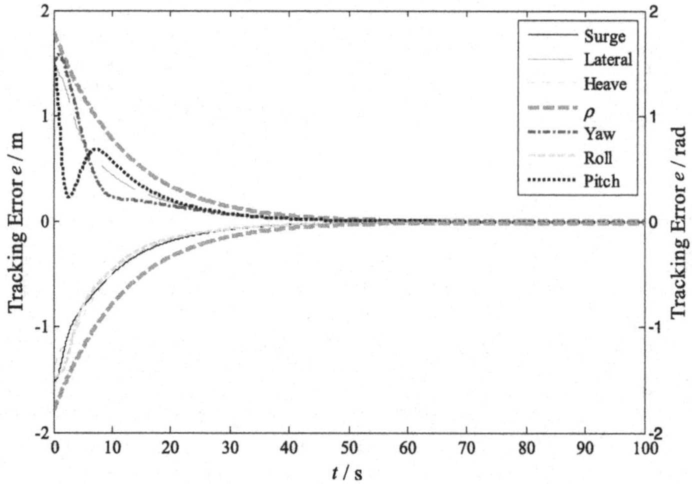

Figure 5.19 Tracking error with thruster abrupt fault.

Conclusion

In this chapter, several autonomous control methods which include the PID control, observer technique, RBFNN, and prescribed performance control are proposed for trajectory tracking control of AUV under the influence of modeling uncertainties, ocean current, and thruster faults. First,

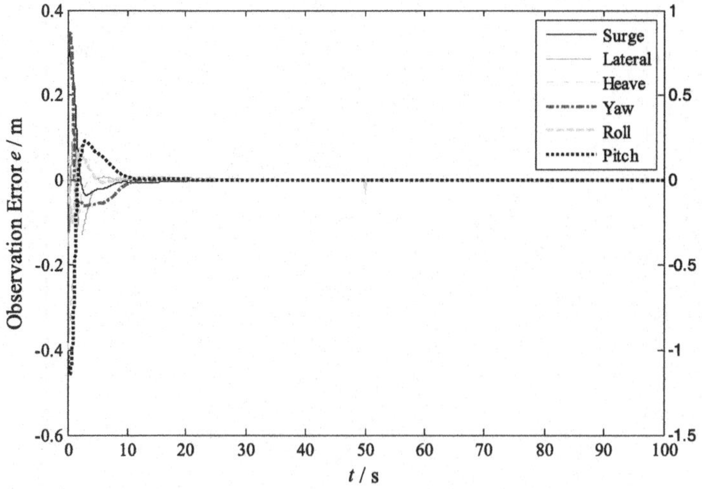

Figure 5.20 Observation error with thruster abrupt fault.

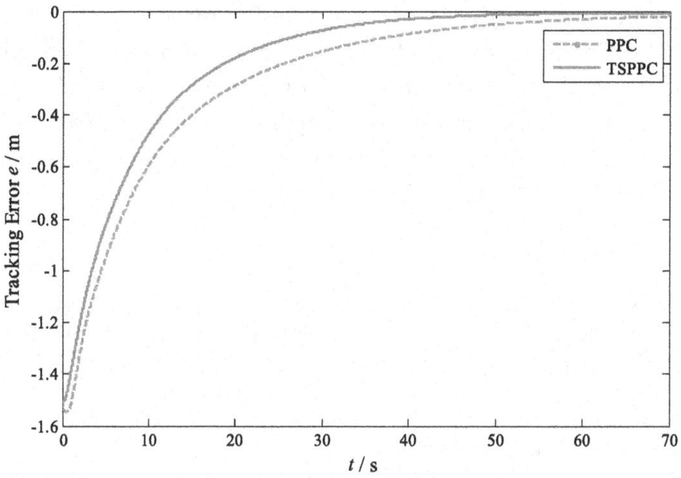

Figure 5.21 Tracking error in surge with thruster saturation.

we established the AUV 6-DOF dynamic model. Then, we provided a brief description and design idea of the above control methods. In case study, the proposed performance function is able to make the tracking errors not only achieve the desired steady precision, but also enforce the convergence time to satisfy the given index. Moreover, an observer is designed to compensate for the uncertainty of AUV. The stability of the closed-loop trajectory

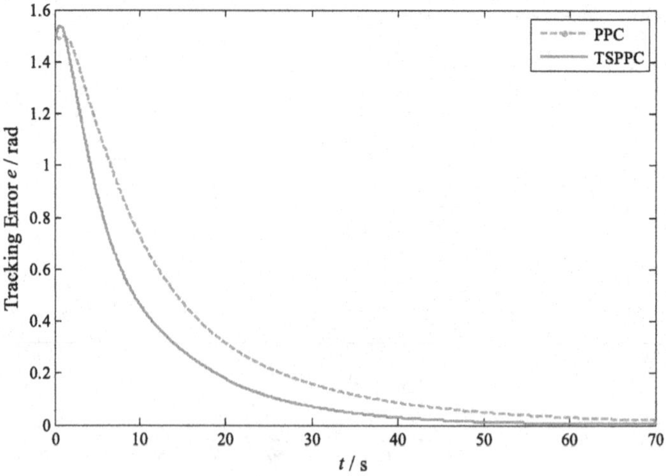

Figure 5.22 Tracking error in yaw with thruster saturation.

tracking control system is analyzed based on Lyapunov theory. Simulations are carried out to illustrate the effectiveness of the proposed method. How to extend the results of this chapter to a more general practical and academic applications may be an interesting issue, such as unknown control directions and finite-time convergence. Another future work might be extending this current chapter to multiple AUV systems.

References

[1] M. Bidoki, M. Mortazavi, M. Sabzehparvar, A new approach in system and tactic design optimization of an autonomous underwater vehicle by using Multidisciplinary Design Optimization, Ocean Engineering 147 (2018) 517–530.

[2] N. Wang, G. Xie, X. Pan, et al., Full-state regulation control of asymmetric under-actuated surface vehicles, IEEE Transactions on Industrial Electronics (2018), https://doi.org/10.1109/TIE.2018.2890500.

[3] R.B. Wynn, V.A.I. Huvenne, T.P.L. Bas, et al., Autonomous underwater vehicles (AUVs): their past, present and future contributions to the advancement of marine geoscience, Marine Geology 352 (2) (2014) 451–468.

[4] A. Shukla, H. Karki, Application of robotics in offshore oil and gas industry – a review. Part II, Robotics and Autonomous Systems 75 (PB) (2015) 508–524.

[5] M. Carreras, J.D. Hernández, E. Vidal, et al., Sparus II AUV—a hovering vehicle for seabed inspection, IEEE Journal of Oceanic Engineering 43 (2) (2018) 344–355.

[6] B. Allotta, R. Costanzi, L. Pugi, et al., Identification of the main hydrodynamic parameters of Typhoon AUV from a reduced experimental dataset, Ocean Engineering 147 (2018) 77–88.

[7] T. Elmokadem, M. Zribi, K. Youcef-Toumi, Trajectory tracking sliding mode control of underactuated AUVs, Nonlinear Dynamics 84 (2) (2016) 1079–1091.

[8] T.I. Fossen, Handbook of Marine Craft Hydrodynamics and Motion Control, 1st ed., John Wiley and Sons, New York, 2011.
[9] E. Omerdic, G. Roberts, Thruster fault diagnosis and accommodation for open-frame underwater vehicles, Control Engineering Practice 12 (12) (2004) 1575–1598.
[10] M. Zhang, X. Liu, F. Wang, Backstepping based adaptive region tracking fault tolerant control for autonomous underwater vehicles, Journal of Navigation 70 (1) (2016) 184–204.
[11] T. Chen, H. Wen, Autonomous assembly with collision avoidance of a fleet of flexible spacecraft based on disturbance observer, Acta Astronautica 147 (Jun. 2018) 86–96.
[12] T. Chen, J. Shan, H. Wen, Distributed adaptive attitude control for networked underactuated flexible spacecraft, IEEE Transactions on Aerospace and Electronic Systems 55 (1) (Jun. 2018) 215–225.
[13] Y. Chen, J. Li, K. Wang, S. Ning, Robust trajectory tracking control of underactuated underwater vehicle subject to uncertainties, Journal of Marine Science and Technology 25 (3) (Jun. 2017) 283–298.
[14] T. Elmokadem, M. Zribi, K. Youcef-Toumi, Terminal sliding mode control for the trajectory tracking of underactuated autonomous underwater vehicles, Ocean Engineering 129 (Jan. 2017) 613–625.
[15] J. Xu, M. Wang, G. Zhang, Trajectory tracking control of an underactuated unmanned underwater vehicle synchronously following mother submarine without velocity measurement, Advances in Mechanical Engineering 7 (7) (Jul. 2015).
[16] H. Joe, M. Kim, S. Yu, Second-order sliding-mode controller for autonomous underwater vehicle in the presence of unknown disturbances, Nonlinear Dynamics 78 (1) (Oct. 2014) 183–196.
[17] L. Zhang, X. Qi, Y. Pang, D. Jiang, Adaptive output feedback control for trajectory tracking of AUV in wave disturbance condition, International Journal of Wavelets, Multiresolution and Information Processing 11 (3) (May 2013).
[18] G.V. Lakhekar, L.M. Waghmare, Robust maneuvering of autonomous underwater vehicle: an adaptive fuzzy PI sliding mode control, Intelligent Service Robotics 10 (3) (Jul. 2017) 195–212.
[19] F. Liu, D. Xu, J. Yu, L. Bai, Fault isolation of thrusters under redundancy in frame-structure unmanned underwater vehicles, International Journal of Advanced Robotic Systems 15 (2) (Apr. 2018).
[20] Z. Wang, P. Shi, C.C. Lim, H-/H∞ fault detection observer in finite frequency domain for linear parameter-varying descriptor systems, Automatica 86 (Dec. 2017) 38–45.
[21] Z.H. Ismail, A.A. Faudzi, M.W. Dunnigan, Fault-tolerant region-based control of an underwater vehicle with kinematically redundant thrusters, Mathematical Problems in Engineering 2014 (Jun. 2014).
[22] Y.S. Sun, Y.M. Li, G.C. Zhang, Y.H. Zhang, H.B. Wu, Actuator fault diagnosis of autonomous underwater vehicle based on improved Elman neural network, Journal of Central South University 23 (4) (Apr. 2016) 808–816.
[23] T. Chen, J. Shan, Rotation-matrix-based attitude tracking for multiple flexible spacecraft with actuator faults, Journal of Guidance, Control, and Dynamics 42 (1) (Sep. 2018) 181–188.
[24] H. Hai, W. Lei, C. Wen-tian, P. Yong-jie, J. Shu-qiang, A fault-tolerable control scheme for an open-frame underwater vehicle, International Journal of Advanced Robotic Systems 11 (May 2014).

[25] M. Zhang, X. Liu, B. Yin, W. Liu, Adaptive terminal sliding mode based thruster fault tolerant control for underwater vehicle in time-varying ocean currents, Journal of the Franklin Institute 352 (11) (Nov. 2015) 4935–4961.
[26] C.P. Bechlioulis, G.A. Rovithakis, Robust adaptive control of feedback linearizable MIMO nonlinear systems with prescribed performance, IEEE Transactions on Automatic Control 53 (9) (Oct. 2008) 2090–2099.
[27] S. Shao, M. Chen, X. Yan, Prescribed performance synchronization for uncertain chaotic systems with input saturation based on neural networks, Neural Computing & Applications 29 (12) (Jun. 2018) 1349–1361.
[28] A. Fan, J. Li, Adaptive neural network prescribed performance matrix projection synchronization for unknown complex dynamical networks with different dimensions, Neurocomputing 281 (May 2018) 55–66.
[29] J. Luo, Z. Yin, C. Wei, J. Yuan, Low-complexity prescribed performance control for spacecraft attitude stabilization and tracking, Aerospace Science and Technology 74 (Oct. 2018) 173–183.
[30] J. Luo, C. Wei, H. Dai, Z. Yin, X. Wei, J. Yuan, Robust inertia-free attitude takeover control of postcapture combined spacecraft with guaranteed prescribed performance, ISA Transactions 74 (Mar. 2018) 28–44.
[31] Q. Guo, Y. Liu, D. Jiang, Q. Wang, W. Xiong, J. Liu, X. Li, Prescribed performance constraint regulation of electrohydraulic control based on backstepping with dynamic surface, Applied Sciences 8 (1) (Jan. 2018) 76.
[32] H. Qin, Z. Wu, Y. Sun, H. Chen, Disturbance-observer-based prescribed performance fault-tolerant trajectory tracking control for ocean bottom flying node, IEEE Access 7 (1) (Dec. 2019) 49004–49013.
[33] N. Wu, C. Wu, T. Ge, D. Yang, R. Yang, Pitch channel control of a REMUS AUV with input saturation and coupling disturbances, Applied Sciences 8 (2) (Feb. 2018) 253.
[34] P. Sarhadi, A.R. Noei, A. Khosravi, Adaptive integral feedback controller for pitch and yaw channels of an AUV with actuator saturations, ISA Transactions 65 (Nov. 2016) 284–295.
[35] P. Sarhadi, A.R. Noei, A. Khosravi, Model reference adaptive PID control with anti-windup compensator for an autonomous underwater vehicle, Robotics and Autonomous Systems 83 (Sep. 2016) 87–93.
[36] F. Rezazadegan, K. Shojaei, F. Sheikholeslam, A. Chatraei, A novel approach to 6-DOF adaptive trajectory tracking control of an AUV in the presence of parameter uncertainties, Ocean Engineering 107 (Oct. 2015) 246–258.
[37] X. Wei, H. Zhang, S. Sun, H.R. Karimi, Composite hierarchical antidisturbance control for a class of discrete-time stochastic systems, International Journal of Robust and Nonlinear Control 28 (9) (Mar. 2018) 3292–3302.
[38] H. Zhang, X. Wei, L. Zhang, M. Tang, Disturbance rejection for nonlinear systems with mismatched disturbances based on disturbance observer, Journal of the Franklin Institute 354 (11) (Jul. 2017) 4404–4424.
[39] H. Qin, Z. Wu, Y. Sun, Y. Sun, Prescribed performance adaptive fault-tolerant trajectory tracking control for an ocean bottom flying node, International Journal of Advanced Robotic Systems 16 (3) (May 2019) 172988141984194.
[40] H. Qin, Z. Wu, Y. Sun, H. Chen, Disturbance-observer-based prescribed performance fault-tolerant trajectory tracking control for ocean bottom flying node, IEEE Access 7 (2019) 49004–49013.

CHAPTER SIX

Development of hybrid control architecture for a small autonomous underwater vehicle

Zhenzhong Chu
Shanghai, China

Contents

Abstract

According to the design scheme of "Beaver III", the hybrid control architecture with the characteristics of hierarchical and subsumption architectures is proposed in this chapter. The workflow of the hybrid control architecture has been described by an example. It indicates the reasonability of the proposed hybrid control architecture. A pool experiment verifies the feasibility and effectiveness of the proposed hybrid control architecture in practical work.

Keywords

Autonomous underwater vehicle, Hybrid control architecture, Target search, Image acquisition, PD sliding mode control

6.1. Introduction

Small autonomous underwater vehicle (SAUV) is an important equipment for oceanographic and resource development [1]. It works in a complex marine environment, and there are typically many intricate relationships between its functional modules [2,3]. Control architecture refers to the time and spatial distribution patterns of intelligent decision-making,

https://doi.org/10.1016/B978-0-12-820271-5.00011-0

information collection, and behavior control of robot systems [4]. The research on the problem of control architecture is one of the most basic and critical for SAUV to efficiently accommodate various functional modules, so that the information flows timely and smoothly between various functional modules, and all functional modules reasonably play their respective roles in time and space [5].

In order to meet the requirements of different control tasks, many distinctive control architectures have been proposed for autonomous underwater vehicles (AUVs) based on different design ideas. These existing control architectures can usually be divided into four categories: hierarchical, behavioral, subsumption, and hybrid [6].

Hierarchical architecture is constructed in accordance with the principle of "intelligent reducing while accuracy increasing", which generally can be divided into three layers: organization, coordination, and execution [7–10]. In general, the organization layer contains reasoning, planning, and decision-making function modules. The coordination layer contains navigation, controlling, and monitoring function modules. The execution layer is the lowest layer of the architecture, which connects directly to sensors and actuators. Hierarchical architecture has good ability of planning and reasoning. Through the task decomposition layer by layer, the work scope of every function module will be gradually narrowed, whereas the accuracy of problem solving will be gradually increased. Due to the features of structure and easy implementation, hierarchical architecture is one of the more commonly used architectures. However, each function module in hierarchical architecture can only exchange information with its vertically adjacent modules. In operation, the function modules in a lower layer have to wait for the programming instructions from the function modules in an upper layer, and the function modules in an upper layer need to wait for the completion of the function modules in a lower layer. Because of the longer reaction time in dealing with external events, the reliability and robustness of hierarchical architecture is relatively poor.

Behavioral architecture is constructed based on the behavior decomposition principle [11,12]. In this architecture, each function module uses a parallel structure distribution pattern, and all of the information from the sensors to actuators is a one-way flow. The advantages of behavioral architecture are that the knowledge and sensor information can be quickly integrated into various function modules with good scalability. However, the behavioral architecture overemphasizes independence and parallelism of various function modules, which will lead to a lack of overall control and coordination. Hence, it is difficult to put expert knowledge into such architecture and coordinate function modules with different control frequencies.

Subsumption and behavioral architectures are similar in that they are constructed on the behavior decomposition principle [5]. But unlike the behavioral architecture, the upper function modules of subsumption architecture can accommodate the lower function modules, and the behavior of upper function modules can suppress the behavior of lower function modules. Compared with hierarchical architecture, subsumption architecture has obvious advantages in terms of real-time. It has higher reliability and robustness by processing sensor information along in each layer. However, as with the increase in the complexity of the system functionality, the output behaviors of function modules will also increase. Since the upper function modules have to interfere with the output behavior of lower function modules, a reasonable arrangement of the level of each function module in subsumption architecture is very complicated.

Since hierarchical, behavioral, and subsumption architectures have these shortcomings and their respective advantages, the hybrid control architecture is constructed by combining different architectures [13]. Currently, there are two common hybrid architectures. One combines hierarchical and behavioral architectures [14,15], and the other combines hierarchical and subsumption architectures [16,17]. These two types of architecture are typically divided into three layers vertically, which is similar with hierarchical architecture. In the middle layer, it is usually designed as behavioral or subsumption architecture. Through the reasonable layer arrangement of various functional modules, the behavioral or subsumption architecture can combine with the reasoning and planning function modules in the upper layer. Hence, the hybrid control architecture has the rapid response capability and robustness in a complex marine environment [18,19].

Considering the characteristics of the above control architectures and the developed SAUV design scheme, a hybrid architecture combined with hierarchical and subsumption architectures is presented in this paper. From the vertical perspective, the proposed hybrid architecture can be divided into management, function, and hardware layers so that it has good planning and reasoning abilities. In the function layer, the subsumption architecture is used and multiple independent function modules are designed. According to different tasks, the management layer can choose appropriate function modules of the function layer to interact with the hardware layer. Thus, a better real-time response can be obtained. In this chapter, an overall design scheme of SAUV is introduced, and then the hybrid control architecture is described in detail. Finally, an example is analyzed to introduce the workflow of the control architecture, and the effectiveness is verified by a pool experiment.

6.2. Design scheme of SAUV

As shown in Fig. 6.1, "Beaver III" is an SAUV. It can perform a variety of tasks, such as target search and image acquisition. It can also be used as a carrier equipped with a manipulator, side-scan sonar and other equipment for autonomous operations. The overall design scheme of "Beaver III" consists of two parts: actuators system scheme and sensors system scheme.

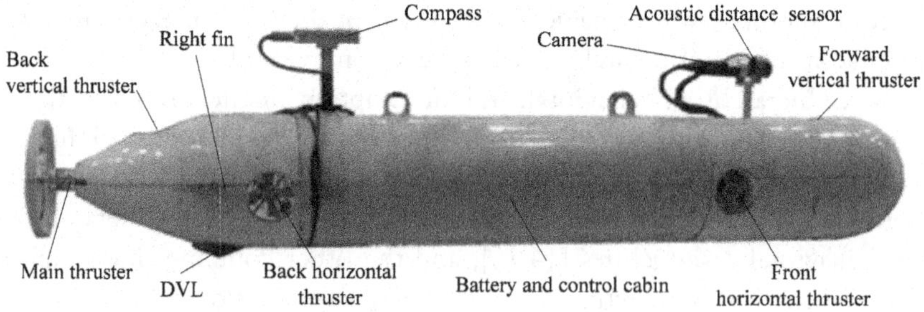

Figure 6.1 Appearance of "Beaver III".

Due to the diversity of tasks, the actuators system scheme has a clear distinction from the traditional SAUV. The main task of traditional SAUV is underwater detection. In order to reduce energy consumption, it is usually designed as an underactuated system, such as a combination of thruster and rudders [20,21]. Such an actuator system makes SAUV difficult to achieve dynamic positioning control or multiple degrees of freedom (DOF) trajectory tracking control. So that it is difficult to complete the task of continuous video transmission for a fixed target and other tasks. Based on the above considerations, "Beaver III" is equipped with five thrusters to control the five degrees of freedom (DOF), that is, surge, sway, vertical, yaw, and pitch. In addition, taking into account the limitations of volume, energy, and other conditions, as well as the efficiency reduction of vertical thrusters while sailing at high speed, two independent fins are used.

For different tasks, different actuators can be used. In the process of long-distance sailing, the main thruster, two horizontal thrusters, and two fins can be used to reduce energy consumption. In the complete image acquisition, manipulator operations, and other tasks, five thrusters can be used for dynamic positioning or trajectory tracking. Meanwhile, if there is current interference in dynamic positioning, two independent fins can be used to reduce the rolling movement.

Considering the dynamic positioning and trajectory tracking, the sensors which can get their own information about position and attitude must be employed. In "Beaver III", the Doppler Velocity Log (DVL), electronic compass, and depth gauge are used for dead reckoning. In order to perform underwater target search, image acquisition, and other tasks, a camera is also included. Considering the complexity of the underwater environment, an ultrasonic distance measuring device is used. Meanwhile, in order to monitor the operation status, "Beaver III" is also equipped with a battery voltage detection sensor and a water leakage detection sensor.

The major technical indicators of "Beaver III" are shown in Table 6.1.

Table 6.1 The major technical indicators of "Beaver III".

Size	Length: 1.95 m, diameter: 0.3 m
Weight	89.6 kg
Actuators	One main thruster, forward thrust: 5.5 kg, reverse thrust: 3.1 kg
	Two horizontal thrusters, forward thrust: 2.6 kg, reverse thrust: 2.4 kg
	Two vertical thrusters, same with horizontal thrusters
	Two fins driven by stepper motor
Drive	High speed sailing: main thruster, two horizontal thrusters, and two fins
	Low speed sailing and dynamic positioning: five thrusters
	Dynamic positioning under current disturbance: five thrusters and two fins
Maximum speed	2 kn
Sensors	Doppler Velocity Log, electronic compass, depth gauge, camera, ultrasonic distance measuring device, battery voltage detection sensor, and water leakage detection sensor
Endurance	4 h (1 k)
Maximum depth	100 m

6.3. Hybrid control architecture

"Beaver III" is driven by thrusters and fins. It is equipped with various types of sensors and can complete a variety of tasks. Hence, a reasonable control architecture is the foundation of intelligent autonomous operations. Compared with traditional SAUV, the control system of "Beaver III" is more complex. By building a control architecture with modular features, it will effectively simplify the design and development of the system, and it also can improve the efficiency of each function module in practice. Although the hierarchical architecture is simple to implement, it has large computational load, poor function expansion, and poor real-time performance. It is difficult to complete the complex underwater tasks. Taking into account that the subsumption architecture has better real-time response and modularization, a hybrid architecture is presented in this paper. It combines with the features of hierarchical and subsumption architectures. From the vertical perspective, the proposed control architecture can be divided into the management, function, and hardware layers. The function layer is designed as subsumption architecture. It has function modules, such as a dead reckoning module, motion control module, optical vision module, and other independent function modules. The proposed hybrid control architecture is shown in Fig. 6.2.

In Fig. 6.2, the management and function layers are the core parts of the hybrid control architecture, and the hardware layer contains the actuator and sensor systems of "Beaver III". Here, we will describe the management and the function layers in detail.

6.3.1 Management layer

The management layer is an important manifestation of intelligence. Through interaction with an onshore monitoring system, the management layer can undertake, decompose, and plan a task. It is also responsible for database maintenance.

The task undertaking module can interact with the onshore monitoring system. By operating this onshore monitoring system, an operator can send instructions to "Beaver III" via Ethernet or wireless serial link.

The task decomposition module is crucial for the successful implementation of a task. After undertaking it, the task decomposition module decomposes the task into several subtasks depending on time, space restrictions, task type and degree of difficulty, and other factors, and then the logical relationship network of subtasks is formed. The logical relationship

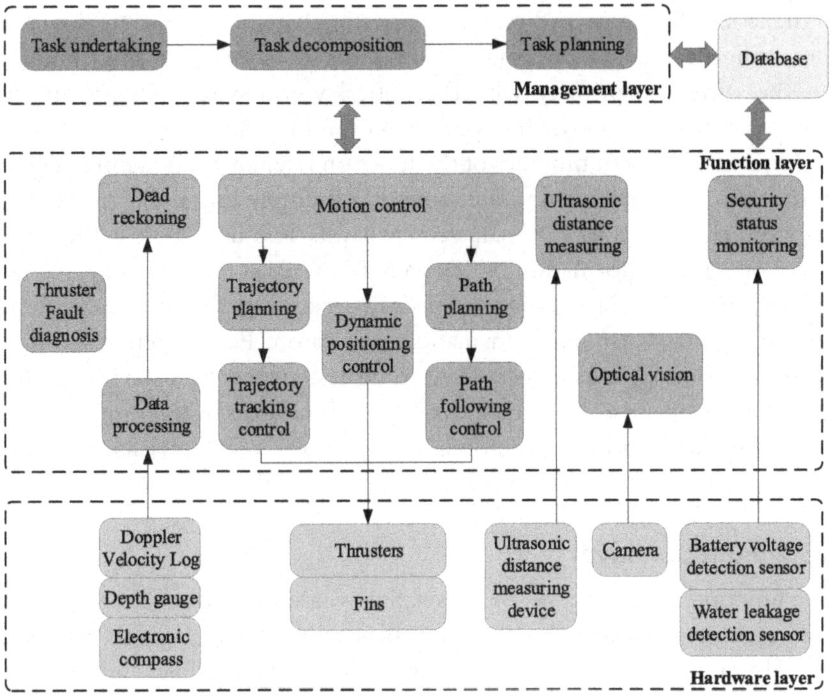

Figure 6.2 Hybrid control architecture.

network describes the serial or parallel sequence of each subtask in time and space. In the operation process, each function module will be executed sequentially in accordance with the relationship network.

The task planning module is used to select the corresponding function modules of the function layer and arrange the priority of function modules. For different tasks, the required function modules of the function layer are not the same. In the task of long-distance sailing, the function modules are needed to be executed, including the ultrasonic distance measuring module, path following control module, fault diagnosis module, and security status monitoring module. In the task of optical vision-based target three-dimensional positioning, the function modules needed include the dead reckoning module, dynamic position control module, optical vision module, fault diagnosis and security status monitoring module. In addition, for different tasks, the task priority of function modules may be different. For example, the ultrasonic distance measuring module has a higher task priority in a long-distance sailing task to achieve obstacle avoidance. However, in the task of target three-dimensional locating based on ultrasonic distance measuring device and camera, the ultrasonic distance measuring module has the same task priority with the optical vision system. Its priority is lower than that of the dynamic position control module.

Since the function modules of the function layer are working in parallel, in order to prevent accessing the same data multiple times, the database is managed uniformly by the management layer. The database has the necessary operation information of "Beaver III", such as the data collected by sensors, trajectory information from dead reckoning, thruster fault information, and task planning information, and so on. Each function module of the function layer submits its data to corresponding local variables, and then the management layer updates the global variables in real time according to the local variables. According to these global variables, each function module can access the data of other function modules.

6.3.2 Function layer

The function layer contains several relatively independent modules, such as dead reckoning module, motion control module, fault diagnosis module, optical vision module, ultrasonic distance measuring module, and security status monitoring module, and each function module can exchange data through the database. In the course of their working, the function modules can run according to priority or run in parallel with the same priority, which produces better real-time performance. In the function layer, the function modules can communicate directly with the hardware devices of the hardware layer, and the security status monitoring module has been given the power to send commands to the thrusters directly. Therefore, the function layer has the same characteristics with subsumption architecture.

Because "Beaver III" is not equipped with the three-dimensional position measurement sensors, such as ultrashort base line (USBL), inertial navigation system (INS), and so on. Thus, the dead reckoning module is designed to get the three-dimensional position in the earth-fixed frame. Considering the complexity of underwater environment, the return data of navigation sensors contains outliers and noise. The data processing must be carried out based on Kalman filter first. Considering that the measurement error of the vertical velocity in DVL is large, and the roll and pitch angles of "Beaver III" are generally small, only the data of longitudinal and transverse velocities of DVL are used. Combined with the data of electronic compass, longitudinal and transverse velocities, dead reckoning can be carried out to get the position in the horizontal plane of the earth-fixed frame.

The vertical position is obtained by the depth gauge [22]. The calculation process of the dead reckoning system is shown in Fig. 6.3, where ν^I is the filtered velocity vector in the DVL-measuring-fixed frame, and its source data is given by DVL; ν^B is the velocity vector in the body-fixed frame; ν^E is the velocity vector in the earth-fixed frame; x^E, y^E, and z^E are the three-dimensional positions in the earth-fixed frame.

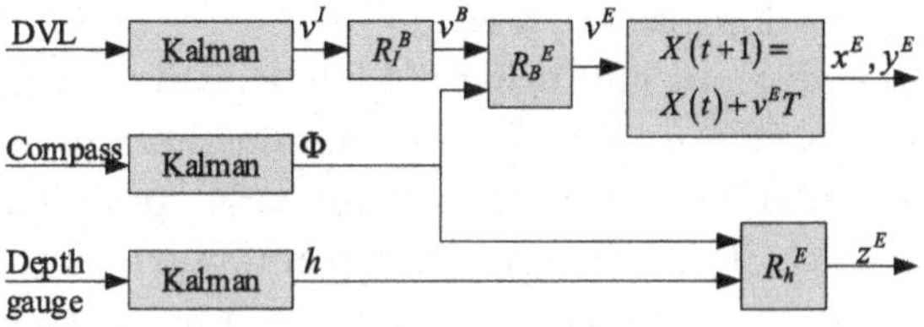

Figure 6.3 Process of dead reckoning.

The motion control module is responsible for generating motion control law to control the thrusters and fins. For different subtasks, the motion control module can choose the dynamic positioning control submodule, path following control submodule, or trajectory tracking control submodule. In this module, the desired trajectory of the trajectory tracking control submodule is generated by the trajectory planning submodule, and the desired path of the path following control submodule is generated by the path planning submodule. Due to the complexity of the marine environment, many methods have been proposed to improve the motion control performance of the SAUVs [23]. However, it is undeniable that PID is still a commonly used method. Compared with many advanced algorithms, the gains are easily set. For the SAUV system discussed in this chapter, a PD sliding mode control method is adopted in trajectory tracking control and dynamics positioning control. It is more focused on the robustness and fast response of the motion control, and its control law is shown as follows [22]:

$$\tau = -K_p J^T(\eta)\tilde{\eta} - K_d \nu - \left(e^{|K_s \tilde{\eta}|} - 1\right) J^T(\eta)\,\mathrm{sgn}(\tilde{\eta}) + J^T(\eta) K_p \dot{\eta}_d \tag{6.1}$$

where $\tilde{\eta} = \eta - \eta_d$ is the tracking error vector, η is the actual position in the body-fixed frame, η_d is the desired position, K_p, K_d, and K_s are all positive definite diagonal matrix, J is the transformation matrix.

Thruster fault diagnosis module is responsible for thruster's fault diagnosis. In order to realize the fault diagnosis, the dynamics model of the SAUV is first identified in a pool environment, then the fault diagnosis method is used based on terminal sliding mode observer and equivalent output injection method to reconstruct the thruster fault signal directly [24]. The required information of the fault diagnosis module can be obtained by accessing the database, including the motion state of SAUV, thrusters' control signal, and so on. If a thruster fault is detected, the fault degree will be further determined. The fault information and fault degree will be submitted to the database. According to the fault information and fault degree, the management layer will replan the task.

The ultrasonic distance measuring module is used to measure the distances of the objects in front of an SAUV. This function makes "Beaver III" have the capabilities of obstacle avoidance and target positioning. In terms of obstacle avoidance, if the ultrasonic distance measuring module finds an obstacle ahead when performing a long-distance sailing task, an instruction will be sent directly to thrusters and fins to move "Beaver III" away from the obstacle. At the same time, the information of the obstacle position will be sent to the path planning module by the database, and the path planning module will regenerate a viable local path. For the target location, "Beaver III" is equipped with only one camera, which is used to find the three-dimensional location of the target with the known dimensions. For the unknown target, the distance information related to the target will be sent to the optical vision module by the ultrasonic distance measuring module to perform target positioning.

The security status monitoring module is responsible for detecting if there is a leakage of power voltage in the electronic cabin. According to the power voltage information, the management layer can determine whether to proceed with the task. Once the water leakage detection sensor detects that the electronic cabin has a leak, the leakage information will be send to the management layer through the database. At the same time, the security status monitoring module will send commands to cut off the power of thrusters. "Beaver III" can float to the surface relying on buoyancy to wait for rescue.

The optical vision module is responsible for collecting and processing the camera image. The function of image processing includes underwater target recognition and target localization [25].

6.4. Case study

In order to illustrate the workflow of the proposed hybrid control architecture, a case is analyzed. In this case, the task of "Beaver III" is to sail from the starting point to the designated area, and to acquire the image of a spherical target, whose size is already known, and then return to the starting point.

The above given task will be undertaken by the task undertaking module first. It will be divided into four subtasks by the task decomposition module, namely, sailing to destination, target search, image acquisition, and return to starting point. The working order of these four subtasks is shown in Fig. 6.4.

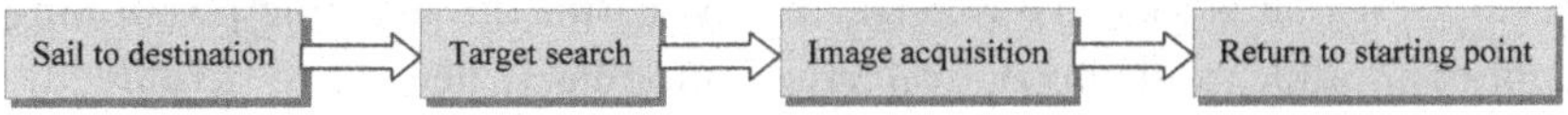

Figure 6.4 Task decomposition.

In the subtask of sailing to destination, the function modules are selected by the task planning module and include dead reckoning module, path planning module, path following module, fault diagnosis module, ultrasonic distance measuring module, and security status monitoring module. In the implementation process, the path from the starting point to the destination is planned by the path planning module first, and then the path following module will calculate the control signals of thrusters and fins to make "Beaver III" follow the desired path. Through the course of this subtask, other function modules keep working. If the ultrasonic distance measuring module detects an obstacle in front of "Beaver III", the main thruster will be controlled to reverse rotation, while the obstacle distance information will be sent to the path planning module through the database, and then the path planning module will generate a local path to make "Beaver III" bypass the obstacle. If the fault diagnosis module detects that there is thruster fault, the fault information will be submitted to the management layer, which will determine whether to continue to work according to the fault degree. If the fault is large, "Beaver III" will float to the surface and wait for the rescue. If the security status monitoring module detects that the power voltage is not normal, the voltage state information will be submitted to the management layer to determine whether to continue work. If the power voltage is too low, then a return to the original location will be undertaken. If the security status monitoring module detects that the electronic cabin is leaking, then the power of thrusters is cut off directly to make "Beaver III" float to the surface. For the subtask of return to the starting point, the workflow is the same as for the subtask of sailing to destination.

In the subtask of a target search, due to the inevitable presence of projection errors of the dead reckoning module in long-time work, "Beaver

III" cannot find its target directly through reaching the destination. Hence, in addition to the same function modules with the subtask of sailing to destination, the optical vision module will be run for underwater target recognition. If the security status monitoring module detects that the power voltage drops below a preset safety value at a time while still searching for the target, "Beaver III" will return to the starting point to give up the task.

The difference between the subtask of image acquisition and the subtask of target search is that "Beaver III" needs to run the dynamic positioning control module, and the optical vision module is needed to acquire the image of a spherical target and then the target positioning is performed. During the process of dynamic positioning control, the management layer needs to adjust the desired position of "Beaver III" real-time according to the position information of the target which is got from the optical vision module, so that the constant relative position of "Beaver III" and the spherical target facilitates image acquisition.

As shown in Fig. 6.5, in order to verify the feasibility and effectiveness of the proposed hybrid control architecture in practical work, a pool experiment is organized for the above tasks. The running trajectory of "Beaver III" in the earth-fixed frame is shown in Fig. 6.6.

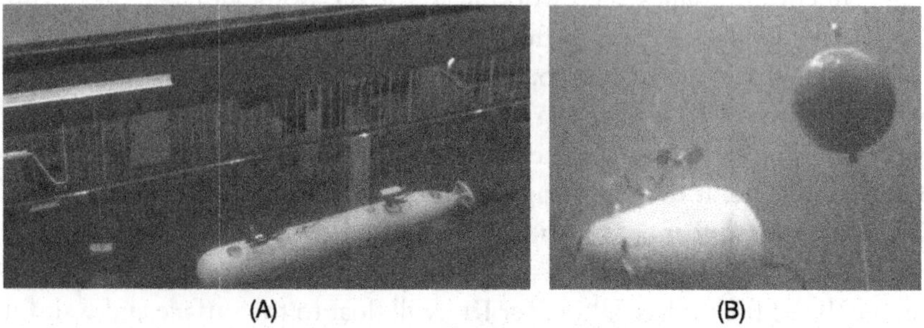

Figure 6.5 Pool environment experiment. (A) Shore image. (B) Underwater image.

As shown in Fig. 6.6, during the time from 0 s to A, "Beaver III" starts the subtask of sailing to the destination and sails to the position of about 3 m from the destination. From time A, "Beaver III" starts the subtask of target search. Due to the restriction of the pool size, the yaw angle of "Beaver III" is set as a constant, and the target searching is carried out by the way of gradually moving forward. From time B, "Beaver III" finds the spherical target and starts the subtask of image acquisition and performs

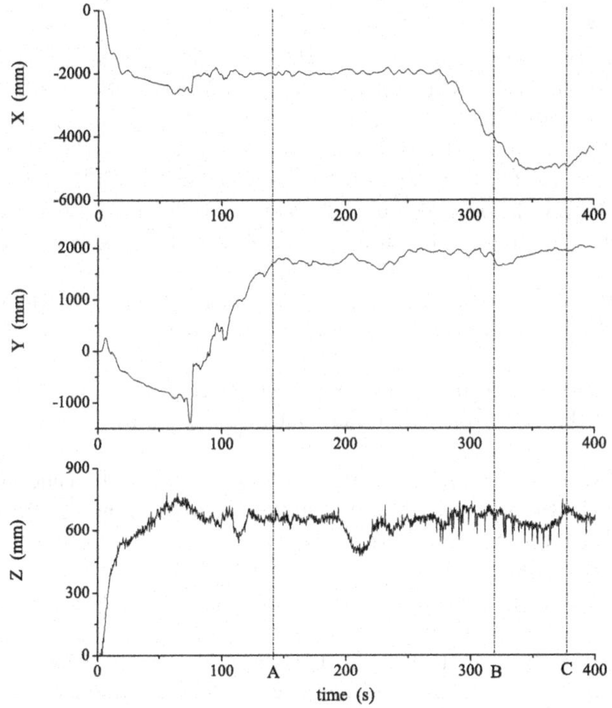

Figure 6.6 The running trajectory of "Beaver III" in earth-fixed frame.

dynamic positioning control. At time C, "Beaver III" completes the image acquisition subtask and starts the subtask of return to the starting point. Experiment results show that the proposed hybrid control architecture can complete the given task well. It verified the feasibility and effectiveness of the proposed hybrid control architecture in practical work.

6.5. Conclusion

According to the design scheme of "Beaver III", the hybrid control architecture with the characteristics of hierarchical and subsumption architectures is proposed in this chapter. The workflow of the hybrid control architecture has been described by an example. It indicates the reasonability of the proposed hybrid control architecture. A pool experiment verifies the feasibility and effectiveness of the proposed hybrid control architecture in practical work.

References

[1] J. Petrich, M.F. Brown, J.L. Pentzer, et al., Side scan sonar based self-localization for small autonomous underwater vehicles, Ocean Engineering 161 (2018) 221–226.

[2] J.K. Yoo, J.H. Kim, Fuzzy integral-based gaze control architecture incorporated with modified-univector field-based navigation for humanoid robots, IEEE Transactions on Systems, Man, and Cybernetics-Part B: Cybernetics 42 (2012) 125–139.

[3] L. Tang, Y. Jiang, J. Lou, Reliability architecture for collaborative robot control systems in complex environments, International Journal of Advanced Robotic Systems (2016) 13.

[4] V.K. Kaliappan, H. Yong, M. Dugki, et al., Reconfigurable intelligent control architecture of a small-scale unmanned helicopter, Journal of Aerospace Engineering 27 (2012) 1–13.

[5] T.V. Arredondo, W. Freund, N. Navarro-Guerrero, et al., Fuzzy motivations in a multiple agent behavior-based architecture, International Journal of Advanced Robotic Systems 10 (2013) 1–13.

[6] K.P. Valavanis, D. Gracanin, M. Matijasevic, et al., Control architectures for autonomous underwater vehicles, IEEE Control Systems Magazine 17 (1997) 48–64.

[7] J. Li, B. Jun, P. Lee, A hierarchical real-time control architecture for a semi-autonomous underwater vehicle, Ocean Engineering 32 (2005) 1631–1641.

[8] J. Gao, W. Yan, Hierarchical control system of underwater robots, Ship Science and Technology 27 (2005) 35–38.

[9] L.D. Marinovici, J. Lian, K. Kalsi, et al., Distributed hierarchical control architecture for transient dynamics improvement in power systems, IEEE Transactions on Power Systems 28 (2013) 3065–3074.

[10] X. Bian, Z. Qin, Z. Yan, Design and evaluation of a hierarchical control architecture for an autonomous underwater vehicle, Journal of Marine Science and Application 7 (2008) 53–58.

[11] C.R. Arkin, Motor schema-based mobile robot navigation, International Journal of Robotics Research 8 (1989) 91–112.

[12] T. Fujii, T. Ura, Autonomous underwater robots with distributed behavior control architecture, in: IEEE International Conference on Robotics and Automation, Nagoya, Japan, 1995, pp. 1868–1873.

[13] E.G. Hernandez-Martinez, S.A. Foyo-Valdes, E.S. Puga-Velazquez, et al., Hybrid architecture for coordination of AGVs in FMS, International Journal of Advance Robotics Systems 11 (2014) 1–12.

[14] G. Marani, K. Song, J. Yuh, Underwater autonomous manipulation for intervention missions AUVs, Ocean Engineering 36 (2009) 15–23.

[15] W. Chang, G. You, Y. Pang, et al., Multiple autonomous underwater vehicles cooperative control system based on hybrid architecture, China Offshore Platform 17 (2002) 12–16.

[16] C.R. Carignan, J.C. Lane, D.L. Akin, Control architecture and operator interface for a free-flying robotic vehicle, IEEE Transactions on Systems, Man, and Cybernetics-Part C: Applications and Reviews 31 (2001) 327–336.

[17] J. Han, J. Ok, W.K. Chung, An ethology-based hybrid control architecture for an autonomous underwater vehicle for performing multiple tasks, IEEE Journal of Oceanic Engineering 38 (2013) 514–521.

[18] C.C. Min, D. Srinivasan, R.L. Cheu, Cooperative hybrid agent architecture for real-time traffic signal control, IEEE Transactions on Systems, Man, and Cybernetics-Part A: Systems and Humans 33 (2003) 597–607.

[19] X. Zhu, W. Zeng, Z. Li, et al., A new hybrid control architecture to attenuate large horizontal wind disturbance for a small-scale unmanned helicopter, International Journal of Advanced Robotics Systems 9 (2012) 1–12.

[20] M.T. Sabet, P. Sarhadi, M. Zarini, Extended and unscented Kalman filters for parameter estimation of an autonomous underwater vehicle, Ocean Engineering 91 (2014) 329–339.

[21] Y. Zhang, M.A. Godin, J.G. Bellingham, et al., Using an autonomous underwater vehicle to tracking a coastal upwelling front, IEEE Journal of Oceanic Engineering 37 (2012) 338–347.

[22] Z.Z. Chu, D.Q. Zhu, B. Sun, et al., Design of a dead reckoning based motion control system for small autonomous underwater vehicle, in: IEEE 28th Canadian Conference on Electrical and Computer Engineering, 2015.

[23] N. Wang, Z. Sun, J. Yin, et al., Finite-time observer based guidance and control of underactuated surface vehicles with unknown sideslip angles and disturbances, IEEE Access 6 (2018) 14059–14070.

[24] Zhenzhong Chu, Mingjun Zhang, Fault reconstruction of thruster for autonomous underwater vehicle based on terminal sliding mode observer, Ocean Engineering 88 (2014) 426–434.

[25] Zhenzhong Chu, Fei Meng, Daqi Zhu, Chaomin Luo, Fault reconstruction using a terminal sliding mode observer for a class of second-order MIMO uncertain nonlinear systems, ISA Transactions 97 (2020) 67–75.

CHAPTER SEVEN

Adaptive sliding mode control based on local recurrent neural networks for an underwater robot

Zhenzhong Chu

Contents

Abstract

The trajectory tracking control problem of an underwater robot is addressed in this chapter. In general, an accurate thrust modeling is very difficult to establish for an underwater robot in practice. Hence, the control voltage of the thruster is designed directly as the input of the system by the controller in this article. First, Taylor's polynomial is used to transform the form of the trajectory tracking error system of an underwater robot to the form of affine nonlinear systems, whose input is the control voltage of the thruster. Then, according to the principle of sliding mode control, and using the local recurrent neural network to estimate the unknown term of an affine system online, an adaptive sliding mode control is proposed. Aiming at the chattering problem which is caused by the sliding mode control term, we propose a switch gain adjustment method based on an exponential function. It is proved that the trajectory tracking error of the underwater robot control system is uniformly ultimately bounded through Lyapunov theory. The feasibility and effectiveness of the proposed approach is demonstrated with trajectory tracking experiments of the experimental prototype of an underwater robot.

Fundamental Design and Automation Technologies in Offshore Robotics
https://doi.org/10.1016/B978-0-12-820271-5.00012-2

Keywords
Underwater robot, Trajectory tracking, Adaptive control, Neural networks, Sliding mode control

7.1. Introduction

An underwater robot works in the complex marine environment, whose movement has obvious nonlinear and cross-coupling characteristics. It is a complex problem how to design the control system of underwater robot motion [20].

In the literature, some adaptive control algorithms for trajectory tracking and dynamic positioning of an underwater robot have been proposed, such as PD adaptive control [5,9,15], neural networks adaptive control [3,21,2], which overcome the relevant control problems caused by the uncertainty of the hydrodynamic coefficients and marine environment disturbance to some extent. However, since the thruster is the main driving device of an underwater robot, the precondition to realize these methods is that a thrust model is already known [12,8]. Underwater robot motion control usually uses a voltage signal as control input of the thruster. As a complex nonlinear mapping relation exists between the thrust, control voltage, and underwater robot state, it is very difficult to get accurate thruster modeling [6]. In a practical control process, ignoring thrust model error will lead to the decline of the trajectory tracking control precision. Based on the above, we mainly study the trajectory tracking control problem using the control voltage of the thruster as system input without knowing the dynamics and thrust model of an underwater robot.

In the view of the nonlinear identification ability, neural networks have been widely used in the study of underwater robot motion controller. Most of these approaches based on forward neural networks approximate online, such as RBF neural network [21,10], BP neural network [1]. However, the forward neural network is very difficult to reflect the time series influence between system input and output variables, the weights of neural network would take a long time to convergence, especially when the underwater robot has a big disturbance or the target trajectory has an abrupt change. This will lead to poor transition process and even uncontrolled output. Due to this disadvantage of the forward neural network, an alternative approach is to use recurrent neural network instead of forward neural network for adaptive control law design. Wang and Lee [18] provide a self-adaptive recurrent neuro-fuzzy controller and a proportional-plus-derivative controller for an autonomous underwater vehicle in an unstructured environment. In [23], an adaptive output feedback controller based on dynamic recurrent fuzzy neural network is studied for trajectory tracking control of an autonomous underwater vehicle. However, a recurrent neural network has low learning efficiency and poor mapping accuracy, so it is very difficult to meet the requirement of the high-precision trajectory tracking control of an underwater robot. Moreover, because of the approximation error of the neural network, the sliding mode control term is used for asymptotic stability of the closed-loop system in [17], while it will cause the chattering problem. Hence, in [1], the saturated function is used to replace the sign function in the sliding mode control term, but it reduces the robustness of the control system near the switch surface of the sliding mode. A fuzzy strategy is used to estimate uncertain terms online, which realizes a switch gain fuzzy adaptive adjustment [24,16]. However, fuzzy rules acquisition depends on the designer's experience and knowledge. A neural network is adopted for adaptive learning of the upper bound of the disturbance to reduce system chattering [13,11]. But it may reduce the neural network learning speed when using it for adaptive learning of the nonlinear unknown term and the upper bound of the disturbance.

For the above considerations, we propose an adaptive sliding mode control method based on the local recurrent neural network for trajectory tracking control of an underwater robot. In the current researches, most controllers are used to design the force/torque on each degree of freedom. Different from these, we use the underwater robot as nonaffine nonlinear system for analysis, and then get an affine nonlinear system whose input is the control voltage of thrusters by an affine transformation. According to the controlled system, the control voltage of thrusters is designed directly based on the sliding mode control principle. For the unknown function term in the controller, an adaptive learning method is proposed based on the local recurrent neural network. Considering the chattering problem caused by the sliding mode control term, a sliding mode switch gain adjustment method is proposed based on an exponential function to ensure that the trajectory tracking error is uniformly ultimately bounded.

This chapter is organized as follows. Section 7.2 describes the trajectory tracking control problem of underwater robot and gives the affine trajectory tracking error system whose input is the control voltage of thrusters. In Section 7.3, according to the sliding mode control theory, the adaptive sliding mode control approach is studied based on a local recurrent neural network. Section 7.4 contains the stability analysis. Then in Section 7.5, the feasibility and effectiveness of the proposed approach is verified by experiments. Finally, we make a brief conclusion in Section 7.6 of this paper.

7.2. Problem formulation

If the control voltage of thrusters is used as system input, the underwater robot motion control system will be analyzed as a nonaffine nonlinear system. Hence, the dynamics equations of underwater robot in 6 DOF space can be described as (7.1) with a nonaffine nonlinear system form. Compared with the models in the literature [3,21,1,2], the thrusters output is expressed as a function associated with the control voltage of thrusters and the speed state of underwater robot:

$$\begin{aligned} \dot{\eta} &= J(\eta)\nu \\ M\dot{\nu} + C(\nu)\nu + D(\nu)\nu + g(\eta) + \tau_d &= B\tau(\nu, u) \end{aligned} \tag{7.1}$$

where η denotes the vector of location and orientation in the earth-fixed frame, ν is the vector of velocity expressed in the body-fixed frame; M is the inertia matrix including extra mass, matrix $C(\nu)$ groups centripetal and Coriolis forces, including the centripetal force and Coriolis force produced by extra mass, $D(\nu)$ is the hydrodynamic damping term, vector $g(\eta)$ is the combined gravitational and buoyancy forces in the body-fixed frame, τ_d is the external disturbances, $J(\eta)$ is the kinematic transformation matrix expressing the transformation from the body-fixed frame to earth-fixed frame, u is the control voltage of thrusters, $\tau(\nu, u)$ is the thruster force, and B is the distribution matrix of thrusters.

For a given desired trajectory $[\eta_d, \dot{\eta}_d]^T$, the trajectory tracking error $\xi = [\xi_1, \xi_2]^T$ is defined as

$$\begin{aligned} \xi_1 &= \eta - \eta_d \\ \xi_2 &= J(\eta)\nu - \dot{\eta}_d \end{aligned} \tag{7.2}$$

According to (7.1) and (7.2), the tracking error system can be written as

$$\begin{aligned} \dot{\xi}_1 &= \xi_2 \\ \dot{\xi}_2 &= \overline{F}(\xi) + \overline{B}\tau(\nu, u) \end{aligned} \tag{7.3}$$

where

$$\begin{aligned}\overline{F}(\xi) =& \dot{J}(\xi_1)J^{-1}(\xi_1)(\xi_2+\dot{\eta}_d)-\ddot{\eta}_d+\\ & J(\xi_1)M^{-1}\left(-C\left(J^{-1}(\xi_1)(\xi_2+\dot{\eta}_d)\right)J^{-1}(\xi_1)(\xi_2+\dot{\eta}_d)-\right.\\ & \left.D\left(J^{-1}(\xi_1)(\xi_2+\dot{\eta}_d)\right)J^{-1}(\xi_1)(\xi_2+\dot{\eta}_d)-g(\xi_1)-\tau_d\right)\\ \overline{B}=& J(\xi_1)M^{-1}B\end{aligned}$$

In (7.3), $\tau(\nu, u)$ is an unknown nonlinear function vector about the control voltage of thrusters. The Taylor expansion of $\tau(\nu, u)$ about u^* is given as

$$\tau_i(\nu, u_i) = \tau_i\left(\nu, u_i^*\right) + \left.\frac{\partial \tau_i(\nu, u_i)}{\partial u_i}\right|_{u_i=u_i^*}\left(u_i - u_i^*\right) + O\left(\left(u_i - u_i^*\right)^2\right) \tag{7.4}$$

where $i = 1, \ldots, n$; n is the number of thrusters.

Let

$$F(\xi) = \overline{F}(\xi) + \tau\left(\nu, u^*\right) - \frac{\partial \tau(\nu, u)}{\partial u}u^* + O\left(\left(u - u^*\right)^2\right), \tag{7.5}$$

$$G(\xi) = \left[\left.\frac{\partial \tau_1(\nu, u_1)}{\partial u_1}\right|_{u_1=u_1^*}; \ldots; \left.\frac{\partial \tau_n(\nu, u_n)}{\partial u_n}\right|_{u_n=u_n^*}\right]. \tag{7.6}$$

From (7.4), (7.5) and (7.6), the tracking error system (7.3) is expressed in the form of an affine nonlinear system whose input is the control voltage of thrusters, which can be rewritten as

$$\begin{aligned}\dot{\xi}_1 &= \xi_2\\ \dot{\xi}_2 &= F(\xi) + \overline{B}diag(G(\xi))u\end{aligned} \tag{7.7}$$

To sum up, the control objective of this paper is to design an adaptive controller u with expected trajectory $[\eta_d, \dot{\eta}_d]^T$ and make the tracking error ξ uniformly ultimately bounded.

7.3. Controller design

For the affine nonlinear tracking error system (7.7), if nonlinear terms $F(\xi)$ and $G(\xi)$ are exactly known, the control law based on sliding mode control theory will be designed as

$$u = -diag(G(\xi))^{-1}\overline{B}^{-1}\left(K_1 S + F(\xi) + l\dot{\xi}_1 + K_2\,\mathrm{sgn}(S)\right) \tag{7.8}$$

where $S = \xi_2 + l\xi_1$ is the vector of the sliding mode switch function; l, K_1 and K_2 are positive definite diagonal matrices, $\overline{B}^{-1} = \overline{B}^T\left(\overline{B}\overline{B}^T\right)^{-1}$ is the generalized inverse matrix of $\overline{B}$, when underwater robots have redundant thruster configuration.

However, there exist uncertainties in the underwater robot dynamics and thruster modeling in practice. Therefore, the nonlinear terms $F(\xi)$ and $G(\xi)$ cannot be obtained exactly. For this reason, we will use a neural network to approximate $F(\xi)$ and $G(\xi)$ online.

7.3.1 Local recurrent neural network

Compared with a forward neural network, recurrent neural network has a recurrent middle layer structure, which can reflect the time series between the input and output variables of the system effectively. When the underwater robot has a big disturbance or the target trajectory has an abrupt change, the adaptive motion control system based on a recurrent neural network has better transition characteristics. However, in the experimental research, we find that the number of recurrent neurons has a great impact on learning efficiency and accuracy. But for underwater robot motion control system, the time series between system input/output variables can be reflected with several recurrent neurons. For the above considerations, we will design an adaptive tracking controller of the underwater robot based on a local recurrent neural network which is shown in Fig. 7.1.

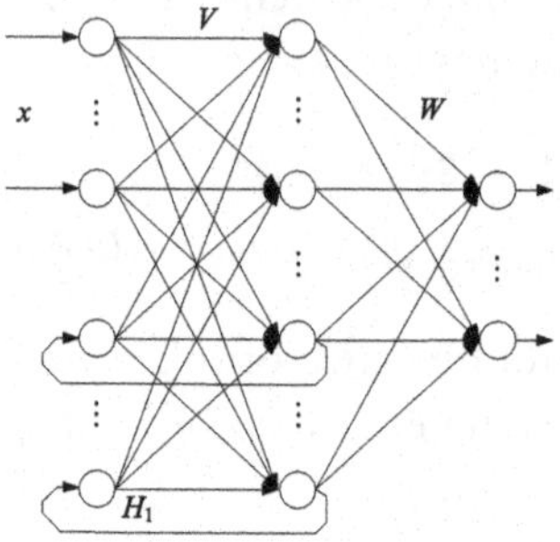

Figure 7.1 Local recurrent neural network architecture.

Different from traditional Elman neural network, the local recurrent neural network in Fig. 7.1 reflects the time series through the return of part of the middle layer neurons to the input layer. In Fig. 7.1, x is the input vector of input layer, H_1 is the output vector of the recurrent layer, W is the network weight matrix between the hidden and output layers, V

is the network weight matrix between input, recurrent, and hidden layers. The output of the local recurrent neural network can be written as

$$f(x) = W\sigma(VH) \tag{7.9}$$

where $\sigma(\cdot)$ is a sigmoid function, $H = [x; H_1]$.

We use the local recurrent neural network (7.9) to approximate $F(\xi)$ and $G(\xi)$ online. Due to the nonlinear mapping capacity of the neural network, we can get the optimal network weights W_G^*, V_G^*, W_F^*, and V_F^*, which make

$$\begin{aligned} G(\xi) &= W_G^*\sigma\left(V_G^* H_G\right) + \varepsilon_g \\ F(\xi) &= W_F^*\sigma\left(V_F^* H_F\right) + \varepsilon_f \end{aligned} \tag{7.10}$$

where $H_F = [x_F; H_{1F}]$, $H_G = [x_G; H_{1G}]$, $x_F = x_G = \overline{\xi}$, $\overline{\xi}$ is a normalization of ξ; ε_g and ε_f are the neural network approximation errors, satisfying $\|\varepsilon_g\| \le \overline{\varepsilon}_g$, $\|\varepsilon_f\| \le \overline{\varepsilon}_f$, where $\overline{\varepsilon}_g$ and $\overline{\varepsilon}_f$ are positive constants; $\sigma(V_F^* H_F)$ and $\sigma(V_G^* H_G)$ are the outputs of the hidden layer.

7.3.2 Adaptive sliding mode controller

Using the local recurrent neural network for control law (7.8), the adaptive sliding mode control law based on the local recurrent neural network in (7.11) can be obtained, where the adaptive law of network weights is shown in (7.12):

$$u = -diag\left(\hat{G}(\xi)\right)^{-1} \overline{B}^{-1}\left(K_1 S + \hat{F}(\xi) + l\dot{\xi}_1 + K_2 \operatorname{sgn}(S)\right), \tag{7.11}$$

$$\begin{aligned} \dot{\hat{W}}_G &= \Gamma_{WG}\left(S^T \overline{B} diag(u)\right)^T \sigma\left(\hat{V}_G H_G\right)^T \\ \dot{\hat{V}}_G &= \Gamma_{VG}\left(S^T \overline{B} diag(u)\, \hat{W}_G \sigma'\left(\hat{V}_G H_G\right)^T\right)^T H_G^T \\ \dot{\hat{W}}_F &= \Gamma_{WF} S \sigma\left(\hat{V}_F H_F\right)^T \\ \dot{\hat{V}}_F &= \Gamma_{VF} \sigma'\left(\hat{V}_F H_F\right) W_F^T S H_F^T \end{aligned} \tag{7.12}$$

where $\hat{G}(\xi) = \hat{W}_G \sigma\left(\hat{V}_G H_G\right)$, $\hat{F}(\xi) = \hat{W}_F \sigma\left(\hat{V}_F H_F\right)$ are the outputs of the neural network; $\hat{W}_G$, $\hat{W}_F$, $\hat{V}_G$, $\hat{V}_F$ are estimated weights; Γ_{WG}, Γ_{WF}, Γ_{VG}, Γ_{VF} are positive definite diagonal matrices; $\sigma'\left(\hat{V}_F H_F\right)$, $\sigma'\left(\hat{V}_G H_G\right)$ are determined by the Taylor expansion of $\sigma\left(V_F^* H_F\right)$, $\sigma\left(V_G^* H_G\right)$ about

$\hat{V}_F H_F$, $\hat{V}_G H_G$, respectively. Also

$$\begin{aligned}\sigma\left(V_F^* H_F\right) &= \sigma\left(\hat{V}_F H_F\right) + \sigma'\left(\hat{V}_F H_F\right)\tilde{V}_F H_F + O\left(\tilde{V}_F H_F\right)^2 \\ \sigma\left(V_G^* H_G\right) &= \sigma\left(\hat{V}_G H_G\right) + \sigma'\left(\hat{V}_G H_G\right)\tilde{V}_G H_G + O\left(\tilde{V}_G H_G\right)^2\end{aligned} \tag{7.13}$$

where $O\left(\tilde{V}_F H_F\right)^2$, $O\left(\tilde{V}_G H_G\right)^2$ are the high-order terms of the Taylor expansion.

When the estimated matrix $diag(\hat{G}(\xi))$ is singular, the adaptive control law defined in (7.11) is not well-defined. For this problem, the parameter constraint method is used to make the network weights in a feasible region to ensure $diag(\hat{G}(\xi))$ is regular in [22,4]. However, it is very difficult to determine the feasible region for network weights in practice. Hence, in order to avoid the singularity, the control law (7.11) is rewritten as follows [14]:

$$\begin{aligned}u = &-diag\left(\hat{G}(\xi)\right)\left[\varepsilon_o I + diag\left(\hat{G}(\xi)\right) diag\left(\hat{G}(\xi)\right)\right]^{-1} \times \\ &\overline{B}^{-1}\left(K_1 S + \hat{F}(\xi) + l\dot{\xi}_1 + K_2 \operatorname{sgn}(S)\right)\end{aligned} \tag{7.14}$$

where ε_o is an arbitrarily small positive constant.

With the control term (7.14), we can get the sliding mode trajectory tracking error system as

$$\begin{aligned}\dot{S} &= F(\xi) + l\dot{\xi}_1 + \overline{B} diag(G(\xi))\, u \\ &= \left(F(\xi) - \hat{F}(\xi)\right) + \overline{B} diag\left(G(\xi) - \hat{G}(\xi)\right) u - K_1 S - K_2 \operatorname{sgn}(S) + \\ &\quad \frac{\varepsilon_o I}{\varepsilon_o I + diag\left(\hat{G}(\xi)\right) diag\left(\hat{G}(\xi)\right)}\left(K_1 S + \hat{F}(\xi) + l\dot{\xi}_1 + K_2 \operatorname{sgn}(S)\right) \\ &= \left(W_F^* \sigma\left(V_F^* H_F\right) - \hat{W}_F \sigma\left(\hat{V}_F H_F\right)\right) + \\ &\quad \overline{B} diag\left(W_G^* \sigma\left(V_G^* H_G\right) - \hat{W}_G \sigma\left(\hat{V}_G H_G\right)\right) u - K_1 S - K_2 \operatorname{sgn}(S) + \varepsilon_f + \\ &\quad \overline{B} diag\left(\varepsilon_g\right) u + \Sigma \\ &= \tilde{W}_F \sigma\left(\hat{V}_F H_F\right) + \overline{B} diag\left(\tilde{W}_G \sigma\left(\hat{V}_G H_G\right)\right) u + \hat{W}_F \sigma'\left(\hat{V}_F H_F\right)\tilde{V}_F H_F + \\ &\quad \overline{B} diag\left(\hat{W}_G \sigma'\left(\hat{V}_G H_G\right)\tilde{V}_G H_G\right) u + w_1 + w_2 + \Sigma - K_1 S - K_2 \operatorname{sgn}(S)\end{aligned} \tag{7.15}$$

$$\Sigma = -\frac{\varepsilon_o I}{\varepsilon_o I + diag\left(\hat{G}(\xi)\right) diag\left(\hat{G}(\xi)\right)} \left(K_1 S + \hat{F}(\xi) + l\dot{\xi}_1 + K_2 \operatorname{sgn}(S)\right) \tag{7.16}$$

where $\Sigma \le \Sigma_M$; the uncertain terms w_1 and w_2 are defined in (7.17), which also meet the constraint condition in (7.18) [7]:

$$\begin{aligned} w_1 &= \tilde{W}_F \sigma'\left(\hat{V}_F H_F\right) \tilde{V}_F H_F + W_F O\left(\tilde{V}_F H_F\right)^2 + \varepsilon_f \\ w_2 &= \overline{B} diag\left(\tilde{W}_G \sigma'\left(\hat{V}_G H_G\right) \tilde{V}_G H_G + W_G O\left(\tilde{V}_G H_G\right)^2 + \varepsilon_g\right) u \end{aligned} \tag{7.17}$$

$$\begin{aligned} \|w_1\| &\le c_{10} + c_{11}\|H_F\| + c_{12}\left\|\tilde{W}_F\right\| \|H_F\| + \overline{\varepsilon}_f \\ \|w_2\| &\le \left\|\overline{B}\right\| \left(c_{20} + c_{21}\|H_G\| + c_{22}\left\|\tilde{W}_G\right\| \|H_G\| + \overline{\varepsilon}_g\right) \|u\| \end{aligned} \tag{7.18}$$

where c_{10}, c_{11}, c_{12}, c_{20}, c_{21}, c_{22} are positive constant.

7.3.3 The improvement of sliding mode switch gain

To ensure the robustness and asymptotic stability of the trajectory tracking control system, the sliding mode switch gain K_2 in the adaptive control law (7.14) should satisfy

$$\begin{aligned} K_2 > {} & c_{10} + c_{11}\|H_F\| + c_{12}\left\|\tilde{W}_F\right\| \|H_F\| + \overline{\varepsilon}_f + \\ & \left\|\overline{B}\right\| \left(c_{20} + c_{21}\|H_G\| + c_{22}\left\|\tilde{W}_G\right\| \|H_G\| + \overline{\varepsilon}_g\right) \|u\| + \Sigma_M. \end{aligned} \tag{7.19}$$

However, in practical control process, it is difficult to select a reasonable value for parameters c_{10}, c_{11}, c_{12}, c_{20}, c_{21}, c_{22}, $\overline{\varepsilon}_f$, $\overline{\varepsilon}_g$, Σ_M, etc., in (7.19), and too conservative estimated values would cause chattering problem easily. Hence, we propose a sliding mode switch gain adjustment method based on an exponential function.

For the sliding mode control, when the system has a pretty large tracking error, the sliding mode switch gain should have a larger value to make the control system approximate quickly. While after the control system has completed the target tracking, the sliding mode switch gain should be decreased in order to reduce the control system chattering. To resolve this, we propose a sliding mode switch gain that changes as an exponential function. The gain is given by

$$K_2 = \exp\left(|kS|\right) - 1 \tag{7.20}$$

where k is a regular diagonal matrix.

From the above analysis, the architecture of an adaptive sliding mode trajectory tracking controller based on local recurrent neural networks is shown in Fig. 7.2.

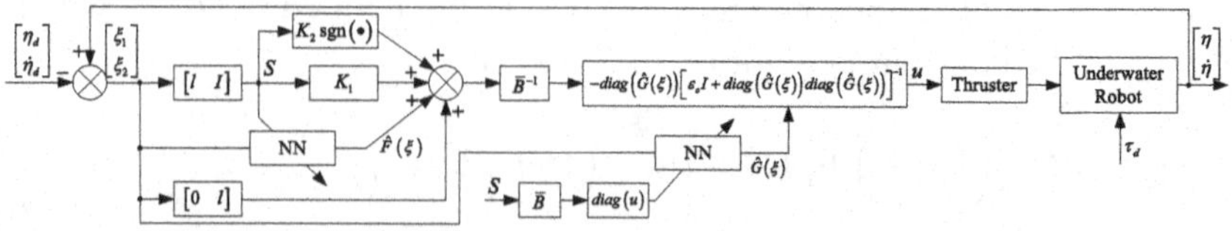

Figure 7.2 Architecture of adaptive sliding mode controller based on local recurrent neural networks.

7.4. Stability analysis

For the adaptive control law (7.14) and adaptive adjusting law of network weights in (7.12), Lyapunov theory is used for stability analysis in this paper.

Consider the Lyapunov function candidate

$$V = \frac{1}{2}S^T S + \frac{1}{2}tr\left(\tilde{W}_G^T \Gamma_{WG}^{-1} \tilde{W}_G\right) + \frac{1}{2}tr\left(\tilde{W}_F^T \Gamma_{WF}^{-1} \tilde{W}_F\right) + \frac{1}{2}tr\left(\tilde{V}_G^T \Gamma_{VG}^{-1} \tilde{V}_G\right) + \frac{1}{2}tr\left(\tilde{V}_F^T \Gamma_{VF}^{-1} \tilde{V}_F\right). \tag{7.21}$$

The derivative of V along (7.15) will be

$$\begin{aligned}
\dot{V} &= S^T \dot{S} + tr\left(\tilde{W}_G^T \Gamma_{WG}^{-1} \dot{\tilde{W}}_G\right) + tr\left(\tilde{W}_F^T \Gamma_{WF}^{-1} \dot{\tilde{W}}_F\right) + tr\left(\tilde{V}_G^T \Gamma_{VG}^{-1} \dot{\tilde{V}}_G\right) + \\
&\quad tr\left(\tilde{V}_F^T \Gamma_{VF}^{-1} \dot{\tilde{V}}_F\right) \\
&= S^T \left(\tilde{W}_F \sigma\left(\hat{V}_F H_F\right) + \overline{B} diag\left(\tilde{W}_G \sigma\left(\hat{V}_G H_G\right)\right) u + \right. \\
&\quad \left. \hat{W}_F \sigma'\left(\hat{V}_F H_F\right) \tilde{V}_F H_F + \overline{B} diag\left(\hat{W}_G \sigma'\left(\hat{V}_G H_G\right) \tilde{V}_G H_G\right) u\right) + \\
&\quad S^T \left(w_1 + w_2 + \Sigma - K_1 \hat{S} - K_2 \operatorname{sgn}\left(\hat{S}\right)\right) + tr\left(\tilde{W}_G^T \Gamma_{WG}^{-1} \dot{\tilde{W}}_G\right) + \\
&\quad tr\left(\tilde{W}_F^T \Gamma_{WF}^{-1} \dot{\tilde{W}}_F\right) + tr\left(\tilde{V}_G^T \Gamma_{VG}^{-1} \dot{\tilde{V}}_G\right) + tr\left(\tilde{V}_F^T \Gamma_{VF}^{-1} \dot{\tilde{V}}_F\right)
\end{aligned} \tag{7.22}$$

From (7.12) and (7.20), it follows

$$\begin{aligned}\dot{V} &= S^T\left(w_1 + w_2 + \Sigma - K_1 S - \left(\exp\left(|kS|\right) - 1\right)\operatorname{sgn}(S)\right)\\ &\le S^T\left(w_1 + w_2 + \Sigma - \left(\exp\left(|kS|\right) - 1\right)\operatorname{sgn}(S)\right)\end{aligned} \tag{7.23}$$

For (7.23), to prove the stability of the trajectory tracking control system, we analyze sliding mode tracking errors in the following three conditions:

① If $S = 0$, then the sliding mode tracking errors will converge to zero and meet the control system requirements.

② If $S < 0$, then $\dot{V} < 0$ does not hold unless the sliding mode tracking errors are such that

$$\begin{aligned}S \in \Big(0, k^{-1}\ln\Big(1 + c_{10} + c_{11}\|H_F\| + c_{12}\left\|\tilde{W}_F\right\|\|H_F\| + \overline{\varepsilon}_f +\\ \|\overline{B}\|\left(c_{20} + c_{21}\|H_G\| + c_{22}\left\|\tilde{W}_G\right\|\|H_G\| + \overline{\varepsilon}_g\right)\|u\| + \Sigma_M\Big)\Big].\end{aligned} \tag{7.24}$$

③ If $S < 0$, then $\dot{V} < 0$ does not hold unless the sliding mode tracking errors are such that

$$\begin{aligned}S \in \Big[-k^{-1}\ln\Big(1 + c_{10} + c_{11}\|H_F\| + c_{12}\left\|\tilde{W}_F\right\|\|H_F\| + \overline{\varepsilon}_f +\\ \|\overline{B}\|\left(c_{20} + c_{21}\|H_G\| + c_{22}\left\|\tilde{W}_G\right\|\|H_G\| + \overline{\varepsilon}_g\right)\|u\| + \Sigma_M\Big), 0\Big).\end{aligned} \tag{7.25}$$

From the above analysis, the sliding trajectory tracking error will finally converge into the region (7.26). Thus, it shows the tracking error is uniformly stably bounded if

$$\begin{aligned}S \in \Big[&-k^{-1}\ln\Big(1 + c_{10} + c_{11}\|H_F\| + c_{12}\left\|\tilde{W}_F\right\|\|H_F\| + \overline{\varepsilon}_f +\\ &\|\overline{B}\|\left(c_{20} + c_{21}\|H_G\| + c_{22}\left\|\tilde{W}_G\right\|\|H_G\| + \overline{\varepsilon}_g\right)\|u\| + \Sigma_M\Big),\\ &k^{-1}\ln\Big(1 + c_{10} + c_{11}\|H_F\| + c_{12}\left\|\tilde{W}_F\right\|\|H_F\| + \overline{\varepsilon}_f +\\ &\|\overline{B}\|\left(c_{20} + c_{21}\|H_G\| + c_{22}\left\|\tilde{W}_G\right\|\|H_G\| + \overline{\varepsilon}_g\right)\|u\| + \Sigma_M\Big)\Big].\end{aligned} \tag{7.26}$$

7.5. Experimental studies

In this section, the adaptive sliding mode control based on local recurrent neural networks is validated. We use an underwater robot experimental prototype for trajectory tracking control experiment in a water tank. The water tank size is 1.2 m × 1.2 m × 1.2 m and the underwater robot experimental prototype is shown in Fig. 7.3. The prototype control system was installed on the shore, which is set up by a multiboard hardware system based on PC104 and uses real-time embedded operating system based on VxWorks for software system development. The prototype weighs 18 kg and has 3 thrusters in the vertical plane. The thrusters use DC motor drive and the drive input is PWM signal. In the control process, the motor voltage is adjusted through adjusting the duty cycle of the PWM signal. Because there is a linear mapping relationship between PWM duty cycle signal and motor voltage, in the experimental analysis, we use voltage signal from −5 V to 5 V instead of duty cycle signal from 0% to 100%. The experiment uses HMR3000 digital compass and CYT-151 depth gauge for the collection of the attitude and depth information. The sensor accuracy is 0.5° and 0.02 m, respectively.

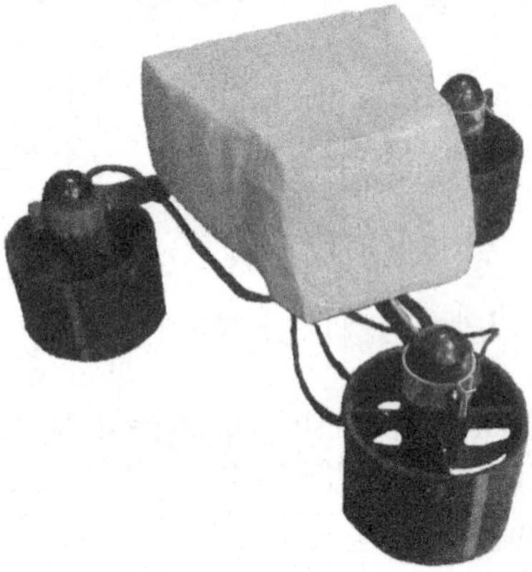

Figure 7.3 Underwater robot experimental prototype.

7.5.1 Trajectory tracking control experiment 1

In order to validate the feasibility and effectiveness of the proposed controller which uses the thruster voltage as system input, contrasting experiments have been carried out with PID control and the proposed control method. In the PID control experiment, the controller was used to design the force/torque on each DOF and then thrust allocation algorithm was used to calculate the thrust force of each thruster, and finally the thruster voltage was calculated by thrust model [19]. In the water tank experiment, the desired pitch and roll angles of the underwater robot are both 0°, and the desired depth trajectory is written as

$$Z = -0.2 \times \cos(2\pi t/20) + 0.3 \text{ (m)} \quad 0 \text{ s} \leq t \leq 60 \text{ s}.$$

The experimental results of tracking control experiments of underwater robot are shown in Fig. 7.4. The thin solid line shows the desired trajectory. The thick solid line illustrates the experimental results from the proposed method. The thin dashed line depicts the experimental results from PID control.

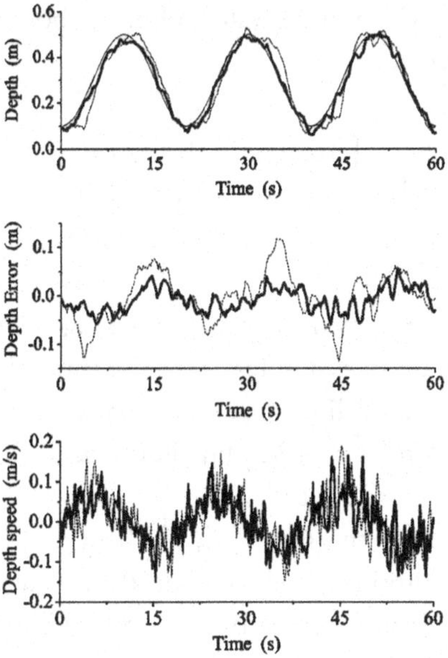

Figure 7.4 Experiment 1: (A) trajectory tracking of depth, (B) trajectory tracking error of depth, and (C) trajectory tracking of depth speed.

From the experimental results, the proposed method can well track the desired trajectory signal through the adaptive learning without knowing the thrust model. And the trajectory tracking error is uniformly ultimately bounded. In depth trajectory tracking control, the maximum absolute depth tracking error produced by the proposed approach is 0.056 m and the mean square deviation of the tracking error is 0.024 m. The maximum absolute depth tracking error produced by the PID control is 0.135 m and the mean square deviation of tracking error is 0.050 m. The results show that the proposed approach has higher tracking control accuracy. And so the feasibility and effectiveness of the control which uses the thruster voltage as input of the underwater robot system is verified.

7.5.2 Trajectory tracking control experiment 2

In order to verify that the proposed method based on local recurrent neural network has better transition characteristics when the target trajectory has an abrupt change, contrasting experiments are carried out with adaptive control based on RBF neural network. Here the center of the RBF is obtained by K clustering method. In the water tank experiment, the desired pitch and roll angles of the underwater robot are both 0°, and the desired depth trajectory is written as

$$Z = \begin{cases} 0.2\,(\mathrm{m}) & t < 20\ \mathrm{s} \\ 0.5\,(\mathrm{m}) & 20\ \mathrm{s} \leq t < 40\ \mathrm{s} \\ 0.3\,(\mathrm{m}) & 40\ \mathrm{s} \leq t \leq 60\ \mathrm{s} \end{cases}$$

The experimental results of trajectory tracking control experiments of an underwater robot are shown in Fig. 7.5. The thin solid line is the desired trajectory. The thick solid line shows the experimental results from the method we address. The thin dashed line illustrates the experimental results from RBF neural network control.

From Fig. 7.5, when the target depth changes, the output trajectory obtained from the control method based on RBF neural network has a big overshoot, while the output trajectory of the proposed method has a smaller overshoot and faster convergence. In the depth trajectory tracking control, the depth adjustment time of our approach is 2.2 s and the depth adjustment time of RBF neural network is 7.0 s. The mean square deviation of depth tracking error produced by the proposed approach is 0.047 m and the mean square deviation of depth tracking error produced by RBF neural network is 0.056 m. The results show that the adaptive sliding mode controller we proposed has better transition characteristics and control accuracy.

7.5.3 Trajectory tracking control experiment 3

For proving the effectiveness of sliding mode switch gain based on an exponential function, contrasting experiments are carried out with fixed sliding mode switch gain. In the water tank experiment, the pitch and roll angles

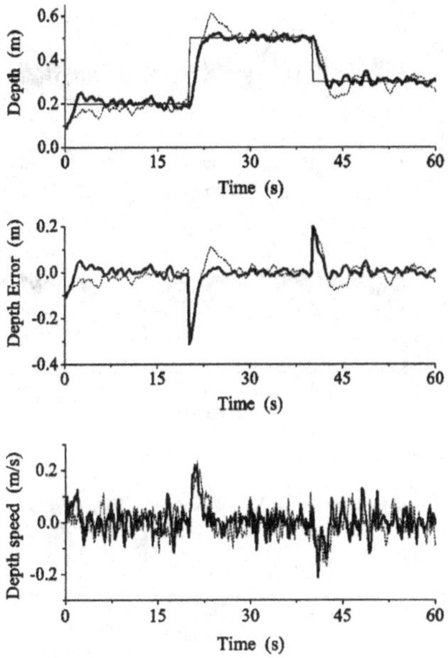

Figure 7.5 Experiment 2: (A) trajectory tracking of depth, (B) trajectory tracking error of depth, and (C) trajectory tracking of depth speed.

and the desired depth trajectory of the underwater robot are the same as in Section 7.5.2.

The control voltage of each thruster is shown in Fig. 7.6. The thick solid line provides the experimental results from the method we address. The thin dashed line shows the experimental results from traditional fixed sliding mode switch gain.

In our approach, the mean square deviation of the control voltage of thrusters is 0.61, 0.31, and 0.35 V, respectively. For fixed sliding mode switch gain, the mean square deviation of the control voltage of thrusters is 1.24, 0.65, and 0.65 V, respectively. Compared to the traditional fixed sliding mode switch gain, the controller we proposed produces lower chattering. From a further analysis of Fig. 7.6, we can find that the proposed method can make the positive and negative rotation switching frequency of the thruster motor smaller in the tracking control process for the same target trajectory, which has a practical significance to improve the thruster system reliability. The experimental results show the effectiveness of the sliding mode switch gain based on an exponential function.

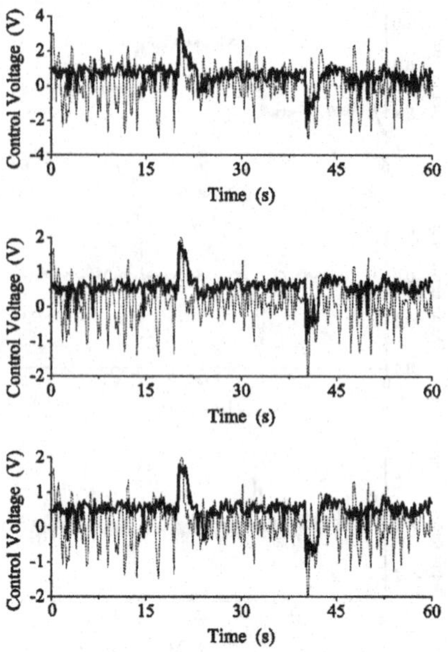

Figure 7.6 Experiment 3: (A) control voltage of thruster 1, (B) control voltage of thruster 2, and (C) control voltage of thruster 3.

7.6. Conclusions

In this chapter, an adaptive sliding mode controller based on a local recurrent neural network has been proposed for an underwater robot system without knowing the dynamics and thrust model, and the trajectory tracking system relying on this approach is proved to be uniformly ultimately bounded. The approach transforms the underwater robot trajectory tracking error system into an affine nonlinear system which uses the thruster control voltage as system input and adopts a local recurrent neural network to approximate unknown terms online for adaptive trajectory tracking control. Experimental results show that it is feasible to use the thrusters control voltage as system input. Compared with PID, the proposed approach has better tracking control accuracy. Compared with adaptive control based on RBF neural network, the approach has better transition characteristics and control accuracy when target trajectory has an abrupt change. Compared with the fixed sliding mode switch gain, the approach produces lower amplitude of chattering. The effectiveness of the approach is proved by theoretical study and experiments.

References

[1] Ahmad Bagheri, Jalal Javadi Moghaddam, Simulation and tracking control for underwater remotely operated vehicle, Neurocomputing 72 (2009) 1934–1950.

[2] A. Bagheri, T. Karimi, N. Amanifard, Tracking performance control of a cable communicated underwater vehicle using adaptive neural network controllers, Applied Soft Computing 10 (2010) 908–918.

[3] T. Chatchanayuenyong, M. Parnichkun, Neural network based-time optimal sliding mode control for an autonomous underwater robot, Mechatronics 16 (2006) 471–478.

[4] Chaio Shiung Chen, Dynamic structure adaptive neural fuzzy control for MIMO uncertain nonlinear systems, Information Sciences 179 (2009) 2676–2688.

[5] Gianluca Antonelli, Stefano Chiaverini, Nilanjan Sarkar, Michael West, Adaptive control of an autonomous underwater vehicle experimental results on ODIN, IEEE Transactions on Control Systems Technology 9 (2001) 756–765.

[6] Jinhyun Kim, WanKyun Chung, Accurate and practical thruster modeling for underwater vehicles, Ocean Engineering 33 (2006) 566–586.

[7] N. Kumar, V. Panwar, N. Sukavanam, Neural network based nonlinear tracking control of kinematically redundant robot manipulators, Mathematical and Computer Modelling 53 (2011) 1889–1901.

[8] R.P. Kumar, C.S. Kumar, D. Sen, A. Dasgupta, Discrete time-delay control of an autonomous underwater vehicle: theory and experimental results, Ocean Engineering 36 (2009) 71–81.

[9] Jihong Li, Panmook Lee, Design of an adaptive nonlinear controller for depth control of an autonomous underwater vehicle, Ocean Engineering 32 (2005) 2165–2181.

[10] Jihong Li, Panmook Lee, A neural network adaptive controller design for free-pitch-angle diving behavior of an autonomous underwater vehicle, Robotics and Autonomous Systems 52 (2005) 132–147.

[11] F.J. Lin, R.J. Wai, Sliding-mode-controlled slider-crank mechanism with fuzzy neural network, IEEE Transactions on Industrial Electronics 48 (2001) 60–70.

[12] Jian Liu, Chuang Yu, Aiming Liu, Research on untethered autonomous underwater vehicle control method, ROBOT 26 (2004) 7–10.

[13] Z.H. Man, X.H. Yu, K. Eshraghian, M. Palaniswami, A robust adaptive sliding mode tracking control using an RBF neural network for robotic manipulators, in: IEEE International Conference on Neural Networks, Perth, WA, Australia, 1995, pp. 2403–2408.

[14] V. Nekoukar, A. Erfanian, Adaptive fuzzy terminal sliding mode control for a class of MIMO uncertain nonlinear systems, Fuzzy Sets and Systems 179 (2011) 34–49.

[15] Quang Hoang Nguyen, Edwin Kreuzer, Adaptive PD-controller for positioning of a remotely operated vehicle close to an underwater structure: theory and experiments, Control Engineering Practice 15 (2007) 411–419.

[16] A. Shahraz, B.R. Boozarjomehry, A fuzzy sliding mode control approach for nonlinear chemical process, Control Engineering Practice 17 (2009) 541–550.

[17] Wallace M.Bessa, Max S. Dutra, Edwin Kreuzer, Depth control of remotely operated underwater vehicles using an adaptive fuzzy sliding mode controller, Robotics and Autonomous Systems 56 (2008) 670–677.

[18] Jeenshing Wang, C.S. George Lee, Self-adaptive recurrent neuro-fuzzy control of an autonomous underwater vehicle, IEEE Transactions on Robotics and Automation 19 (2003) 283–295.

[19] Jianan Xu, Yujia Wang, Mingjun Zhang, Development and experiment of an autonomous underwater vehicle test-bed, Journal of Harbin Engineering University 28 (2007) 212–217.
[20] Yuru Xu, Kun Xiao, Technology development of autonomous ocean vehicle, Acta Autonomous Sinica 33 (2007) 518–521.
[21] Jiancheng Yu, Qiang Li, Aiqun Zhang, Neural network adaptive control for underwater vehicles, Control Theory & Applications 25 (2008) 9–13.
[22] Jiancheng Yu, Aiqun Zhang, Xiaohui Wang, Direct adaptive control of underwater vehicles based on fuzzy neural networks, Acta Autonomous Sinica 33 (2007) 840–846.
[23] Lijun Zhang, Xue Qi, Yongjie Pang, Adaptive output feedback control based on DRFNN, Ocean Engineering 36 (2009) 716–722.
[24] K.Y. Zhuang, H.Y. Su, J. Chu, K.Q. Zhang, Globally stable robust tracking of uncertain systems via fuzzy integral sliding mode control, in: Proceedings of the 3th World Congress on Intelligent Control and Automation, Hefei, P. R. China, 2000, pp. 1821–1831.

CHAPTER EIGHT

Thruster fault reconstruction for autonomous underwater vehicle based on terminal sliding mode observer

Zhenzhong Chu
Shanghai, China

Contents

Abstract

In marine environment with high-pressure and poor-visibility, autonomous underwater vehicles' (AUVs) task could not be achieved, or even the AUV sometimes could not be reclaimed, when an accident happens in AUV. As the important technology of AUV, fault diagnosis has great significance on AUV's safety.

Keywords

Truster fault reconstruction, Autonomous underwater vehicle, RBF neural network, Affine nonlinear system, Terminal sliding observer, Finite time convergence

8.1. Introduction

In marine environment with high-pressure and poor-visibility, autonomous underwater vehicles' (AUVs) task could not be achieved, or even the AUV sometimes could not be reclaimed, when an accident happens in AUV. As the important technology of AUV, fault diagnosis has great significance on AUV's safety [1,2].

Fundamental Design and Automation Technologies in Offshore Robotics
https://doi.org/10.1016/B978-0-12-820271-5.00013-4

Among the researches about thruster fault diagnosis, state-estimation method has been widely used in AUV field, where the inconsistencies can be captured by using the residual signal between the estimated and observed states [3]. The existing state estimation methods include the Kalman filter [4,5], sliding mode observer [6,7], etc. However, an exact mathematical analytical model is always required in the state-estimation method [8]. Due to the complexity of the marine environment, together with the serious nonlinearity and intercoupling features among the AUV 6-DOF motion, the AUV motion model based on the dynamics analysis method has great uncertainty [9]. If fault diagnosis is completed based on this model with great uncertainty, it is relatively easy to result in faulty or missed diagnosis, due to the lack of the prior knowledge of the system uncertainty in practice. Due to the great ability in terms of nonlinear approximation, neural networks are widely used to approximate AUV motion model and provide estimated values for the AUV states. The AUV thruster fault was monitored based on fuzzy neural networks in Wang et al. [10]. In [11], AUV motion model was established based on wavelet neural networks and then thruster fault diagnosis was performed. However, the internal knowledge representation of neural networks is like a black–box, which makes it difficult to explicitly reflect the relationship between input and output, and brings difficulties for the thruster fault reconstruction.

Due to the good robustness against the system uncertainty and external disturbances, a sliding mode observer is used to reconstruct the fault through the introduction of the equivalent output injection. Edwards et al. [12] studied fault reconstruction problems of an actuator and sensor for a class of linear systems. Based on this, the fault diagnosis problems of nonlinear uncertain systems were further explored in [13–16]. A remarkable feature of this method for fault reconstruction is that the sliding mode motion still needs to be maintained when the actuator faults. That is, the state estimation error should be still zero. In general, the occurrence of thruster fault would damage the sliding mode motion in the sliding mode observer. Hence, it requires the state estimation error converge to zero in a finite time for fault reconstruction. However, Chee et al. [17] pointed out that the sliding mode observer given in [12–16] can only guarantee the finite-time convergence of the state which can be measured directly by the sensor, but asymptotic convergence was obtained for the unmeasurable state. This results in that the thruster fault cannot be reconstructed timely based on the sliding mode observer. A modeling method of AUV motion model with the affine form is given in this chapter based on neural networks, where the position and orientation state with respect to the earth-fixed frame can be measured by sensors while the linear and angular velocity states are not measurable. In this model, the relationship between the AUV system input and output can be reflected clearly. Based on the established AUV motion model, a terminal sliding mode observer is developed to guarantee the finite-time convergence for system states, and then the equivalent output injection method is used to reconstruct thruster fault online.

This chapter is organized as follows. Section 8.2 describes the AUV thruster fault reconstruction problem. In Section 8.3, the motion modeling method is presented based on RBF neural networks. And in Section 8.4, the thruster fault is reconstructed based on a terminal sliding mode observer. Then Section 8.5 verifies the effectiveness of the established motion model and fault reconstruction approach. Finally, we make a brief conclusion in Section 8.6.

8.2. Problem formulation

The AUV mathematical model in 6 DOF can be described as

$$\begin{aligned} \dot{\eta} &= J(\eta)\, \nu, \\ M\dot{\nu} + C(\nu)\,\nu + D(\nu)\,\nu + g(\eta) &= B\tau(\nu, u), \\ y_{out} &= \eta \end{aligned} \tag{8.1}$$

where $\eta \in \mathbb{R}^6$ denotes the vector of position and orientation in the earth-fixed frame, $\nu \in \mathbb{R}^6$ is the vector of linear and angular velocity expressed in the body-fixed frame; $M \in \mathbb{R}^{6\times6}$ is the inertia matrix including extra mass, the matrix $C(\nu) \in \mathbb{R}^{6\times6}$ is centripetal and Coriolis forces, including the centripetal force and Coriolis force produced by extra mass, $D(\nu) \in \mathbb{R}^{6\times6}$ is the hydrodynamic resistance and lift matrix, $g(\eta) \in \mathbb{R}^6$ is restoring force and torque vector, $\tau(\nu, u) \in \mathbb{R}^n$ is the thruster force, $u = [u_1, \ldots, u_n]^T$ is the vector of thruster control signal from the controller output which satisfies $\|u\| \le u_d$, n is the number of the thrusters, $y_{out} \in \mathbb{R}^6$ are the state variables which can be measured directly by sensors; $J(\eta) \in \mathbb{R}^{6\times6}$ is the transformation matrix; $B \in \mathbb{R}^{6\times n}$ is the thruster distribution matrix. In this chapter, we will not consider the situation of redundant thrusters' configuration, namely matrix B will be of full column rank.

There exists a linear change of coordinates $w = Tv$ such that

$$\begin{gathered}\dot{\eta} = J(\eta)\, T^{-1} w,\\ TMT^{-1}\dot{w} + TC\left(T^{-1}w\right) T^{-1}w + TD\left(T^{-1}w\right) T^{-1}w + Tg(\eta) = \underline{B}\tau\left(T^{-1}w, u\right),\\ y_{out} = \eta,\end{gathered} \tag{8.2}$$

where $\underline{B} = [0_{6-n} \quad I_n]^T$.

Define

$$\begin{aligned}\bar{\xi}_1 &= \eta,\\ \bar{\xi}_2 &= J(\eta)\, T^{-1} w.\end{aligned} \tag{8.3}$$

Then (8.2) can be expressed as a strict feedback nonlinear system shown as follows:

$$\begin{aligned}\dot{\bar{\xi}}_1 &= \bar{\xi}_2,\\ \dot{\bar{\xi}}_2 &= h\left(\bar{\xi}_1, \bar{\xi}_2\right) + \bar{B}\tau\left(J^{-1}\left(\bar{\xi}_1\right)\bar{\xi}_2, u\right),\\ y_{out} &= \bar{\xi}_1\end{aligned} \tag{8.4}$$

where

$$\begin{aligned}h\left(\bar{\xi}_1, \bar{\xi}_2\right) = J\left(\bar{\xi}_1\right) M^{-1}\left\{-C\left(T^{-1}w\right) T^{-1}w - D\left(T^{-1}w\right) T^{-1}w - g\left(\bar{\xi}_1\right)\right\}\\ + \dot{J}\left(\bar{\xi}_1\right) T^{-1}w,\\ \bar{B} = J\left(\bar{\xi}_1\right) M^{-1} T^{-1} \underline{B}.\end{aligned}$$

With respect to AUV motion modeling, the thrust model, describing the mapping relationship between control signal and thrust, should be at first established for thrusters in most of the existing methods [18,19]. But the thrust is related with the propeller speed, AUV state, and other factors, which leads to be difficult to establish accurate thrust model [18]. An inaccurate thruster model will cause the AUV motion model to have modeling error, and also affect the accuracy of fault diagnosis. Considering that the purpose of AUV motion modeling is to reconstruct thruster fault, in this chapter, thus, we put the thruster control signal u as the input of AUV system directly without considering the thrust modeling.

From (8.4), if u is treated as system input, then the AUV is a nonaffine nonlinear system. For the subsequent research for AUV thruster fault reconstruction, affine transformation is applied to the equation shown in (8.4) and the affine nonlinear system with system input u is obtained.

The Taylor expansion of $\tau\left(J^{-1}(\bar{\xi}_1)\bar{\xi}_2, u\right)$ about u^* is given by

$$\tau\left(J^{-1}(\bar{\xi}_1)\bar{\xi}_2, u\right) = \tau\left(J^{-1}(\bar{\xi}_1)\bar{\xi}_2, u^*\right) + \left.\frac{\partial\tau\left(J^{-1}(\bar{\xi}_1)\bar{\xi}_2, u\right)}{\partial u}\right|_{u=u^*}(u-u^*) + O\left((u-u^*)^2\right) \tag{8.5}$$

where $O\left((u-u^*)^2\right)$ is the higher-order term of the Taylor expansion.

Let

$$\begin{aligned}\bar{F}(\bar{\xi}_1, \bar{\xi}_2) &= a\bar{\xi}_2 + h(\bar{\xi}_1, \bar{\xi}_2) + \tau\left(J^{-1}(\bar{\xi}_1)\bar{\xi}_2, u^*\right) \\ &\quad - \left.\frac{\partial\tau\left(J^{-1}(\bar{\xi}_1)\bar{\xi}_2, u\right)}{\partial u}\right|_{u=u^*} u^* + O\left((u-u^*)^2\right), \\ \bar{G}(\bar{\xi}_1, \bar{\xi}_2) &= \mathrm{diag}\left(\left.\frac{\partial\tau\left(J^{-1}(\bar{\xi}_1)\bar{\xi}_2, u\right)}{\partial u}\right|_{u=u^*}\right).\end{aligned}$$

Then (8.4) can be rewritten in the form of an affine nonlinear system where the thruster control signal u is used as input, shown as follows:

$$\begin{aligned}\dot{\bar{\xi}}_1 &= \bar{\xi}_2, \\ \dot{\bar{\xi}}_2 &= -a\bar{\xi}_2 + \bar{F}(\bar{\xi}_1, \bar{\xi}_2) + \bar{B}\bar{G}(\bar{\xi}_1, \bar{\xi}_2)u\end{aligned} \tag{8.6}$$

where matrix a is chosen such that the eigenvalues of matrix $\bar{A} = [0, I_6; 0, -a]$ are nonpositive.

In practice, when a thruster fault occurs, the actual thrust of the thruster will be less than the theoretical thrust. There is a direct proportion between thrust output and control signal. Therefore, in this chapter, the AUV motion model with thruster fault is described as

$$\begin{aligned}\dot{\bar{\xi}} &= \bar{A}\bar{\xi} + b\left[\bar{F}(\bar{\xi}_1, \bar{\xi}_2) + \bar{B}\bar{G}(\bar{\xi}_1, \bar{\xi}_2)(u-f)\right], \\ y_{out} &= C\bar{\xi}\end{aligned} \tag{8.7}$$

where $\bar{\xi} = [\bar{\xi}_1^T \quad \bar{\xi}_2^T]^T$, $b = \begin{bmatrix} 0 \\ I_6 \end{bmatrix}$, $C = [I_6 \quad 0]$, u is the control signal of controller output; $f = [f_1, \ldots, f_n]^T$ is the vector of thruster fault such that $f_i u_i \geq 0$ $(i = 1, \ldots, n)$; $f_i = 0$ means that there is no thruster fault; $0 < |f_i| \leq |u_i|$ means that there is a thruster fault.

In summary, the research in this chapter can be expressed as follows: for a class of AUV system with only position and orientation measurements, at first the RBF neural networks are used to identify nonlinear functions

$\bar{F}(\bar{\xi}_1, \bar{\xi}_2)$ and $\bar{G}(\bar{\xi}_1, \bar{\xi}_2)$ and the AUV motion model is established in the form of an affine nonlinear system; furthermore, a terminal sliding mode observer is constructed for thrust fault reconstruction based on the model.

8.3. AUV motion modeling

When there is no thruster fault, (8.7) can be written as

$$\dot{\bar{\xi}} = \bar{A}\bar{\xi} + b\left[\bar{F}(\bar{\xi}_1, \bar{\xi}_2) + \bar{B}\bar{G}(\bar{\xi}_1, \bar{\xi}_2)\, u\right],$$
$$y_{out} = C\bar{\xi}. \tag{8.8}$$

For AUV motion modeling, RBF neural networks are used to identify the unknown functions $\bar{F}(\bar{\xi}_1, \bar{\xi}_2)$ and $\bar{G}(\bar{\xi}_1, \bar{\xi}_2)$. From the analysis in Section 8.2, both $\bar{F}(\bar{\xi}_1, \bar{\xi}_2)$ and $\bar{G}(\bar{\xi}_1, \bar{\xi}_2)$ are functions of the states $\bar{\xi}_1$ and $\bar{\xi}_2$. However, since the velocity and angular velocity of AUV cannot be measured, the state $\bar{\xi}_2$ is unavailable. But for the second-order nonlinear system shown in (8.8), the AUV velocity and angular velocity information can be indirectly reflected by the difference between the two position or orientation states at adjacent moments. For this reason, we use $x = \left[\bar{\xi}_1^t; \bar{\xi}_1^t - \bar{\xi}_1^{t-1}\right]$ as the input of RBF neural networks in this chapter, where $\bar{\xi}_1^t$ is the position and orientation state vector at time t and $\bar{\xi}_1^{t-1}$ is the position and orientation state vector at time $t-1$.

From the nonlinear identification ability of RBF neural networks, there are network weights $W_F^* \in \mathbb{R}^{6 \times H_F}$ and $W_G^* \in \mathbb{R}^{6 \times H_G}$, centers of radial basis function $c_F^* \in \mathbb{R}^{h_{Fc} \times H_F}$ and $c_G^* \in \mathbb{R}^{h_{Gc} \times H_G}$, the variances of radial basis function $b_F^* \in \mathbb{R}$ and $b_G^* \in \mathbb{R}$, such that the following equations hold:

$$\bar{F}(\bar{\xi}_1, \bar{\xi}_2) = W_F^* \Phi\left(x, c_F^*, b_F^*\right) + \varepsilon_F,$$
$$\bar{G}(\bar{\xi}_1, \bar{\xi}_2) = \operatorname{diag}\left(W_G^* \Phi\left(x, c_G^*, b_G^*\right) + \varepsilon_G\right) \tag{8.9}$$

where $\Phi(\cdot)$ is a radial basis function, ε_F, ε_G are approximation errors, H_F, H_G are the numbers of neurons in the hidden layer, h_{Fc}, h_{Gc} are the numbers of neural network inputs.

The output of RBF neural networks can be written as

$$F(x) = W_F \Phi\left(x, c_F, b_F\right),$$
$$G(x) = \operatorname{diag}\left(W_G \Phi\left(x, c_G, b_G\right)\right) \tag{8.10}$$

where W_F, W_G are estimated network weights, b_F, b_G are estimated radial basis function variances, $c_F = [c_{F1}, \ldots, c_{FH_F}]$, $c_G = [c_{G1}, \ldots, c_{GH_G}]$ are estimated centers of the radial basis function.

From (8.9) and (8.10), the neural networks estimation errors can be written as

$$\begin{aligned} F_e &= F(x) - \bar{F}(\bar{\xi}_1, \bar{\xi}_2) \\ &= \widetilde{W}_F \Phi(x, c_F, b_F) + W_F Q_F - \widetilde{W}_F Q_F - \varepsilon_F, \end{aligned} \tag{8.11}$$

$$\begin{aligned} G_e &= G(x) - \bar{G}(\bar{\xi}_1, \bar{\xi}_2) \\ &= \operatorname{diag}\left(\widetilde{W}_G \Phi(x, c_G, b_G) + W_G Q_G - \widetilde{W}_G Q_G - \varepsilon_G\right) \end{aligned} \tag{8.12}$$

where $\widetilde{W}_F = W_F - W_F^*$ and $\widetilde{W}_G = W_G - W_G^*$ are the estimation errors of network weights; Q_F and Q_G are as follows:

$$\begin{aligned} Q_F &= \Phi(x, c_F, b_F) - \Phi(x, c_F^*, b_F^*), \\ Q_G &= \Phi(x, c_G, b_G) - \Phi(x, c_G^*, b_G^*). \end{aligned} \tag{8.13}$$

To obtain the explicit expressions of c_F, c_G, b_F, and b_G, the Taylor expansions of function Q_F and Q_G are written respectively as

$$\begin{aligned} Q_F &= \sum_{i=1}^{H_F} \left.\frac{\partial \Phi(x, c_F^*, b_F^*)}{\partial c_{Fi}^*}\right|_{c_{Fi}^* = c_{Fi}} \tilde{c}_{Fi} + \left.\frac{\partial \Phi(x, c_F^*, b_F^*)}{\partial b_F^*}\right|_{b_F^* = b_F} \tilde{b}_F + o_F\left(\tilde{c}_F^2, \tilde{b}_F^2\right) \\ &= \sum_{i=1}^{H_F} \sigma_{Fci} \tilde{c}_{Fi} + \sigma_{Fb} \tilde{b}_F + o_F\left(\tilde{c}_F^2, \tilde{b}_F^2\right), \end{aligned} \tag{8.14a}$$

$$\begin{aligned} Q_G &= \sum_{j=1}^{H_G} \left.\frac{\partial \Phi(x, c_G^*, b_G^*)}{\partial c_{Gj}^*}\right|_{c_{Gj}^* = c_{Gj}} \tilde{c}_{Gj} + \left.\frac{\partial \Phi(x, c_G^*, b_G^*)}{\partial b_G^*}\right|_{b_G^* = b_G} \tilde{b}_G + o_G\left(\tilde{c}_G^2, \tilde{b}_G^2\right) \\ &= \sum_{j=1}^{H_G} \sigma_{Gcj} \tilde{c}_{Gj} + \sigma_{Gb} \tilde{b}_G + o_G\left(\tilde{c}_G^2, \tilde{b}_G^2\right) \end{aligned} \tag{8.14b}$$

where $\tilde{c}_{Fi} = c_{Fi} - c_{Fi}^*$ is the estimation error for the center of the radial basis function in the ith hidden layer unit of $F(x)$, $\tilde{c}_{Gj} = c_{Gj} - c_{Gj}^*$ is the estimation error for the center of the radial basis function in the jth hidden layer unit of $G(x)$, $\tilde{b}_F = b_F - b_F^*$ and $\tilde{b}_G = b_G - b_G^*$ are variance estimation errors of the radial basis functions, respectively; $o_F\left(\tilde{c}_F^2, \tilde{b}_F^2\right)$ and $o_G\left(\tilde{c}_G^2, \tilde{b}_G^2\right)$ are the higher-order terms of the Taylor expansion.

From (8.11) to (8.14), it follows that

$$F_e = \widetilde{W}_F \Phi\left(x, c_F, b_F\right) + W_F \left(\sum_{i=1}^{H_F} \sigma_{Fci} \tilde{c}_{Fi} + \sigma_{Fb} \tilde{b}_F \right) + \omega_F, \tag{8.15}$$

$$G_e = \operatorname{diag}\left(\widetilde{W}_G \Phi\left(x, c_G, b_G\right) + W_G \left(\sum_{j=1}^{H_G} \sigma_{Gcj} \tilde{c}_{Gj} + \sigma_{Gb} \tilde{b}_G \right) + \omega_G \right) \tag{8.16}$$

where w_F and w_G are as follows:

$$w_F = W_F o_F\left(\tilde{c}_F^2, \tilde{b}_F^2\right) + \widetilde{W}_F Q_F - \varepsilon_F,$$
$$w_G = \operatorname{diag}\left(W_G o_G\left(\tilde{c}_G^2, \tilde{b}_G^2\right) + \widetilde{W}_G Q_G - \varepsilon_G \right).$$

In Bessa et al. [20] and Yu et al. [21], the estimation errors of neural networks are assumed to be bounded with known constants, which is a strong restricting condition in the neural networks learning process. Hence, the following assumption is made for the estimation errors of neural networks [22,23]:

Assumption 8.1. Assume

$$F_e \le C_F^* \varphi_F(x) \quad \text{and} \quad G_e \le C_G^* \varphi_G(x) \tag{8.17}$$

where C_F^* and C_G^* are unknown and positive vectors, $\varphi_F(x)$ and $\varphi_G(x)$ are known, bounded and nonnegative functions of x, and there are known and positive constants l_F and l_G such that

$$|\varphi_F(x_1) - \varphi_F(x_2)| \le l_F |x_1 - x_2|,$$
$$|\varphi_G(x_1) - \varphi_G(x_2)| \le l_G |x_1 - x_2|. \tag{8.18}$$

According to the output of RBF neural networks as in (8.10), the AUV motion model can be expressed as

$$\dot{\xi} = \bar{A}\xi + b\left[F(x) + \bar{B}G(x)u - C_F\varphi_F - \bar{B}C_G\varphi_G u\right] \tag{8.19}$$

where C_F and C_G are the estimates of unknown vectors C_F^* and C_G^*, respectively.

Letting $e = \xi - \bar{\xi}$, and according to (8.8) and (8.19), the estimation error state equation of the model is

$$\dot{e} = \bar{A}e + b\left[F_e + \bar{B}G_e u - C_F\varphi_F - \bar{B}C_G\varphi_G u\right],$$
$$g = Ce. \tag{8.20}$$

In the process of the neural networks training, we need to know the estimation errors state e. Considering that the AUV velocity and angular velocity state cannot be measured directly, a high-gain observer is constructed as follows:

$$\dot{\hat{e}} = \bar{A}\hat{e} - k_1 C\tilde{e},$$
$$\hat{g} = C\hat{e} \tag{8.21}$$

where $\tilde{e} = \hat{e} - e$ and matrix k_1 is chosen such that $\tilde{A} = A - k_1 C$ is asymptotically stable.

Form (8.19) and (8.21), the observer error dynamics can be written as

$$\dot{\tilde{e}} = \bar{A}\tilde{e} - b\left[F_e + \bar{B}G_e u - C_F\varphi_F - \bar{B}C_G\varphi_G u\right]. \tag{8.22}$$

Based on Lyapunov theory, the adaptive law of the parameters W_F, W_G, c_F, c_G, b_F, b_G, C_F, C_G is designed as follows:

$$\dot{W}_F = \Gamma_{WF}\left[-\left(\hat{e}^T b\right)^T \Phi\left(x, c_F, b_F\right)^T - \lambda_{WF}\left(W_F - W_{F0}\right)\right],$$
$$\dot{W}_G = \Gamma_{WG}\left[-\left(\hat{e}^T b\bar{B}\mathrm{diag}\left(u\right)\right)^T \Phi\left(x, c_G, b_G\right)^T - \lambda_{WG}\left(W_G - W_{G0}\right)\right], \tag{8.23}$$

$$\dot{c}_{Fi} = \Gamma_{Fci}\left[-\left(\hat{e}^T b W_F \sigma_{Fci}\right)^T - \lambda_{Fc}\left(c_{Fi} - c_{Fi0}\right)\right],$$
$$\dot{c}_{Gj} = \Gamma_{Gcj}\left[-\left(\hat{e}^T b\bar{B} W_G \sigma_{Gcj}\right)^T \mathrm{diag}\left(u\right) - \lambda_{Gc}\left(c_{Gj} - c_{Gj0}\right)\right], \tag{8.24}$$

$$\dot{b}_F = \Gamma_{Fb}\left[-\hat{e}^T b W_F \sigma_{Fb} - \lambda_{Fb}\left(b_F - b_{F0}\right)\right],$$
$$\dot{b}_G = \Gamma_{Gb}\left[-\hat{e}^T b\bar{B} W_G \sigma_{Gb}\mathrm{diag}\left(u\right) - \lambda_{Gb}\left(b_G - b_{G0}\right)\right], \tag{8.25}$$

$$\dot{C}_F = \Gamma_{CF}\left[\left(\hat{e}^T b\right)^T \varphi_F^T - \lambda_{CF}\left(C_F - C_{F0}\right)\right],$$
$$\dot{C}_G = \Gamma_{CG}\left[\left(u\hat{e}^T b\bar{B}\right)^T \varphi_G^T - \lambda_{CG}\left(C_G - C_{G0}\right)\right] \tag{8.26}$$

where W_{F0}, W_{G0}, c_{Fi0}, c_{Gj0}, b_{F0}, b_{G0}, C_{F0}, C_{G0} are the best guesses of corresponding parameters, $\lambda_{(\cdot)} > 0$, $\lambda_{CF} > 2\left\|\varphi_F\right\|^2$, $\lambda_{CG} > 2\left\|\varphi_G\right\|^2$, and $\Gamma_{(\cdot)}$ is the gain matrix.

Proposition 8.1. *Consider the AUV motion model as in (8.8). Under Assumption 8.1 and with the adaptive law defined by (8.23)–(8.26), model (8.19) will eventually converge to AUV actual dynamics.*

Proof. Define $\tilde{C}_F = C_F - C_F^*$, $\tilde{C}_G = C_G - C_G^*$.

Choose Lyapunov function candidate as

$$
\begin{aligned}
V_1 = {} & \frac{1}{2}e^T e + \frac{1}{2}\tilde{e}^T\tilde{e} + \frac{1}{2}\operatorname{tr}\left[\widetilde{W}_F^T \Gamma_{WF}^{-1}\widetilde{W}_F\right] + \frac{1}{2}\operatorname{tr}\left[\widetilde{W}_G^T \Gamma_{WG}^{-1}\widetilde{W}_G\right] \\
& + \frac{1}{2}\sum_{i=1}^{H_F}\Gamma_{Fci}^{-1}\tilde{C}_{Fi}^T\tilde{C}_{Fi} + \frac{1}{2}\sum_{J=1}^{H_G}\Gamma_{Gcj}^{-1}\tilde{C}_{cj}^T\tilde{C}_{Gj} + \frac{1}{2}\Gamma_{Fb}^{-1}\tilde{b}_F^2 + \frac{1}{2}\Gamma_{Gb}^{-1}\tilde{b}_G^2 \\
& + \frac{1}{2}\Gamma_{CF}^{-1}\tilde{C}_F^T\tilde{C}_F + \frac{1}{2}\Gamma_{CG}^{-1}\tilde{C}_G^T\tilde{C}_G. \qquad (8.27)
\end{aligned}
$$

The time derivative of (8.27) is

$$
\begin{aligned}
\dot{V}_1 = {} & e^T\bar{A}e + \tilde{e}^T\tilde{A}\tilde{e} + \hat{e}^T b\left[F_e + \bar{B}G_e u - C_F\varphi_F - \bar{B}C_G\varphi_G u\right] \\
& - 2\tilde{e}^T b\left[F_e + \bar{B}G_e u - C_F\varphi_F - \bar{B}C_G\varphi_G u\right] + tr\left[\widetilde{W}_F^T\Gamma_{WF}^{-1}\dot{\widetilde{W}}_F\right] \\
& + \operatorname{tr}\left[\widetilde{W}_G^T\Gamma_{WG}^{-1}\dot{\widetilde{W}}_G\right] + \sum_{i=1}^{H_F}\Gamma_{Fci}^{-1}\tilde{C}_{Fi}^T\dot{\tilde{C}}_{Fi} + \sum_{J=1}^{H_G}\Gamma_{Gcj}^{-1}\tilde{C}_{Gj}^T\dot{\tilde{C}}_{Gj} \\
& + \Gamma_{Fb}^{-1}\tilde{b}_F\dot{\tilde{b}}_F + \Gamma_{Gb}^{-1}\tilde{b}_G\dot{\tilde{b}}_G + \Gamma_{CF}^{-1}\tilde{C}_F\dot{\tilde{C}}_F + \Gamma_{CG}^{-1}\tilde{C}_G\dot{\tilde{C}}_G. \qquad (8.28)
\end{aligned}
$$

Substituting (8.15) and (8.16) into (8.28), one has

$$
\begin{aligned}
\dot{V}_1 = {} & e^T\bar{A}e + \tilde{e}^T\tilde{A}\tilde{e} + \hat{e}^T b\Bigg[\widetilde{W}_F\Phi\left(X, C_F, b_F\right) + W_F\left(\sum_{i=1}^{H_F}\sigma_{Fci}\tilde{C}_{Fi} + \sigma_{Fb}\tilde{b}_F\right) \\
& + \omega_F - C_F\varphi_F\Bigg] - 2\tilde{e}^T b\left[F_e + \bar{B}G_e u - C_F\varphi_F - \bar{B}C_G\varphi_G u\right] \\
& + \hat{e}^T b[\bar{B}\mathrm{diag}\left(\widetilde{W}_F\Phi\left(X, C_F, b_F\right) + W_G\left(\sum_{j=1}^{G}\sigma_{Gcj}\tilde{C}_{Gj} + \sigma_{Gb}\tilde{b}_G\right) + \omega_G\right)u \\
& - \bar{B}C_G\varphi_G u] + \operatorname{tr}\left[\widetilde{W}_F^T\Gamma_{WF}^{-1}\dot{\widetilde{W}}_F\right] + tr\left[\widetilde{W}_G^T\Gamma_{WG}^{-1}\dot{\widetilde{W}}_G\right] \\
& + \sum_{i=1}^{H_F}\Gamma_{Fci}^{-1}\tilde{C}_{Fi}^T\dot{\tilde{C}}_{Fi} + \sum_{J=1}^{H_G}\Gamma_{Gcj}^{-1}\tilde{C}_{Gj}^T\dot{\tilde{C}}_{Gj} + \Gamma_{Fb}^{-1}\tilde{b}_F\dot{\tilde{b}}_F + \Gamma_{Gb}^{-1}\tilde{b}_G\dot{\tilde{b}}_G \\
& + \Gamma_{CF}^{-1}\tilde{C}_F\dot{\tilde{C}}_F + \Gamma_{CG}^{-1}\tilde{C}_G\dot{\tilde{C}}_G]. \qquad (8.29)
\end{aligned}
$$

Substituting (8.23), (8.24) and (8.25) into (8.29), one obtains

$$
\begin{aligned}
\dot{V}_1 = {} & e^T\bar{A}e + \tilde{e}^T\tilde{A}\tilde{e} + \hat{e}^T b\left(\omega_F + \bar{B}\mathrm{diag}\left(\omega_G\right)u - C_F\varphi_F - \bar{B}C_G\varphi_G u\right) \\
& - 2\tilde{e}^T b\left[F_e + \bar{B}G_e u - C_F\varphi_F - \bar{B}C_G\varphi_G u\right]
\end{aligned}
$$

$$
\begin{aligned}
&+\Gamma_{CF}^{-1}\tilde{C}_F^T\dot{\tilde{C}}_F+\Gamma_{CG}^{-1}\tilde{C}_G^T\dot{\tilde{C}}_G-\operatorname{tr}\left[\lambda_{WF}\widetilde{W}_F^T\left(W_F-W_{F0}\right)\right]\\
&-\operatorname{tr}\left[\lambda_{WG}\widetilde{W}_G^T\left(W_G-W_{G0}\right)\right]-\operatorname{tr}\left[\lambda_{Fc}\tilde{c}_F^T\left(c_F-c_{F0}\right)\right]\\
&-\operatorname{tr}\left[\lambda_{Gc}\tilde{c}_G^T\left(c_G-c_{G0}\right)\right]-\operatorname{tr}\left[\lambda_{Fb}\tilde{b}_F^T\left(b_F-b_{F0}\right)\right]-\operatorname{tr}\left[\lambda_{Gb}\tilde{b}_G^T\left(b_G-b_{G0}\right)\right],
\end{aligned}
\tag{8.30}
$$

$$
\begin{aligned}
\dot{V}_1\le{}&\lambda_{\bar{A}}\|e\|^2+\lambda_{\tilde{A}}\|\tilde{e}\|^2+\hat{e}^Tb\left(-\tilde{C}_F\varphi_F-\bar{B}\tilde{C}_G\varphi_G u\right)\\
&+\left\|\hat{e}^Tb\right\|\left\|\widetilde{W}_F\Phi\left(x,c_F,b_F\right)+W_F\left(\sum_{i=1}^{H_F}\sigma_{Fci}\tilde{c}_{Fi}+\sigma_{Fb}\tilde{b}_F\right)\right.\\
&\left.+\operatorname{diag}\left(\bar{B}\widetilde{W}_G\Phi\left(x,c_G,b_G\right)u+\bar{B}W_G\left(\sum_{j=1}^{H_G}\sigma_{Gcj}\tilde{c}_{Gj}+\sigma_{Gb}\tilde{b}_G\right)u\right)\right\|\\
&+2\left\|\tilde{e}^Tb\right\|\left(\left\|\tilde{C}_F\varphi_F\right\|+\left\|\bar{B}u\tilde{C}_G\varphi_G\right\|\right)\\
&+\Gamma_{CF}^{-1}\tilde{C}_F^T\dot{\tilde{C}}_F+\Gamma_{CG}^{-1}\tilde{C}_G^T\dot{\tilde{C}}_G-\operatorname{tr}\left[\lambda_{WF}\widetilde{W}_F^T\left(W_F-W_{F0}\right)\right]\\
&-\operatorname{tr}\left[\lambda_{WG}\widetilde{W}_G^T\left(W_G-W_{G0}\right)\right]-\operatorname{tr}\left[\lambda_{Fc}\tilde{c}_F^T\left(c_F-c_{F0}\right)\right]\\
&-\operatorname{tr}\left[\lambda_{Gc}\tilde{c}_G^T\left(c_G-c_{G0}\right)\right]-\operatorname{tr}\left[\lambda_{Fb}\tilde{b}_F^T\left(b_F-b_{F0}\right)\right]-\operatorname{tr}\left[\lambda_{Gb}\tilde{b}_G^T\left(b_G-b_{G0}\right)\right]
\end{aligned}
\tag{8.31}
$$

where $\lambda_{\bar{A}}$ and $\lambda_{\tilde{A}}$ are the largest eigenvalues of $\bar{A}$ and $\tilde{A}$, respectively.

Substituting (8.26) into (8.31) yields

$$
\begin{aligned}
\dot{V}_1\le{}&\lambda_{\bar{A}}\|e\|^2+\lambda_{\tilde{A}}\|\tilde{e}\|^2\\
&+\left\|\hat{e}^Tb\right\|\left\|\widetilde{W}_F\Phi\left(x,c_F,b_F\right)+W_F\left(\sum_{i=1}^{H_F}\sigma_{Fci}\tilde{c}_{Fi}+\sigma_{Fb}\tilde{b}_F\right)\right.\\
&\left.+\operatorname{diag}\left(\bar{B}\widetilde{W}_G\Phi\left(x,c_G,b_G\right)u+\bar{B}W_G\left(\sum_{j=1}^{H_G}\sigma_{Gcj}\tilde{c}_{Gj}+\sigma_{Gb}\tilde{b}_G\right)u\right)\right\|\\
&+2\left\|\tilde{e}^Tb\right\|\left(\left\|\tilde{C}_F\varphi_F\right\|+\left\|\bar{B}u\tilde{C}_G\varphi_G\right\|\right)-\operatorname{tr}\left[\lambda_{WF}\widetilde{W}_F^T\left(W_F-W_{F0}\right)\right]\\
&-\operatorname{tr}\left[\lambda_{WG}\widetilde{W}_G^T\left(W_G-W_{G0}\right)\right]-\operatorname{tr}\left[\lambda_{Fc}\tilde{c}_F^T\left(c_F-c_{F0}\right)\right]\\
&-\operatorname{tr}\left[\lambda_{Gc}\tilde{c}_G^T\left(c_G-c_{G0}\right)\right]-\lambda_{CF}\operatorname{tr}\left[\tilde{C}_F^T\left(C_F-C_{F0}\right)\right]\\
&-\lambda_{CG}\operatorname{tr}\left[\tilde{C}_G^T\left(C_G-C_{G0}\right)\right]-\operatorname{tr}\left[\lambda_{Fb}\tilde{b}_F^T\left(b_F-b_{F0}\right)\right]\\
&-\operatorname{tr}\left[\lambda_{Gb}\tilde{b}_G^T\left(b_G-b_{G0}\right)\right].
\end{aligned}
\tag{8.32}
$$

Since

$$
\begin{aligned}
2\left\|\tilde{e}^T b\right\|\left\|\tilde{C}_F\varphi_F\right\| &\le \|b\|^2\|\tilde{e}\|^2+\left\|\tilde{C}_F\right\|^2\|\varphi_F\|^2,\\
2\left\|\tilde{e}^T b\right\|\left\|\bar{B}u\tilde{C}_G\varphi_G\right\| &\le \|b\|^2\left\|\bar{B}\right\|^2\|u_d\|^2\|\tilde{e}\|^2+\left\|\tilde{C}_G\right\|^2\|\varphi_G\|^2,\\
-\operatorname{tr}\left[\lambda_{WF}\widetilde{W}_F^T\left(W_F-W_{F0}\right)\right] &= -\operatorname{tr}\left[\lambda_{WF}\widetilde{W}_F^T\left(\widetilde{W}_F+W_F^*-W_{F0}\right)\right]\\
&\le -\frac{\lambda_{WF}}{2}\left\|\widetilde{W}_F\right\|^2+\frac{\lambda_{WF}}{2}\left\|W_F^*-W_{F0}\right\|^2,\\
-\operatorname{tr}\left[\lambda_{WG}\widetilde{W}_G^T\left(W_G-W_{G0}\right)\right] &= -\operatorname{tr}\left[\lambda_{WG}\widetilde{W}_G^T\left(\widetilde{W}_G+W_G^*-W_{G0}\right)\right]\\
&\le -\frac{\lambda_{WG}}{2}\left\|\widetilde{W}_G\right\|^2+\frac{\lambda_{WG}}{2}\left\|W_G^*-W_{G0}\right\|^2,\\
-\operatorname{tr}\left[\lambda_{Fc}\tilde{c}_F^T\left(c_F-c_{F0}\right)\right] &= -\operatorname{tr}\left[\lambda_{Fc}\tilde{c}_F^T\left(\tilde{c}_F+c_F^*-c_{F0}\right)\right]\\
&\le -\frac{\lambda_{Fc}}{2}\|\tilde{c}_F\|^2+\frac{\lambda_{Fc}}{2}\left\|c_F^*-c_{F0}\right\|^2,\\
-\operatorname{tr}\left[\lambda_{Gc}\tilde{c}_G^T\left(c_G-c_{G0}\right)\right] &= -\operatorname{tr}\left[\lambda_{Gc}\tilde{c}_G^T\left(\tilde{c}_G+c_G^*-c_{G0}\right)\right]\\
&\le -\frac{\lambda_{Gc}}{2}\|\tilde{c}_G\|^2+\frac{\lambda_{Gc}}{2}\left\|c_G^*-c_{G0}\right\|^2,\\
-\operatorname{tr}\left[\lambda_{Fb}\tilde{b}_F^T\left(b_F-b_{F0}\right)\right] &= -\operatorname{tr}\left[\lambda_{Fb}\tilde{b}_F^T\left(\tilde{b}_F+b_F^*-b_{F0}\right)\right]\\
&\le -\frac{\lambda_{Fb}}{2}\left\|\tilde{b}_F\right\|^2+\frac{\lambda_{Fb}}{2}\left\|b_F^*-b_{F0}\right\|^2,\\
-\operatorname{tr}\left[\lambda_{Gb}\tilde{b}_G^T\left(b_G-b_{G0}\right)\right] &= -\operatorname{tr}\left[\lambda_{Gb}\tilde{b}_G^T\left(\tilde{b}_G+b_G^*-b_{G0}\right)\right]\\
&\le -\frac{\lambda_{Gb}}{2}\left\|\tilde{b}_G\right\|^2+\frac{\lambda_{Gb}}{2}\left\|b_G^*-b_{G0}\right\|^2,\\
-\lambda_{CF}\operatorname{tr}\left[\tilde{C}_F^T\left(C_F-C_{F0}\right)\right] &= -\lambda_{CF}\operatorname{tr}\left[\tilde{C}_F^T\left(\tilde{C}_F+C_F^*-C_{F0}\right)\right]\\
&\le -\frac{\lambda_{CF}}{2}\left\|\tilde{C}_F\right\|^2+\frac{\lambda_{CF}}{2}\left\|C_F^*-C_{F0}\right\|^2,\\
-\lambda_{CG}\operatorname{tr}\left[\tilde{C}_G^T\left(C_G-C_{G0}\right)\right] &= -\lambda_{CG}\operatorname{tr}\left[\tilde{C}_G^T\left(\tilde{C}_G+C_G^*-C_{G0}\right)\right]\\
&\le -\frac{\lambda_{CG}}{2}\left\|\tilde{C}_G\right\|^2+\frac{\lambda_{CG}}{2}\left\|C_G^*-C_{G0}\right\|^2,
\end{aligned}
$$

we get

$$
\begin{aligned}
\dot{V}_1 \le \lambda_{\bar{A}}\|e\|^2 &+ \lambda_{\tilde{A}}\|\tilde{e}\|^2+\left\|\hat{e}^T b\right\|\left[\left\|\widetilde{W}_F\right\|\left\|\Phi\left(x,c_F,b_F\right)\right\|\right.\\
&+\|W_F\|\left(\|\sigma_{Fc}\|\|\tilde{c}_F\|+\|\sigma_{Fb}\|\left\|\tilde{b}_F\right\|\right)+\left\|\bar{B}\right\|\left\|\Phi\left(x,c_G,b_G\right)\right\|\|u_d\|\left\|\widetilde{W}_G\right\|\\
&\left.+\left\|\bar{B}\right\|\|W_G\|\left(\|\sigma_{Gc}\|\|\tilde{c}_G\|+\|\sigma_{Gb}\|\left\|\tilde{b}_G\right\|\right)\|u_d\|\right]+\|b\|^2\|\tilde{e}\|^2
\end{aligned}
$$

$$
\begin{aligned}
&+\left\|\tilde{C}_F\right\|^2\|\varphi_F\|^2+\|b\|^2\left\|\bar{B}\right\|^2\|u_d\|^2\|\tilde{e}\|^2+\left\|\tilde{C}_G\right\|^2\|\varphi_G\|^2\\
&-\frac{\lambda_{WF}}{2}\left\|\widetilde{W}_F\right\|^2+\frac{\lambda_{WF}}{2}\left\|W_F^*-W_{F0}\right\|^2-\frac{\lambda_{WG}}{2}\left\|\widetilde{W}_G\right\|^2\\
&+\frac{\lambda_{WG}}{2}\left\|W_G^*-W_{G0}\right\|^2-\frac{\lambda_{Fc}}{2}\|\tilde{c}_F\|^2+\frac{\lambda_{Fc}}{2}\left\|c_F^*-c_{F0}\right\|^2-\frac{\lambda_{Gc}}{2}\|\tilde{c}_G\|^2\\
&+\frac{\lambda_{Gc}}{2}\left\|c_G^*-c_{G0}\right\|^2-\frac{\lambda_{Fb}}{2}\left\|\tilde{b}_F\right\|^2+\frac{\lambda_{Fb}}{2}\left\|b_F^*-b_{F0}\right\|^2-\frac{\lambda_{Gb}}{2}\left\|\tilde{b}_G\right\|^2\\
&+\frac{\lambda_{Gb}}{2}\left\|b_G^*-b_{G0}\right\|^2-\frac{\lambda_{CF}}{2}\left\|\tilde{C}_F\right\|^2+\frac{\lambda_{CF}}{2}\left\|C_F^*-C_{F0}\right\|^2\\
&-\frac{\lambda_{CG}}{2}\left\|\tilde{C}_G\right\|^2+\frac{\lambda_{CG}}{2}\left\|C_G^*-C_{G0}\right\|^2.
\end{aligned}
\tag{8.33}
$$

Since

$$
\begin{aligned}
&\left\|\hat{e}^Tb\right\|\left\|\widetilde{W}_F\right\|\left\|\Phi\left(x,c_F,b_F\right)\right\|-\frac{\lambda_{WF}}{2}\left\|\widetilde{W}_F\right\|^2\\
&\quad\le\frac{\frac{1}{2}\left\|\hat{e}^Tb\right\|^2\left\|\Phi\left(x,c_F,b_F\right)\right\|^2}{\lambda_{WF}},\\
&\left\|\hat{e}^Tb\right\|\left\|\bar{B}\right\|\left\|\Phi\left(x,c_G,b_G\right)\right\|\|u_d\|\left\|\widetilde{W}_G\right\|-\frac{\lambda_{WG}}{2}\left\|\widetilde{W}_G\right\|^2\\
&\quad\le\frac{\frac{1}{2}\left\|\hat{e}^Tb\right\|^2\left\|\bar{B}\right\|^2\left\|\Phi\left(x,c_G,b_G\right)\right\|^2\|u_d\|^2}{\lambda_{WG}},\\
&\left\|\hat{e}^Tb\right\|\|W_F\|\|\sigma_{Fc}\|\|\tilde{c}_F\|-\frac{\lambda_{Fc}}{2}\|\tilde{c}_F\|^2\\
&\quad\le\frac{\frac{1}{2}\left\|\hat{e}^Tb\right\|^2\|W_F\|^2\|\sigma_{Fc}\|^2}{\lambda_{Fc}},\\
&\left\|\hat{e}^Tb\right\|\|W_F\|\|\sigma_{Fb}\|\left\|\tilde{b}_F\right\|-\frac{\lambda_{Fb}}{2}\left\|\tilde{b}_F\right\|^2\\
&\quad\le\frac{\frac{1}{2}\left\|\hat{e}^Tb\right\|^2\|W_F\|^2\|\sigma_{Fb}\|^2}{\lambda_{Fb}},\\
&\left\|\hat{e}^Tb\right\|\left\|\bar{B}\right\|\|W_G\|\|\sigma_{Gc}\|\|\tilde{c}_G\|\|u_d\|-\frac{\lambda_{Gc}}{2}\|\tilde{c}_G\|^2\\
&\quad\le\frac{\frac{1}{2}\left\|\hat{e}^Tb\right\|^2\left\|\bar{B}\right\|^2\|W_G\|^2\|\sigma_{Gc}\|^2\|u_d\|^2}{\lambda_{Gc}},\\
&\left\|\hat{e}^Tb\right\|\left\|\bar{B}\right\|\|W_G\|\|\sigma_{Gb}\|\left\|\tilde{b}_G\right\|\|u_d\|-\frac{\lambda_{Gb}}{2}\left\|\tilde{b}_G\right\|^2\\
&\quad\le\frac{\frac{1}{2}\left\|\hat{e}^Tb\right\|^2\left\|\bar{B}\right\|^2\|W_G\|^2\|\sigma_{Gb}\|^2\|u_d\|^2}{\lambda_{Gb}},
\end{aligned}
$$

one gets

$$\begin{aligned}
\dot{V}_1 \le\ & \lambda_{\bar{A}} \|e\|^2 + \left(\lambda_{\bar{A}} + \|b\|^2 + \|b\|^2 \|\bar{B}\|^2 \|u_d\|^2\right) \|\tilde{e}\|^2 \\
& + \frac{\frac{1}{2}\|\hat{e}^T b\|^2 \|\Phi(x, c_F, b_F)\|^2}{\lambda_{WF}} + \frac{\frac{1}{2}\|\hat{e}^T b\|^2 \|\bar{B}\|^2 \|\Phi(x, c_G, b_G)\|^2 \|u_d\|^2}{\lambda_{WG}} \\
& + \frac{\frac{1}{2}\|\hat{e}^T b\|^2 \|W_F\|^2 \|\sigma_{Fc}\|^2}{\lambda_{Fc}} + \frac{\frac{1}{2}\|\hat{e}^T b\|^2 \|W_F\|^2 \|\sigma_{Fb}\|^2}{\lambda_{Fb}} \\
& + \frac{\frac{1}{2}\|\hat{e}^T b\|^2 \|\bar{B}\|^2 \|W_G\|^2 \|\sigma_{Gc}\|^2 \|u_d\|^2}{\lambda_{Gc}} \\
& + \frac{\frac{1}{2}\|\hat{e}^T b\|^2 \|\bar{B}\|^2 \|W_G\|^2 \|\sigma_{Gb}\|^2 \|u_d\|^2}{\lambda_{Gb}} - \left(\frac{\lambda_{CF}}{2} - \|\varphi_F\|^2\right) \|\tilde{C}_F\|^2 \\
& - \left(\frac{\lambda_{CG}}{2} - \|\varphi_G\|^2\right) \|\tilde{C}_G\|^2 + \frac{\lambda_{WF}}{2} \|W_F^* - W_{F0}\|^2 \\
& + \frac{\lambda_{VF}}{2} \|V_F^* - V_{F0}\|^2 + \frac{\lambda_{WG}}{2} \|W_G^* - W_{G0}\|^2 \\
& + \frac{\lambda_{VG}}{2} \|V_G^* - V_{G0}\|^2 + \frac{\lambda_{CF}}{2} \|C_F^* - C_{F0}\|^2 + \frac{\lambda_{CG}}{2} \|C_G^* - C_{G0}\|^2 \\
& \le -\chi V_1 + \rho
\end{aligned} \tag{8.34}$$

where

$$\begin{aligned}
\chi = \min\Bigg\{ & -\lambda_{\bar{A}}, -\lambda_{\bar{A}} - \|b\|^2 - \|b\|^2 \|\bar{B}\|^2 \|u_d\|^2, \\
& \Gamma_{CF}\left(\frac{\lambda_{CF}}{2} - \|\varphi_F\|^2\right) + \Gamma_{CG}\left(\frac{\lambda_{CG}}{2} - \|\varphi_G\|^2\right)\Bigg\},
\end{aligned} \tag{8.35}$$

$$\begin{aligned}
\rho = & \frac{\frac{1}{2}\|\hat{e}^T b\|^2 \|\Phi(x, c_F, b_F)\|^2}{\lambda_{WF}} + \frac{\frac{1}{2}\|\hat{e}^T b\|^2 \|\bar{B}\|^2 \|\Phi(x, c_G, b_G)\|^2 \|u_d\|^2}{\lambda_{WG}} \\
& + \frac{\frac{1}{2}\|\hat{e}^T b\|^2 \|W_F\|^2 \|\sigma_{Fc}\|^2}{\lambda_{Fc}} + \frac{\frac{1}{2}\|\hat{e}^T b\|^2 \|W_F\|^2 \|\sigma_{Fb}\|^2}{\lambda_{Fb}} \\
& + \frac{\frac{1}{2}\|\hat{e}^T b\|^2 \|\bar{B}\|^2 \|W_G\|^2 \|\sigma_{Gc}\|^2 \|u_d\|^2}{\lambda_{Gc}} \\
& + \frac{\frac{1}{2}\|\hat{e}^T b\|^2 \|\bar{B}\|^2 \|W_G\|^2 \|\sigma_{Gc}\|^2 \|u_d\|^2}{\lambda_{Gc}} + \frac{\lambda_{WF}}{2} \|W_F^* - W_{F0}\|^2 \\
& + \frac{\lambda_{VF}}{2} \|V_F^* - V_{F0}\|^2 + \frac{\lambda_{WG}}{2} \|W_G^* - W_{G0}\|^2 + \frac{\lambda_{VG}}{2} \|V_G^* - V_{G0}\|^2 \\
& + \frac{\lambda_{CF}}{2} \|C_F^* - C_{F0}\|^2 + \frac{\lambda_{CG}}{2} \|C_G^* - C_{G0}\|^2 .
\end{aligned} \tag{8.36}$$

From (8.36), it follows that

$$0 \leq V_1 \leq \frac{\rho}{\chi} + \left(V_1(0) - \frac{\rho}{\chi}\right)\exp(-\chi t). \tag{8.37}$$

When $t \to \infty$, V_1 is bounded by $\frac{\rho}{\chi}$. It also indicates that the parameters $e, \tilde{e}, \widetilde{W}_F, \widetilde{W}_G, \tilde{C}_F, \tilde{C}_G, b_F, b_G$ are all uniformly ultimately bounded, from which it follows that the motion model (8.19) will eventually converge to AUV actual dynamics. □

8.4. Fault reconstruction

In comparison with the traditional sliding mode control, terminal sliding mode control can achieve finite-time convergence of the tracking error. Due to this advantage, Feng et al. [24] and Chee et al. [17] have used terminal sliding mode for state observer design. Terminal sliding mode observer has been also applied to speed estimation in Permanent Magnet Synchronous Motor in [24], but it was only for a first-order linear system. Chee et al. [17] have presented a terminal sliding mode observer for nonlinear systems, but the system uncertainties have not been considered. In general, the convergence of the observer cannot be guaranteed, if the actuator fault of the controlled system occurs. On the basis of the work in [17], this chapter will further investigate thruster fault reconstruction based on terminal sliding mode observer for AUV.

From Section 8.3, the AUV motion model can be expressed as an affine nonlinear system as in (8.19). Here, we will give the following assumptions about (8.19).

Assumption 8.2. There exist known positive constants $\vartheta_{\bar{B}}$ and ϑ_G such that the following inequalities hold:

$$\left\|\bar{B}\right\| \leq \vartheta_{\bar{B}}, \tag{8.38}$$

$$\|G(x)\| \leq \vartheta_G. \tag{8.39}$$

Assumption 8.3. There exist known positive constants γ_F and γ_G such that the following inequalities hold:

$$\begin{aligned} \|F(x_1) - F(x_2)\| &\leq \gamma_F \|x_1 - x_2\|, \\ \|G(x_1) - G(x_2)\| &\leq \gamma_G \|x_1 - x_2\|. \end{aligned} \tag{8.40}$$

From (8.19), a terminal sliding mode observer is designed as follows:

$$\dot{\hat{\xi}} = \bar{A}\hat{\xi} + b\left[F(\hat{x}) + \bar{B}G(\hat{x})u - C_F\hat{\varphi}_F - \bar{B}C_G\hat{\varphi}_G u\right] - L\tilde{\xi}_1 + \zeta,$$
$$\hat{y}_{out} = C\hat{\xi} \tag{8.41}$$

where $\hat{x} = \left[\hat{\xi}_1^t; \hat{\xi}_1^t - \hat{\xi}_1^{t-1}\right]$, $\hat{\xi} = \left[\hat{\xi}_1^T, \hat{\xi}_2^T\right]^T$, $\tilde{\xi}_1 = \hat{\xi}_1 - \xi_1$, $\zeta = \left[\zeta_1^T, \zeta_2^T\right]^T$, $\zeta_1 = \left[\zeta_{11}, \ \ldots, \ \zeta_{16}\right]^T$, $\zeta_2 = \left[\zeta_{21}, \ \ldots, \ \zeta_{26}\right]^T$,

$$\zeta_1 = -\alpha\,\mathrm{sgn}\left(\tilde{\xi}_1\right),$$
$$\zeta_2 = \beta\zeta_1^{\frac{q}{p}} \tag{8.42}$$

where α and β are positive constants, p and q are odd integers with $p > q$; $L = [L_1^T, L_2^T]^T$ is a gain matrix, and L should be chosen such that $A_o = \bar{A} - LC$ is asymptotically stable.

In (8.41), sliding mode strategy (8.42) has the same form as the sliding mode strategy in [17], but the work in [17] does not consider the convergence problem when there exist uncertainties in the system. In the subsequent analysis, the selection method about parameters α and β is given such that the observer state estimation errors can converge to zero in finite time when a thruster fault occurs.

Define $\tilde{F} = F(\hat{x}) - F(x)$, $\tilde{G} = G(\hat{x}) - G(x)$, $\tilde{\varphi}_F = \hat{\varphi}_F - \varphi_F$, $\tilde{\varphi}_G = \hat{\varphi}_G - \varphi_G$, and $\tilde{\xi} = \hat{\xi} - \xi$. From (8.19) and (8.41), we have

$$\dot{\tilde{\xi}} = A_o\tilde{\xi} + b\left[\tilde{F} + \bar{B}\tilde{G}u - C_F\tilde{\varphi}_F - \bar{B}C_G\tilde{\varphi}_G u + \bar{B}G(x)f\right] + \zeta. \tag{8.43}$$

Then (8.43) can be partitioned into

$$\dot{\tilde{\xi}}_1 = -L_1\tilde{\xi}_1 + \tilde{\xi}_2 + \zeta_1,$$
$$\dot{\tilde{\xi}}_2 = -(a + L_2)\tilde{\xi}_1 + \tilde{F} + \bar{B}\tilde{G}u - C_F\tilde{\varphi}_F - \bar{B}C_G\tilde{\varphi}_G u + \bar{B}G(x)f + \zeta_2. \tag{8.44}$$

Proposition 8.2. *Define*

$$P_oA_o + A_o^TP_o + \frac{4}{\varepsilon}P_oP_o + \varepsilon\psi = -Q_o \tag{8.45}$$

where $P_o = \mathrm{diag}(P_1, P_2) > 0$, $\psi = \mathrm{diag}\left(\left(\gamma_F^2\|b\|^2 + \gamma_G^2\|b\bar{B}\|^2\|u_d\|^2 + l_F^2\|b\|^2C_F^2 + l_G^2\|b\bar{B}\|^2\|u_d\|^2C_G^2\right)I_6, 0_6\right)$, $P_1 = \mathrm{diag}(p_{11}, \ldots, p_{16})$, $P_2 = \mathrm{diag}(p_{21}, \ldots, p_{26})$.

If P_o, L and ε are chosen such that $Q_o > 0$ and

$$\alpha > \left(\frac{2}{\mu}\left(\left\|P_o b\right\| \vartheta_{\bar{B}}\vartheta_G u_d + \alpha^{\frac{q}{p}}\beta\left\|P_2\right\|\right) + \sigma\right)^{\frac{p}{p-q}} \tag{8.46}$$

then $\tilde{\xi}_1 = 0$, $\dot{\tilde{\xi}}_1 = 0$, and $\dot{\tilde{\xi}}_2 \approx 0$ will take place in finite time, where $\mu = -\lambda(Q_o)_{\max}$ is the maximum eigenvalue of matrix Q_o, σ is a positive constant, while β will be given later.

Proof. Choose a candidate Lyapunov function as

$$V_2 = \tilde{\xi}^T P_o \tilde{\xi}. \tag{8.47}$$

The time derivative of (8.47) is

$$\begin{aligned}\dot{V}_2 = {} & \tilde{\xi}^T\left(P_o A_o + A_o^T P_o\right)\tilde{\xi} + 2\tilde{\xi}^T P_o b\left[\tilde{F} + \bar{B}\tilde{G}u - C_F\tilde{\varphi}_F - \bar{B}C_G\tilde{\varphi}_G u\right] \\ & + 2\tilde{\xi}^T P_o b\bar{B}G(x)f + 2\tilde{\xi}^T P_o\zeta.\end{aligned} \tag{8.48}$$

Since

$$\begin{aligned}2\tilde{\xi}^T P_o b\tilde{F} &\le \frac{1}{\varepsilon}\tilde{\xi}^T P_o P_o\tilde{\xi} + \varepsilon\left\|b\right\|^2\tilde{F}^T\tilde{F} \le \frac{1}{\varepsilon}\tilde{\xi}^T P_o P_o\tilde{\xi} + \varepsilon\gamma_F^2\left\|b\right\|^2\tilde{\xi}_1^T\tilde{\xi}_1, \\ 2\tilde{\xi}^T P_o b\bar{B}\tilde{G}u &\le \frac{1}{\varepsilon}\tilde{\xi}^T P_o P_o\tilde{\xi} + \varepsilon\left\|b\bar{B}\right\|^2\left\|u_d\right\|^2\tilde{G}^T\tilde{G} \\ &\le \frac{1}{\varepsilon}\tilde{\xi}^T P_o P_o\tilde{\xi} + \varepsilon\gamma_G^2\left\|b\bar{B}\right\|^2\left\|u_d\right\|^2\tilde{\xi}_1^T\tilde{\xi}_1, \\ -2\tilde{\xi}^T P_o b C_F\tilde{\varphi}_F &\le \frac{1}{\varepsilon}\tilde{\xi}^T P_o P_o\tilde{\xi} + \varepsilon\left\|b\right\|^2 C_F^2\tilde{\varphi}_F^T\tilde{\varphi}_F \\ &\le \frac{1}{\varepsilon}\tilde{\xi}^T P_o P_o\tilde{\xi} + \varepsilon l_F^2\left\|b\right\|^2 C_F^2\tilde{\xi}_1^T\tilde{\xi}_1, \\ -2\tilde{\xi}^T P_o b\bar{B}C_G\tilde{\varphi}_G u &\le \frac{1}{\varepsilon}\tilde{\xi}^T P_o P_o\tilde{\xi} + \varepsilon\left\|b\bar{B}\right\|^2\left\|u_d\right\|^2\tilde{\varphi}_G^T\tilde{\varphi}_G \\ &\le \frac{1}{\varepsilon}\tilde{\xi}^T P_o P_o\tilde{\xi} + \varepsilon l_G^2\left\|b\bar{B}\right\|^2\left\|u_d\right\|^2 C_G^2\tilde{\xi}_1^T\tilde{\xi}_1,\end{aligned}$$

we get

$$\begin{aligned}\dot{V}_2 \le {} & \tilde{\xi}^T\left(P_o A_o + A_o^T P_o + \frac{4}{\varepsilon}P_o P_o + \varepsilon\psi\right)\tilde{\xi} + 2\tilde{\xi}^T P_o b\bar{B}G(x)f + 2\tilde{\xi}_1^T P_1\zeta_1 \\ & + 2\tilde{\xi}_2^T P_2\zeta_2 \\ \le {} & -\tilde{\xi}^T Q_o\tilde{\xi} + 2\tilde{\xi}^T P_o b\bar{B}G(x)f - 2\tilde{\xi}_1^T P_1\alpha sgn\left(\tilde{\xi}_1\right) - 2\alpha^{\frac{q}{p}}\beta P_\beta\tilde{\xi}\end{aligned}$$

$$\begin{aligned}&\leq -\mu\left\|\tilde{\xi}\right\|^2+2\left\|\tilde{\xi}\right\|\left\|P_o b\right\|\vartheta_{\bar{B}}\vartheta_G u_d-2\alpha^{\frac{q}{p}}\beta P_\beta\tilde{\xi}\\&\leq -\left\|\tilde{\xi}\right\|\left(\mu\left\|\tilde{\xi}\right\|-2\left\|P_o b\right\|\vartheta_{\bar{B}}\vartheta_G u_d-2\alpha^{\frac{q}{p}}\beta\left\|P_\beta\right\|\right)\end{aligned}\tag{8.49}$$

where

$$P_\beta=\left[0_{1\times 6}\quad p_{21}\,\mathrm{sgn}\left(\tilde{\xi}_{11}\right)\quad \ldots\quad p_{26}\,\mathrm{sgn}\left(\tilde{\xi}_{16}\right)\right].\tag{8.50}$$

From (8.49), we can know $\dot{V}_2<0$ when $\|\tilde{\xi}\|\geq\frac{\Omega=\left(2\|P_o b\|\vartheta_{\bar{B}}\vartheta_G u_d+2\alpha^{\frac{q}{p}}\beta\|P_\beta\|\right)}{\mu}$. It indicates that the state estimation errors will eventually converge to the region Ω. Proposition 2 in [17] shows that $\tilde{\xi}_1$ and $\dot{\tilde{\xi}}_1$ will converge to zero in a finite time when the value of α satisfies (8.46).

Next, we will prove that $\dot{\tilde{\xi}}_2\approx 0$ takes place in finite time.

When sliding mode motion $\tilde{\xi}_1=0$ and $\dot{\tilde{\xi}}_1=0$ have been attained, from (8.18) and the first equation of (8.44), one can get

$$\dot{\tilde{\xi}}_2=\bar{B}G(\hat{x})f-\beta\tilde{\xi}_2^{\frac{q}{p}}.\tag{8.51}$$

Choosing a candidate Lyapunov function as $V_3=\tilde{\xi}_2^T\tilde{\xi}_2$ and differentiating it, we obtain

$$\begin{aligned}\dot{V}_3&=\tilde{\xi}_2^T\dot{\tilde{\xi}}_2\\&=\tilde{\xi}_2^T\left(\bar{B}G(\hat{x})f-\beta\tilde{\xi}_2^{\frac{q}{p}}\right)\\&=\left(\tilde{\xi}_2^{\frac{p+q}{p}}\right)^T\left(\frac{\bar{B}G(\hat{x})f}{\tilde{\xi}_2^{\frac{q}{p}}}-\beta\right)\\&\leq\left(\tilde{\xi}_2^{\frac{p+q}{p}}\right)^T\left(\frac{\vartheta_{\bar{B}}\vartheta_G u_d}{\|\tilde{\xi}_2\|^{\frac{q}{p}}}-\beta\right).\end{aligned}\tag{8.52}$$

Since $(p+q)$ is an even integer, and if β is chosen to satisfy $\beta\geq\frac{\vartheta_{\bar{B}}\vartheta_G u_d}{\|\tilde{\xi}_2\|^{\frac{q}{p}}}$, then $\dot{V}_3\leq 0$. Hence, the vector $\tilde{\xi}_2$ will converge into the region Δ: $\|\tilde{\xi}_2\|\leq\left(\frac{\vartheta_{\bar{B}}\vartheta_G u_d}{\beta}\right)^{\frac{p}{q}}$ in finite time. Then an appropriate β can be found such that $\tilde{\xi}_2\approx 0$ and $\dot{\tilde{\xi}}_2\approx 0$.

When $\tilde{\xi}_1=0$, $\tilde{\xi}_2\approx 0$, and $\dot{\tilde{\xi}}_2\approx 0$, (8.51) can be expressed as $\bar{B}G(\hat{x})f=\beta\alpha\,\mathrm{sgn}(\tilde{\xi}_1)^{\frac{q}{p}}$. Based on the equation, one can reconstruct the thruster fault as follows:

$$f = G^{-1}(\hat{x})\left[\alpha\beta TMJ^{-1}\left(\hat{\xi}_1\right)\operatorname{sgn}\left(\tilde{\xi}_1\right)^{\frac{q}{p}}\right]_n \tag{8.53}$$

where $\left[\alpha\beta TMJ^{-1}(\hat{\xi}_1)\operatorname{sgn}(\tilde{\xi}_1)^{\frac{q}{p}}\right]_n$ denotes the latter n rows in column vector $\alpha\beta TMJ^{-1}(\hat{\xi}_1)\operatorname{sgn}(\tilde{\xi}_1)^{\frac{q}{p}}$.

Since sgn(·) is a discontinuous function, in order to obtain the continuous reconstruction signal of thruster fault, equivalent output injection (8.54) is used to replace sgn(·):

$$\Theta = \frac{\tilde{\xi}_1}{\|\tilde{\xi}_1\| + \kappa_2\|\tilde{\xi}_1\| + \kappa_1} \tag{8.54}$$

where κ_1 and κ_2 are positive constants.

Form (8.53) and (8.54), the reconstructed thruster fault is given by

$$f = G^{-1}(\hat{x})\left[\alpha\beta TMJ^{-1}\left(\hat{\xi}_1\right)\Theta^{\frac{q}{p}}\right]_n. \tag{8.55}$$

□

8.5. Experimental verification

In order to verify the effectiveness of the proposed fault reconstruction method, pool experiments are performed on BEAVER, an underwater vehicle experimental prototype. BEAVER is an open-frame and low-power experimental platform. And it has two horizontal thrusters and two vertical thrusters. To measure the position of BEAVER in the earth-fixed frame, a three-dimensional position measurement sensor system consisting of three distance sensors is developed for the pool environment. The BEAVER and the sensor system are shown in Fig. 8.1.

The maximum longitude speed of BEAVER is 0.5 m/s and the center of gravity is much lower than the center of buoyancy. Due to these facts, the motion of BEAVER can be decoupled to horizontal and vertical plane motions. In order to verify the effectiveness of the proposed method, this paper only discusses the problems of motion modeling and thruster fault reconstruction in the horizontal plane.

Large number of motion data are required to train the developed neural network based motion model given in Section 8.3. These training samples come from open-loop control, trajectory tracking control, and other motion control experiments. In the process of motion modeling, the input of neural networks is $x = \left[\psi^t; X^t - X^{t-1}; Y^t - Y^{t-1}; \psi^t - \psi^{t-1}\right]$,

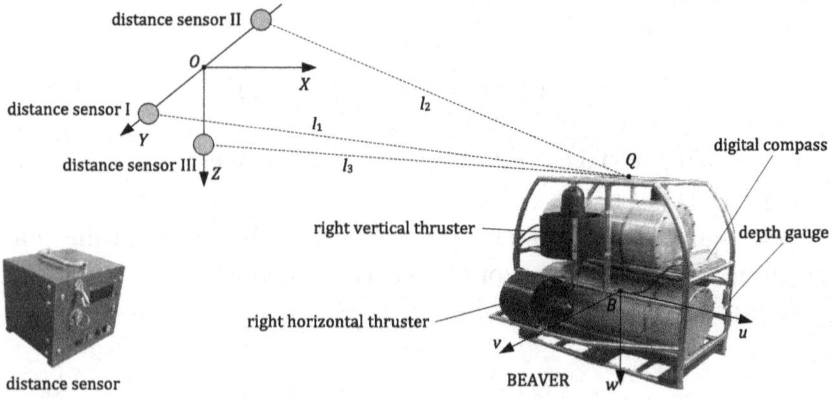

Figure 8.1 The BEAVER and the sensor system.

where X, Y provide the coordinate position in the earth-fixed frame, and ψ is the yaw angle. The other designed parameters are selected as: $\varphi_F(x) = 0.1$, $\varphi_G(x) = 0.1$, $a = \mathrm{diag}([4, 4, 1])$, $\Gamma_{WF} = \Gamma_{Fci} = \Gamma_{Fb} = 0.005$, $\Gamma_{CF} = 0.001$, $\Gamma_{WG} = \Gamma_{Gcj} = \Gamma_{Gb} = 0.0005$, $\Gamma_{CG} = 0.0006$, $\lambda_{CF} = \lambda_{CG} = 0.1$, and $\lambda_{WF} = \lambda_{WG} = \lambda_{Fci} = \lambda_{Gcj} = \lambda_{Fb} = \lambda_{Gb} = 0.02$.

In experimental verification of thruster fault reconstruction, the thruster fault is simulated by reducing the control output of the thruster. In experiments, the fault occurs in the right horizontal thruster and the real control output is only 75% of the desired controller output signal, namely $u_1' = 0.75u_1$, where signal u_1' is loaded into the thruster, and signal u_1 is the desired controller output. The design parameters in fault reconstruction are selected as: $\kappa_1 = 0.2$, $\kappa_2 = 0.1$, $p = 3$, $q = 1$, $\alpha = 0.25$ and $\beta = 0.02$. Figs. 8.2–8.4 respectively show the estimated state based on terminal sliding mode observer and the estimation errors about the position and orientation of BEAVER in the earth-fixed frame. In these figures, the solid line shows the actual measured values, and the dotted line gives the estimated values of the observer. The fault reconstruction signals of left and right horizontal thrusters are shown in Figs. 8.5 and 8.6, respectively. The thin dotted line is the actual fault signal, and the thick solid line is the fault reconstruction signal.

From 0 to 50 s, the thrusters in BEAVER are healthy. And during this period, both the position and orientation estimation errors converge to zero under the action of the developed terminal sliding mode observer. From 0 to 50 s, the mean values of fault reconstruction signals of left and right horizontal thrusters are 0.006 and 0.030 V, respectively, and the mean

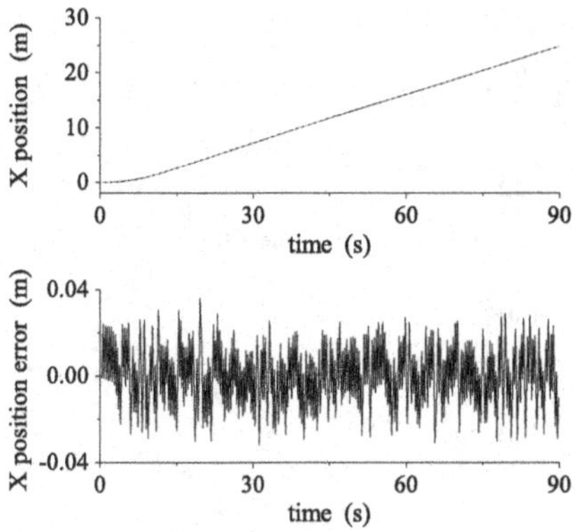

Figure 8.2 *X* position and estimation error.

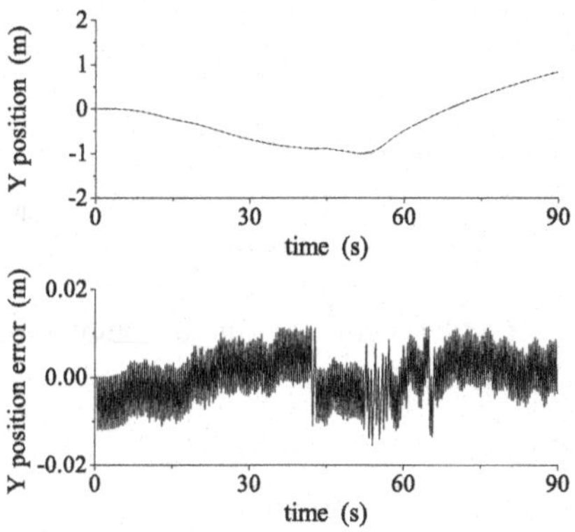

Figure 8.3 *Y* position and estimation error.

square errors are 0.016 and 0.056 V, respectively. Compared with the range of thruster control voltages [−5 V, 5 V], these values of the fault reconstruction can be ignored. The experimental result shows that the proposed motion model can well describe the actual dynamic of BEAVER, which also indicates the effectiveness of developed neural network based motion model.

In experiments, the right horizontal occurs fault from the 50th second. Fig. 8.5 depicts the fault reconstruction signal and the real fault signal. After 52 s, the mean values and the mean square error of the right horizontal thrusters are −0.021 and 0.046 V, respectively. The experimental result illustrates the effectiveness of the proposed fault reconstruction method.

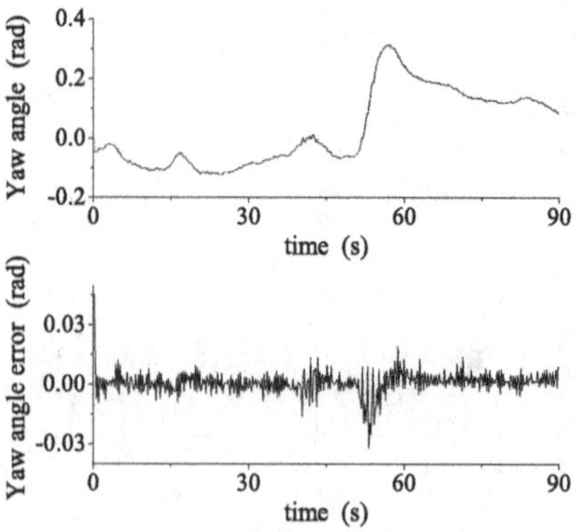

Figure 8.4 Yaw angle and estimation error.

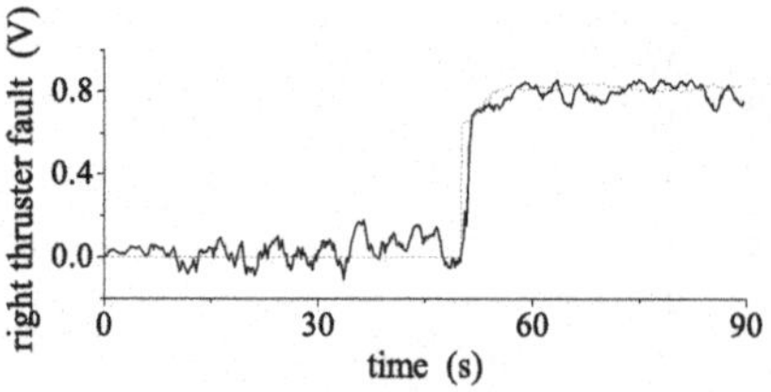

Figure 8.5 The fault reconstruction signal of right horizontal thruster.

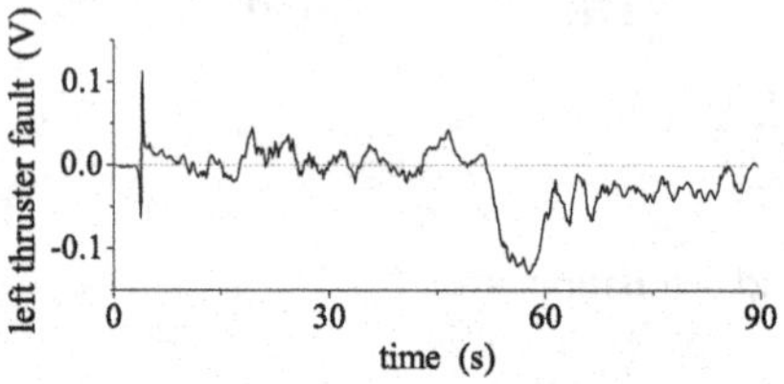

Figure 8.6 The fault reconstruction signal of left horizontal thruster.

Although there is no fault in the left horizontal thruster, the fault reconstruction signal in Fig. 8.6 significantly deviates from zero. After the 50th second, the mean values and the mean square error of the left horizontal thruster are −0.041 and 0.034 V, respectively. In practice, the purpose of thruster fault reconstruction is to provide basic information for the task planning system, which can decide whether the AUV continues to work or how to accommodate the fault. Since the fault reconstruction signal of the left horizontal thruster is smaller, it does not affect the decision-making of the task planning system.

8.6. Conclusions

This chapter investigates thruster fault reconstruction for AUV with thruster fault. Since it is difficult to establish an accurate AUV dynamic model, a motion model with an affine form is developed based on RBF neural network, whose input is the thruster control signal. Thus, the relationship between system input and output can be reflected clearly. The experiment results show that the developed motion model can well describe the AUV dynamics. For AUV thruster fault reconstruction problem, a fault reconstruction method is developed based on a terminal sliding observer. Under the action of the developed method, the estimation errors of all states converge to zero in finite time and the thruster fault can be reconstructed quickly. The experiment results show the effectiveness of the developed method in terms of thruster fault reconstruction.

References

[1] M. Takai, T. Ura, Development of a system to diagnose autonomous underwater vehicles, International Journal of Systems Science 30 (1999) 981–988.

[2] Y.R. Xu, K. Xiao, Technology development of autonomous ocean vehicle, Acta Automatica Sinica 33 (2007) 518–521.

[3] V. Venkatasubramanian, R. Rengaswamy, K. Yin, S.N. Kavuri, A review of process fault detection and diagnosis part I: quantitative model-based methods, Computers and Chemical Engineering 27 (2003) 293–311.

[4] A. Alessangri, M. Caccia, G. Veruggio, Fault detection of actuator faults in unmanned underwater vehicles, Control Engineering Practice 7 (1999) 357–368.

[5] D. Loebis, R. Suttor, J. Chudley, Adaptive tuning of Kalman filter via fuzzy logic for an intelligent AUV navigation system, Control Engineering Practice 12 (2004) 1531–1539.

[6] A. Alessandri, A. Gibbons, J. Healey, G. Veruggio, Robust model-based fault diagnosis for unmanned underwater vehicles using sliding mode-observers, in: Proceedings International Symposium Unmanned Underwater Submersible Technology, Durham, New Hampshire, 1999, pp. 352–359.

[7] L.R. Wang, Y. Gan, Y.R. Xu, L. Wan, Sliding mode observers used in thruster fault diagnosis of an autonomous underwater vehicle, Journal of Harbin Engineering University 26 (2005) 425–429.

[8] S.L. Xiao, X.M. Fang, X.M. Liu, Research on actuator fault diagnosis method for underwater vehicle, Ship Building of China 51 (2010) 132–138.

[9] B. Ahmad, J.M. Jalal, Simulation and tracking control based on neural-network strategy and sliding mode control for underwater remotely operated vehicle, Neurocomputing 72 (2009) 1934–1950.

[10] Y.J. Wang, M.J. Zhang, Z.X. Jin, Condition monitoring system for sensors and thrusters of AUV, Chinese Journal of Mechanical Engineering 42 (2006) 214–218.

[11] L.N. Wang, K. Ding, Actuator fault diagnosis of autonomous underwater vehicle based on wavelet neural network, Journal of System Simulation 19 (2007) 206–209.

[12] C. Edwards, S.K. Spurgeon, R.J. Patton, Sliding mode observers for fault detection and isolation, Automatica 36 (2000) 541–553.
[13] P.T. Chee, C. Edwards, An LMI approach for design sliding mode observers, International Journal of Control 74 (2001) 1559–1568.
[14] P.T. Chee, C. Edwards, Sliding mode observers for robust detection and reconstruction of actuator and sensor faults, International Journal of Robust and Nonlinear Control 13 (2003) 443–463.
[15] X.G. Yan, C. Edwards, Nonlinear robust fault reconstruction and estimation using a sliding mode observer, Automatica 43 (2007) 1605–1614.
[16] W. Chen, M. Saif, Actuator fault diagnosis for uncertain linear systems using a high-order sliding mode robust differentiator, International Journal of Robust and Nonlinear Control 18 (2008) 413–426.
[17] P.T. Chee, X.H. Yu, Z.H. Man, Terminal sliding mode observers for a class of nonlinear systems, Automatica 46 (2010) 1401–1404.
[18] J. Kim, W.K. Chung, Accurate and practical thruster modeling for underwater vehicles, Ocean Engineering 33 (2006) 566–586.
[19] E. Omerdic, G. Roberts, Thruster fault diagnosis and accommodation for open-frame underwater vehicles, Control Engineering Practice. 12 (2004) 1575–1598.
[20] W.M. Bessa, M.S. Dutra, E. Kreuzer, Depth control of remotely operated underwater vehicles using an adaptive fuzzy sliding mode controller, Robotics and Autonomous Systems 56 (2008) 670–677.
[21] J.C. Yu, A.Q. Zhang, X.H. Wang, Neural network adaptive control of underwater vehicles, Control Theory and Applications 25 (2008) 9–13.
[22] J.H. Li, P.M. Lee, A neural network adaptive controller design for free-pitch-angle diving behavior of an autonomous underwater vehicle, Robotics and Autonomous Systems 52 (2005) 132–147.
[23] L.J. Zhang, X. Qi, Y.J. Pang, Adaptive output feedback control based on DRFNN for AUV, Ocean Engineering 36 (2009) 716–722.
[24] Y. Feng, J.F. Zheng, X.H. Yu, N.V. Truong, Hybrid terminal sliding mode observer design method for a permanent magnet synchronous motor control system, IEEE Transactions on Industrial Electronics 56 (2009) 3424–3431.

CHAPTER NINE

Robust sampled-data control for dynamic positioning ships based on T–S fuzzy model

Yueying Wang[a], Minjie Zheng[b]
[a]School of Electrical Information and Electrical Engineering, Shanghai University, Shanghai, China
[b]Navigation College, Jimei University, Xiamen, Fujian, China

Contents

Abstract

This note deals with the robust stabilization problem for nonlinear sampled-data dynamic positioning (DP) of ships based on Takagi–Sugeno (T–S) fuzzy model. Firstly, the T–S fuzzy model of the nonlinear sampled-data DP ship system is established. Next, based on Lyapunov method, sufficient condition is given to guarantee the stability of the system and improve H_∞ performance. Then, the fuzzy sampled-data controller is designed by analyzing the stabilization condition. Finally, a simulation example for DP of ships is given to validate the effectiveness and performance of the proposed system.

Keywords

Dynamic positioning of ships, Sampled-data control, Lyapunov–Krasovskii functional, Takagi–Sugeno (T–S) fuzzy model

9.0. Introduction

Dynamic positioning (DP), a system which is controlled by computer and can keep a ship's desired position and heading by using active thrusters and propellers [1], has been widely employed in various vessel types, such as oceanographic research vessels, cruise ships, and mobile offshore drilling

https://doi.org/10.1016/B978-0-12-820271-5.00014-6

units. With the expansion of ocean activities to deeper waters, the DP ship technology has attracted considerable attention and became a hot topic. In [2], the control problem for DP of ships was presented to guarantee robustness with respect to environmental disturbances based on a discrete-time structure. In [3], a recursive vectorial backstepping controller was design to guarantee DP of a ship to maintain the position and heading under external disturbances. Hassani et al. [4] proposed multiple models and adaptive wave filtering techniques for DP of ships to adapt variable disturbances. Du et al. [5] designed a robust nonlinear controller for the DP of ships using an auxiliary dynamic system and using the dynamic surface control technique.

It has been shown in [6] that Takagi–Sugeno (T–S) fuzzy model can be applied for many complex nonlinear systems. Over the past several years, a rapidly growing interest has been witnessed for DP ship systems employing T–S fuzzy models. Xu and Liu [7] designed a T–S fuzzy PID controller for the DP of ships systems with random noises. Ngongi et al. [8] discussed the robust fuzzy control problem for a DP of ships system based on optimal H_∞ control methods. In [9], both hybrid genetic algorithm and orthogonal function methods were adopted to deal with quadratic optimal control problem for a DP of ships system based on T–S fuzzy models. However, few literature sources were found to consider fuzzy sampled-data control for a DP of ships system.

Sampled-data systems have received considerable attention in the past few years. In such systems, digital computers are widely used for controlling the continuous-time systems. In the DP ship system, a variety of sensors such as Differential Global Positioning System (DGPS) for obtaining higher accuracy and reliability position of ships, gyrocompasses for determining heading of ships, Vertical Reference Unit (VRU) or Motion Reference Unit (MRU) or Vertical Reference System (VRS) for determining the ship's roll pitch and heave, and wind sensors for anticipating wind gusts are used by digital computers to determine the ship's motion state. The sensors produce continuous-time signals, then the signals are transformed into discrete-time control signals by being sampled and quantized with a microcontroller, and finally the discrete-time control signals are transformed back into continuous-time signals again by the zero-order holder. Hence, both discrete- and continuous-time signals exist in one system within a continuous-time framework. So, the sampled-data control systems are more practical and important than the continuous-time systems. Until now, considerable research methods have been adopted for analyzing the sampled-data systems [10]. Input delay is one of the main approaches

[11]. It can convert the sampling period to a bounded time-varying delay and apply to the system with nonuniform uncertain sampling [12]. Its important advantage is that the sampling distance need not be constant. In the past years, the approach has been adopted for several systems such as fuzzy system [13,14], chaotic system [15,16], complex dynamic system [17–19], and neural networks system [20–22]. However, almost all the existing studies have not considered the sampled-data control for nonlinear DP of ships by T–S fuzzy models, which has significance both in practice and theory.

In this paper, the issue of H_∞ sampled-data control for DP of ships in a T–S fuzzy model is discussed. By using input delay approach, the sampled-data DP control system is converted to a system with time-varying delay. By using Lyapunov stability theorems, adequate conditions are established to guarantee the system's asymptotical stability and achieve the H_∞ performance. Then, the fuzzy H_∞ sampled-data controller is obtained by means of the Linear Matrix Inequality (LMI) method. Simulation results of a DP ship system are provided to show the effectiveness of the proposed approaches.

9.1. Problem formulation

The low-speed motion model of a DP ship system is described as follows:

$$\begin{aligned} M\dot{\nu} + D\nu &= u + w, \\ \dot{\eta} &= J(\psi)\nu, \end{aligned} \tag{9.1}$$

where $\eta = [x\ y\ \psi]^{\mathrm{T}}$ represents the x, y position and heading ψ of the ship related to an earth-fixed frame; $\nu = [p\ \upsilon\ r]^{\mathrm{T}}$ represents the ship's velocity vector related to the body-fixed frame, where p, υ, r represent the velocities of surge, sway, and yaw, respectively (the body-fixed coordinate system is shown in Fig. 9.1); $J(\psi)$ represents a transformation matrix of two frames mentioned above,

$$J(\psi) = \begin{bmatrix} \cos\psi & -\sin\psi & 0 \\ 0 & \cos\psi & 0 \\ 0 & 0 & 1 \end{bmatrix}, \tag{9.2}$$

u represents the vector of control force and moment; w represents the system's disturbance input; M is the inertia matrix while D is a linear damping

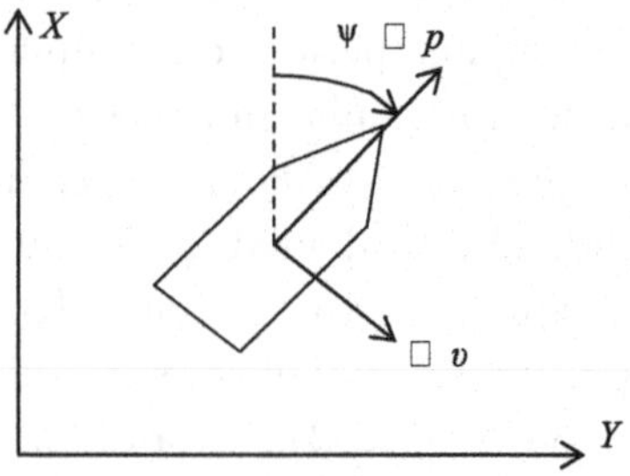

Figure 9.1 The coordinate systems of DP of ships.

matrix, which are given as

$$\boldsymbol{M}=\begin{bmatrix} m-X_{\dot{u}} & 0 & 0 \\ 0 & m-Y_{\dot{v}} & mx_G-Y_{\dot{r}} \\ 0 & mx_G-Y_{\dot{r}} & I_z-N_{\dot{r}} \end{bmatrix}, \quad \boldsymbol{D}=\begin{bmatrix} -X_{\dot{u}} & 0 & 0 \\ 0 & -Y_{\dot{v}} & -Y_{\dot{r}} \\ 0 & -Y_{\dot{r}} & -N_{\dot{r}} \end{bmatrix}, \tag{9.3}$$

where m represents the mass of the ship; I_z represents the inertia moment; x_G is the distance between the center of gravity and the origin of the body-fixed frame, $X_{\dot{u}}$, $Y_{\dot{v}}$, $Y_{\dot{r}}$, $N_{\dot{r}}$ are the accessional mass variables.

Denote the variables as follows:

$$\begin{aligned} & x(t)=\begin{bmatrix} \eta & \upsilon \end{bmatrix}^{\mathrm{T}}=\begin{bmatrix} x_1 & x_2 & x_3 & x_4 & x_5 & x_6 \end{bmatrix}^{\mathrm{T}}, \\ & u(t)=u=\begin{bmatrix} u_1 & u_2 & u_3 \end{bmatrix}^{\mathrm{T}}, \\ & w(t)=w=\begin{bmatrix} w_1 & w_2 & w_3 \end{bmatrix}^{\mathrm{T}}, \\ & \boldsymbol{A}=\begin{bmatrix} 0 & \boldsymbol{J}(\psi) \\ 0 & -\boldsymbol{M}^{-1}\boldsymbol{D} \end{bmatrix}, \quad \boldsymbol{B}=\begin{bmatrix} 0 \\ \boldsymbol{M}^{-1} \end{bmatrix}, \quad \boldsymbol{E}=\begin{bmatrix} 0 \\ \boldsymbol{M}^{-1} \end{bmatrix}, \\ & \boldsymbol{C}=\begin{bmatrix} \boldsymbol{I} & 0 \end{bmatrix}. \end{aligned} \tag{9.4}$$

Here $\boldsymbol{I}$ is identity matrix. Rewriting Eq. (9.1) yields

$$\begin{aligned} \dot{x}(t)&=\boldsymbol{A}x(t)+\boldsymbol{B}u(t)+\boldsymbol{E}w(t), \\ y(t)&=\boldsymbol{C}x(t), \end{aligned} \tag{9.5}$$

where $y(t)$ represents the control system output. Then we calculate and define the matrix parameters as follows:

$$\boldsymbol{M}^{-1} = \begin{bmatrix} 1 & 0 & 0 \\ 0 & m_{22} & m_{23} \\ 0 & m_{32} & m_{33} \end{bmatrix}, \quad -\boldsymbol{M}^{-1}\boldsymbol{D} = \begin{bmatrix} a_{11} & 0 & 0 \\ 0 & a_{22} & a_{23} \\ 0 & a_{32} & a_{33} \end{bmatrix}. \tag{9.6}$$

Substituting Eqs. (9.2) and (9.6) into Eq. (9.4) yields

$$\boldsymbol{A} = \begin{bmatrix} 0 & 0 & 0 & \cos x_3 & -\sin x_3 & 0 \\ 0 & 0 & 0 & \sin x_3 & \cos x_3 & 0 \\ 0 & 0 & 0 & 0 & 0 & 1 \\ 0 & 0 & 0 & a_{11} & 0 & 0 \\ 0 & 0 & 0 & 0 & a_{22} & a_{23} \\ 0 & 0 & 0 & 0 & a_{32} & a_{33} \end{bmatrix}, \quad \boldsymbol{B} = \boldsymbol{E} = \begin{bmatrix} 0 & 0 & 0 \\ 0 & 0 & 0 \\ 0 & 0 & 0 \\ 1 & 0 & 0 \\ 0 & m_{22} & m_{23} \\ 0 & m_{32} & m_{33} \end{bmatrix}. \tag{9.7}$$

It is known from [7–9] that the T–S fuzzy model can be expressed for a nonlinear DP ship motion as follows:

Mode Rule i:

Let $z_1(t)$ be $f_{i1}, \cdots$, and $z_n(t)$ be f_{in}, where $i = 1, 2, \ldots, n$, f_{ij} denotes the fuzzy set, n denotes the number of the rule, and $z_1(t), z_2(t), \ldots, z_n(t)$ are premise variables.

Then set

$$\begin{aligned} \dot{x}(t) &= \boldsymbol{A}_i x(t) + \boldsymbol{B}_i u(t) + \boldsymbol{E}_i w(t), \\ y(t) &= \mathbf{C}_i x(t). \end{aligned} \tag{9.8}$$

Like in Eq. (9.8), the yaw angle x_3 is assumed to be changing between $-\pi/2$ and $\pi/2$, so the exact T–S fuzzy model for the DP system Eq. (9.1) is represented by using the following three rules:

Model Rule 1:

$$\text{IF } x_3(t) = 0, \text{ THEN } \dot{x}(t) = \boldsymbol{A}_1 x(t) + \boldsymbol{B}_1 u(t) + \boldsymbol{E}_1 w(t). \tag{9.9}$$

Model Rule 2:

$$\text{IF } x_3(t) = \frac{\pi}{2}, \text{ THEN } \dot{x}(t) = \boldsymbol{A}_2 x(t) + \boldsymbol{B}_2 u(t) + \boldsymbol{E}_2 w(t). \tag{9.10}$$

Model Rule 3:

$$\text{IF } x_3(t) = -\frac{\pi}{2}, \text{ THEN } \dot{x}(t) = \boldsymbol{A}_3 x(t) + \boldsymbol{B}_3 u(t) + \boldsymbol{E}_3 w(t). \tag{9.11}$$

Above we use

$$\boldsymbol{A}_1=\begin{bmatrix}0&0&0&1&-\alpha&0\\0&0&0&\alpha&1&0\\0&0&0&0&0&1\\0&0&0&a_{11}&0&0\\0&0&0&0&a_{22}&a_{23}\\0&0&0&0&a_{32}&a_{33}\end{bmatrix},\quad \boldsymbol{A}_2=\begin{bmatrix}0&0&0&\beta&-1&0\\0&0&0&1&\beta&0\\0&0&0&0&0&1\\0&0&0&a_{11}&0&0\\0&0&0&0&a_{22}&a_{23}\\0&0&0&0&a_{32}&a_{33}\end{bmatrix},$$

$$\boldsymbol{A}_3=\begin{bmatrix}0&0&0&\beta&1&0\\0&0&0&-1&\beta&0\\0&0&0&0&0&1\\0&0&0&a_{11}&0&0\\0&0&0&0&a_{22}&a_{23}\\0&0&0&0&a_{32}&a_{33}\end{bmatrix},\quad \boldsymbol{B}_1=\boldsymbol{B}_2=\boldsymbol{B}_3=\begin{bmatrix}0&0&0\\0&0&0\\0&0&0\\1&0&0\\0&m_{22}&m_{23}\\0&m_{32}&m_{33}\end{bmatrix}. \tag{9.12}$$

The overall fuzzy model inferred from Eqs. (9.9)–(9.11) is represented as

$$\begin{aligned}\dot{x}(t)&=\frac{\sum_{i=1}^{3}\omega_i(z(t))\left[\boldsymbol{A}_i x(t)+\boldsymbol{B}_i u(t)+\boldsymbol{E}_i w(t)\right]}{\sum_{i=1}^{3}\omega_i(z(t))}\\&=\sum_{i=1}^{3}\mu_i(z(t))\left[\boldsymbol{A}_i x(t)+\boldsymbol{B}_i u(t)+\boldsymbol{E}_i w(t)\right],\end{aligned} \tag{9.13}$$

where

$$\begin{aligned}&\omega_i(z(t))=\prod_{j=1}^{n}f_{ij}(z_j(t))\\&\mu_i(z(t))=\frac{\omega_i(z(t))}{\sum_{i=1}^{3}\omega_i(z(t))}\geq 0,\quad i=1,2,3,\\&\sum_{i=1}^{3}u_i(z(t))=1,\\&z(t)=[z_1(t),z_2(t),\ldots,z_n(t)],\end{aligned} \tag{9.14}$$

with $f_{ij}(z_j(t))$ representing the membership grade of $z_j(t)$ in the fuzzy set f_{ij}.

In this paper, we assume that the state variables of DP of ships are generated at the sampling instants $0 = t_0 < t_1 < t_2 < \cdots < t_k < \ldots$, that is, in the interval $t_k \leq t \leq t_{k+1}$, only $x(t_k)$ is available. The sampling period follows the assumption that it is bounded by a constant d, that is,

$$t_{k+1} - t_k \leq d, \quad \forall k \geq 0,\ d > 0. \tag{9.15}$$

Then, we consider the fuzzy sampled-data controller for the DP system Eq. (9.3) as follows:

Controller Rule 1:

$$\text{IF } x_3(t) = 0, \text{ THEN } u(t) = \boldsymbol{K}_1 x(t_k). \tag{9.16}$$

Controller Rule 2:

$$\text{IF } x_3(t) = \frac{\pi}{2}, \text{ THEN } u(t) = \boldsymbol{K}_2 x(t_k). \tag{9.17}$$

Controller Rule 3:

$$\text{IF } x_3(t) = -\frac{\pi}{2}, \text{ THEN } u(t) = \boldsymbol{K}_3 x(t_k), \tag{9.18}$$

where $\boldsymbol{K}_1, \boldsymbol{K}_2, \boldsymbol{K}_3$ are the local gain matrices. Hence, the overall fuzzy controller is described in the form

$$u(t) = \sum_{j=1}^{3} \mu_j(z(t)) \boldsymbol{K}_j x(t_k), \quad t_k \leq t < t_{k+1}, \quad k = 0, 1, 2, \ldots \tag{9.19}$$

Substituting Eq. (9.4) into Eq. (9.1), we obtain

$$\dot{x}(t) = \sum_{i=1}^{3} \sum_{j=1}^{3} \mu_i(z(t)) \mu_j(z(t)) \left[\boldsymbol{A}_i x(t) + \boldsymbol{B}_i \boldsymbol{K}_j x(t_k) + \boldsymbol{E} w(t) \right]. \tag{9.20}$$

Remark 9.1. Note that both discrete and continuous signals exist in Eq. (9.20), which has more practical significance than the existing continuous-time control method for a DP ship system. Besides, because the parameter uncertainties exist in this system, it is a difficult problem for traditional lifting technique to be dealt with.

By defining $\tau(t) = t - t_k$, $t_k \leq t < t_{k+1}$, the fuzzy sampled-data controller is rewritten in the following form through input delay approach:

$$u(t)=\sum_{j=1}^{3}\mu_{\mathrm{j}}(z(t))\boldsymbol{K}_{\mathrm{j}}x(t_{\mathrm{k}})=\sum_{j=1}^{3}\mu_{\mathrm{j}}(z(t))\boldsymbol{K}_{\mathrm{j}}x(t-(t-t_{\mathrm{k}}))$$
$$=\sum_{j=1}^{3}\mu_{\mathrm{j}}(z(t))\boldsymbol{K}_{\mathrm{j}}x(t-\tau(t)). \tag{9.21}$$

The time-varying delay $\tau(t)$ satisfies

$$0\leq\tau(t)\leq d,\quad \dot{\tau}(t)=1,\quad t\neq t_k. \tag{9.22}$$

Thus, the sampled-data system in Eqs. (9.16)–(9.18) can be converted to a time-varying delay system as follows:

$$\begin{cases}\dot{x}(t)=\sum_{i=1}^{3}\sum_{j=1}^{3}\mu_{\mathrm{i}}(z(t))\mu_{\mathrm{j}}(z(t))\left[\boldsymbol{A}_{\mathrm{i}}x(t)+\boldsymbol{B}_{\mathrm{i}}\boldsymbol{K}_{\mathrm{j}}x(t-\tau(t))+\boldsymbol{E}_{\mathrm{i}}w(t)\right],\\ y(t)=\sum_{i=1}^{3}\sum_{j=1}^{3}\mu_{\mathrm{i}}(z(t))\mu_{\mathrm{j}}(z(t))\boldsymbol{C}_{\mathrm{i}}x(t).\end{cases} \tag{9.23}$$

The object is to find the controller gains $\boldsymbol{K}_1,\boldsymbol{K}_2,\boldsymbol{K}_3$, which satisfy the requirements as follows:

1) The closed-loop system Eq. (9.23) with $w(t)=\mathbf{0}$ is asymptotically stable;

2) To reject the external disturbance like waves, wind, and ocean currents, the closed-loop system is assumed to satisfy $\|y(t)\|_2\leq\gamma\|w(t)\|_2$ for all nonzero $w(t)\in L_2[0,\infty)$ under the zero initial condition, where $\gamma>0$, while L_2 denotes the set of square integrable functions.

9.2. Main results

The robust fuzzy control problem for sampled-data DP of ships is studied in this section. By constructing Lyapunov–Krasovskii functional, an asymptotical stability condition is derived for the system and the H_∞ performance is achieved. Then, the fuzzy sampled-data controller is obtained by analyzing the stabilization condition.

Theorem 9.1. *For given scales $d>0$ and $\gamma>0$, the system in Eq. (9.23) with $w(t)=\mathbf{0}$ is asymptotically stable with H_∞ performance γ if there exist matrices* $\boldsymbol{P}=\boldsymbol{P}^{\mathrm{T}}>0,\boldsymbol{Q}=\boldsymbol{Q}^{\mathrm{T}}>0,\boldsymbol{Z}=\boldsymbol{Z}^{\mathrm{T}}>0,\boldsymbol{M}_{\mathrm{ij}}=\begin{bmatrix}\boldsymbol{M}_{1\mathrm{ij}} & \boldsymbol{M}_{2\mathrm{ij}} & \boldsymbol{M}_{3\mathrm{ij}} & \boldsymbol{M}_{4\mathrm{ij}}\end{bmatrix}^{\mathrm{T}}$

and $\boldsymbol{N}_{ij} = \begin{bmatrix} \boldsymbol{N}_{1ij} & \boldsymbol{N}_{2ij} & \boldsymbol{N}_{3ij} & \boldsymbol{N}_{4ij} \end{bmatrix}^{\mathrm{T}}$ *such that*

$$\begin{bmatrix} \boldsymbol{\Xi}_{ij}^{11} & \boldsymbol{\Xi}_{ij}^{12} & \boldsymbol{\Xi}_{ij}^{13} & \boldsymbol{\Xi}_{ij}^{14} & \boldsymbol{A}_i^{\mathrm{T}} & \sqrt{d}\boldsymbol{M}_{1ij} & \sqrt{d}\boldsymbol{N}_{1ij} \\ * & \boldsymbol{\Xi}_{ij}^{22} & \boldsymbol{\Xi}_{ij}^{23} & \boldsymbol{\Xi}_{ij}^{24} & \boldsymbol{K}_j^{\mathrm{T}}\boldsymbol{B}_i^{\mathrm{T}} & \sqrt{d}\boldsymbol{M}_{2ij} & \sqrt{d}\boldsymbol{N}_{2ij} \\ * & * & \boldsymbol{\Xi}_{ij}^{44} & -\boldsymbol{N}_{4ij}^{T} & \boldsymbol{0} & \sqrt{d}\boldsymbol{M}_{3ij} & \sqrt{d}\boldsymbol{N}_{3ij} \\ * & * & * & -\gamma^2\boldsymbol{I} & \boldsymbol{E}_i^{\mathrm{T}} & \sqrt{d}\boldsymbol{M}_{4ij} & \sqrt{d}\boldsymbol{N}_{4ij} \\ * & * & * & * & -\boldsymbol{Z}^{-1} & \boldsymbol{0} & \boldsymbol{0} \\ * & * & * & * & * & -\boldsymbol{Z} & \boldsymbol{0} \\ * & * & * & * & * & * & -\boldsymbol{Z} \end{bmatrix} < \boldsymbol{0}, \tag{9.24}$$

where

$$\begin{aligned}
\boldsymbol{\Xi}_{ij}^{11} &= \boldsymbol{P}\boldsymbol{A}_i + \boldsymbol{A}_i^{\mathrm{T}}\boldsymbol{P} + \boldsymbol{Q} + \boldsymbol{M}_{1ij} + \boldsymbol{M}_{1ij}^{\mathrm{T}} + \boldsymbol{C}_i^{\mathrm{T}}\boldsymbol{C}_i, \\
\boldsymbol{\Xi}_{ij}^{12} &= \boldsymbol{P}\boldsymbol{B}_i\boldsymbol{K}_j - \boldsymbol{M}_{1ij} + \boldsymbol{M}_{2ij}^{\mathrm{T}} + \boldsymbol{N}_{1ij}, \\
\boldsymbol{\Xi}_{ij}^{13} &= \boldsymbol{M}_{3ij}^{\mathrm{T}} - \boldsymbol{N}_{1ij}, \\
\boldsymbol{\Xi}_{ij}^{14} &= \boldsymbol{P}\boldsymbol{E}_i + \boldsymbol{M}_{4ij}^{\mathrm{T}}, \\
\boldsymbol{\Xi}_{ij}^{22} &= -\boldsymbol{M}_{2ij} - \boldsymbol{M}_{2ij}^{\mathrm{T}} + \boldsymbol{N}_{2ij} + \boldsymbol{N}_{2ij}^{\mathrm{T}}, \\
\boldsymbol{\Xi}_{ij}^{23} &= -\boldsymbol{M}_{3ij}^{\mathrm{T}} - \boldsymbol{N}_{2ij} + \boldsymbol{N}_{3ij}^{\mathrm{T}}, \\
\boldsymbol{\Xi}_{ij}^{24} &= -\boldsymbol{M}_{4ij}^{\mathrm{T}} + \boldsymbol{N}_{4ij}^{\mathrm{T}}, \\
\boldsymbol{\Xi}_{ij}^{33} &= -\boldsymbol{Q} - \boldsymbol{N}_{3ij} + \boldsymbol{N}_{3ij}^{\mathrm{T}}.
\end{aligned}$$

Proof. We choose the Lyapunov functional for system Eq. (9.23) as follows:

$$\begin{aligned}
V(t) &= \sum_{i=1}^{3} V_i(t), \qquad t \in [t_k, t_{k+1}), \\
V_1(t) &= x(t)^{\mathrm{T}}\boldsymbol{P}x(t), \\
V_2(t) &= \int_{t-d}^{t} x(s)^{\mathrm{T}}\boldsymbol{B}x(s)\mathrm{d}s, \\
V_3(t) &= \int_{-d}^{0}\int_{t+\theta}^{t} \dot{x}(s)^{\mathrm{T}}\boldsymbol{Z}\dot{x}(s)\mathrm{d}s\mathrm{d}\theta.
\end{aligned} \tag{9.25}$$

Calculating the derivative of $V(t)$ yields

$$\begin{aligned}
\dot{V}_1(t) &= 2x(t)^{\mathrm{T}}\boldsymbol{P}\dot{x}(t), \\
\dot{V}_2(t) &= x(t)^{\mathrm{T}}\boldsymbol{Q}x(t) - x(t-d)^{\mathrm{T}}\boldsymbol{Q}x(t-d),
\end{aligned}$$

$$\dot{V}_3(t) = d\dot{x}(t)^{\mathrm{T}}\boldsymbol{Z}\dot{x}(t) - \int_{t-d}^{t} \dot{x}(s)^{\mathrm{T}}\boldsymbol{Z}\dot{x}(s)\mathrm{d}s. \tag{9.26}$$

By using the Leibniz–Newton formula, for any matrices of proper dimension $N_{\mathrm{ij}}, M_{\mathrm{ij}}, i = 1, 2, 3$, the equations can be obtained as follows:

$$\begin{aligned}
&2\sum_{i=1}^{3}\sum_{j=1}^{3}\mu_{\mathrm{i}}(z(t))\mu_{\mathrm{j}}(z(t))[x^{\mathrm{T}}(t)\boldsymbol{M}_{1\mathrm{ij}} + x^{\mathrm{T}}(t-\tau(t))\boldsymbol{M}_{2\mathrm{ij}} + x^{\mathrm{T}}(t-d)\boldsymbol{M}_{3\mathrm{ij}}]\\
&\times [x(t) - x(t-\tau(t)) - \int_{t-\tau(t)}^{t} \dot{x}(s)\mathrm{d}s] = 0,
\end{aligned} \tag{9.27}$$

$$\begin{aligned}
&2\sum_{i=1}^{3}\sum_{j=1}^{3}\mu_{\mathrm{i}}(z(t))\mu_{\mathrm{j}}(z(t))[x^{\mathrm{T}}(t)\boldsymbol{N}_{1\mathrm{ij}} + x^{\mathrm{T}}(t-\tau(t))\boldsymbol{N}_{2\mathrm{ij}} + x^{\mathrm{T}}(t-d)\boldsymbol{N}_{3\mathrm{ij}}]\\
&\times [x(t-\tau(t)) - x(t-d) - \int_{t-d}^{t-\tau(t)} \dot{x}(s)\mathrm{d}s] = 0.
\end{aligned} \tag{9.28}$$

Similarly, it can be obtained that

$$\int_{t-d}^{t} \dot{x}(s)\boldsymbol{Z}\dot{x}(s)\mathrm{d}s = \int_{t-\tau(t)}^{t} \dot{x}(s)\boldsymbol{Z}\dot{x}(s)\mathrm{d}s + \int_{t-d}^{t-\tau(t)} \dot{x}(s)\boldsymbol{Z}\dot{x}(s)\mathrm{d}s. \tag{9.29}$$

Substituting Eqs. (9.27)–(9.29) into Eq. (9.26) yields

$$\begin{aligned}
\dot{V}(t) &= 2x^{\mathrm{T}}(t)\boldsymbol{P}\dot{x}(t) + x(t)^{\mathrm{T}}\boldsymbol{Q}x(t) - x(t-d)^{\mathrm{T}}\boldsymbol{Q}x(t-d)\\
&\quad + d\dot{x}(t)^{\mathrm{T}}\boldsymbol{Z}\dot{x}(t) - \int_{t-d}^{t} \dot{x}(s)^{\mathrm{T}}\boldsymbol{Z}\dot{x}(s)\mathrm{d}s\\
&\le 2x^{\mathrm{T}}(t)\boldsymbol{P}\dot{x}(t) + x(t)^{\mathrm{T}}\boldsymbol{Q}x(t) - x(t-d)^{\mathrm{T}}\boldsymbol{Q}x(t-d)\\
&\quad + d\dot{x}(t)^{\mathrm{T}}\boldsymbol{Z}\dot{x}(t) - \int_{t-d}^{t-\tau(t)} \dot{x}(s)\boldsymbol{Z}\dot{x}(s)\mathrm{d}s - \int_{t-\tau(t)}^{t} \dot{x}(s)\boldsymbol{Z}\dot{x}(s)\mathrm{d}s\\
&\quad + 2\sum_{i=1}^{n}\sum_{j=1}^{n}\mu_{\mathrm{i}}(z(t))\mu_{\mathrm{j}}(z(t))[x^{\mathrm{T}}(t)\boldsymbol{M}_{1\mathrm{ij}} + x^{\mathrm{T}}(t-\tau(t))\boldsymbol{M}_{2\mathrm{ij}}\\
&\quad + x^{\mathrm{T}}(t-d)\boldsymbol{M}_{3\mathrm{ij}}][x(t) - x(t-\tau(t)) - \int_{t-\tau(t)}^{t} \dot{x}(s)\mathrm{d}s]\\
&\quad + 2\sum_{i=1}^{n}\sum_{j=1}^{n}\mu_{\mathrm{i}}(z(t))\mu_{\mathrm{j}}(z(t))[x^{\mathrm{T}}(t)\boldsymbol{N}_{1\mathrm{ij}} + x^{\mathrm{T}}(t-\tau(t))\boldsymbol{N}_{2\mathrm{ij}}\\
&\quad + x^{\mathrm{T}}(t-d)\boldsymbol{N}_{3\mathrm{ij}}][x(t-\tau(t)) - x(t-d) - \int_{t-d}^{t-\tau(t)} \dot{x}(s)\mathrm{d}s]
\end{aligned}$$

$$
\begin{aligned}
&\leq \sum_{i=1}^{3}\sum_{j=1}^{3}\mu_{\mathrm{i}}(z(t))\mu_{\mathrm{j}}(z(t))\Big(\zeta^{\mathrm{T}}(t)(\boldsymbol{\Theta}+[\boldsymbol{A}_{\mathrm{i}} \quad \boldsymbol{B}_{\mathrm{i}}\boldsymbol{K}_{\mathrm{j}} \quad 0]^{\mathrm{T}} \\
&\cdot \boldsymbol{Z}[\boldsymbol{A}_{\mathrm{i}} \quad \boldsymbol{B}_{\mathrm{i}}\boldsymbol{K}_{\mathrm{j}} \quad 0] + d\boldsymbol{M}_{\mathrm{ij}}\boldsymbol{Z}^{-1}\boldsymbol{M}_{\mathrm{ij}}^{\mathrm{T}} + d\boldsymbol{N}_{\mathrm{ij}}\boldsymbol{Z}^{-1}\boldsymbol{N}_{\mathrm{ij}}^{\mathrm{T}})\zeta(t)\Big) \\
&-\sum_{i=1}^{3}\sum_{j=1}^{3}\mu_{\mathrm{i}}(z(t))\mu_{\mathrm{j}}(z(t)) \\
&\times \int_{t-\tau(t)}^{t}[\zeta^{\mathrm{T}}(t)\boldsymbol{M}_{ij}+\dot{x}^{\mathrm{T}}(s)\boldsymbol{Z}]\boldsymbol{Z}^{-1}[\boldsymbol{M}_{\mathrm{ij}}^{\mathrm{T}}\zeta(t)+\boldsymbol{Z}\dot{x}(s)]\mathrm{d}s \\
&-\sum_{i=1}^{3}\sum_{j=1}^{3}\mu_{\mathrm{i}}(z(t))\mu_{\mathrm{j}}(z(t)) \\
&\times \int_{t-d}^{t-\tau(t)}[\zeta^{\mathrm{T}}(t)\boldsymbol{N}_{\mathrm{ij}}+\dot{x}^{\mathrm{T}}(s)\boldsymbol{Z}]\boldsymbol{Z}^{-1}[\boldsymbol{N}_{\mathrm{ij}}^{\mathrm{T}}\zeta(t)+\boldsymbol{Z}\dot{x}(s)]\mathrm{d}s, \qquad (9.30)
\end{aligned}
$$

where

$$
\zeta(t) = [x^{\mathrm{T}}(t) \quad x^{\mathrm{T}}(t-\tau(t)) \quad x^{\mathrm{T}}(t-d)]^{\mathrm{T}}, \qquad (9.31)
$$

$$
\boldsymbol{M}_{\mathrm{ij}} = \begin{bmatrix}\boldsymbol{M}_{1\mathrm{ij}}\\ \boldsymbol{M}_{2\mathrm{ij}}\\ \boldsymbol{M}_{3\mathrm{ij}}\end{bmatrix}, \qquad \boldsymbol{N}_{\mathrm{ij}} = \begin{bmatrix}\boldsymbol{N}_{1\mathrm{ij}}\\ \boldsymbol{N}_{2\mathrm{ij}}\\ \boldsymbol{N}_{3\mathrm{ij}}\end{bmatrix},
$$

$$
\boldsymbol{\Theta} = \begin{bmatrix}\boldsymbol{\Xi}_{\mathrm{ij}}^{11}-\boldsymbol{C}_{\mathrm{i}}^{\mathrm{T}}\boldsymbol{C}_{\mathrm{i}} & \boldsymbol{\Xi}_{\mathrm{ij}}^{12} & \boldsymbol{M}_{3\mathrm{ij}}^{\mathrm{T}}-\boldsymbol{N}_{1\mathrm{ij}}\\ * & \boldsymbol{\Xi}_{\mathrm{ij}}^{22} & -\boldsymbol{M}_{3\mathrm{ij}}^{\mathrm{T}}-\boldsymbol{N}_{2\mathrm{ij}}+\boldsymbol{N}_{3\mathrm{ij}}^{\mathrm{T}}\\ * & * & -\boldsymbol{Q}-\boldsymbol{N}_{3\mathrm{ij}}-\boldsymbol{N}_{3\mathrm{ij}}^{\mathrm{T}}\end{bmatrix},
$$

with

$$
\begin{aligned}
\boldsymbol{\Xi}_{\mathrm{ij}}^{11} &= \boldsymbol{P}\boldsymbol{A}_{\mathrm{i}}+\boldsymbol{A}_{\mathrm{i}}^{\mathrm{T}}\boldsymbol{P}+\boldsymbol{Q}+\boldsymbol{M}_{1\mathrm{ij}}+\boldsymbol{M}_{1\mathrm{ij}}^{\mathrm{T}}+\boldsymbol{C}_{\mathrm{i}}^{\mathrm{T}}\boldsymbol{C}_{\mathrm{i}},\\
\boldsymbol{\Xi}_{\mathrm{ij}}^{12} &= \boldsymbol{P}\boldsymbol{B}_{\mathrm{i}}\boldsymbol{K}_{\mathrm{j}}-\boldsymbol{M}_{1\mathrm{ij}}+\boldsymbol{M}_{2\mathrm{ij}}^{\mathrm{T}}+\boldsymbol{N}_{1\mathrm{ij}}, \qquad (9.32)\\
\boldsymbol{\Xi}_{\mathrm{ij}}^{22} &= -\boldsymbol{M}_{2\mathrm{ij}}-\boldsymbol{M}_{2\mathrm{ij}}^{\mathrm{T}}+\boldsymbol{N}_{2\mathrm{ij}}+\boldsymbol{N}_{2\mathrm{ij}}^{\mathrm{T}}.
\end{aligned}
$$

Since $\boldsymbol{Z} > \boldsymbol{0}$, then the last two parts in Eq. (9.30) are all less than $\boldsymbol{0}$. By Schur complement, we have the inequality from Eq. (9.24) as follow

$$
\boldsymbol{\Theta}+[\boldsymbol{A}_{\mathrm{i}} \quad \boldsymbol{B}_{\mathrm{i}}\boldsymbol{K}_{\mathrm{j}} \quad 0]^{\mathrm{T}}\boldsymbol{Z}[\boldsymbol{A}_{\mathrm{i}} \quad \boldsymbol{B}_{\mathrm{i}}\boldsymbol{K}_{\mathrm{j}} \quad 0]+d\boldsymbol{M}_{\mathrm{ij}}\boldsymbol{Z}^{-1}\boldsymbol{M}_{\mathrm{ij}}^{\mathrm{T}}+d\boldsymbol{N}_{\mathrm{ij}}\boldsymbol{Z}^{-1}\boldsymbol{N}_{\mathrm{ij}}^{\mathrm{T}} < 0. \qquad (9.33)
$$

It shows from Eq. (9.33) that $\dot{V}(x_t) < \boldsymbol{0}$. The asymptotical stability is established.

Now, we will establish the H_∞ performance of the system in Eq. (9.23). Based on the same Lyapunov–Krasovskii functional given in Eq. (9.25) and a similar proof, it can be shown that

$$
\begin{aligned}
&y^T(t)y(t) - \gamma^2 w^T(t)w(t) + \dot{V}(t) \leq \\
&\sum_{i=1}^{3}\sum_{j=1}^{3}\mu_i(z(t))\mu_j(z(t))\begin{bmatrix}\zeta(t)\\ w(t)\end{bmatrix}^T \left\{\boldsymbol{\Theta} + [\boldsymbol{A}_i \quad \boldsymbol{B}_i\boldsymbol{K}_j \quad 0 \quad \boldsymbol{E}_i]^T \cdot\right. \\
&\left.\boldsymbol{Z}[\boldsymbol{A}_i \quad \boldsymbol{B}_i\boldsymbol{K}_j \quad 0 \quad \boldsymbol{E}_i] + d\boldsymbol{M}_{ij}\boldsymbol{Z}^{-1}\boldsymbol{M}_{ij}^T + d\boldsymbol{N}_{ij}\boldsymbol{Z}^{-1}\boldsymbol{N}_{ij}^T\right\}\begin{bmatrix}\zeta(t)\\ w(t)\end{bmatrix},
\end{aligned}
\tag{9.34}
$$

where

$$
\boldsymbol{M}_{ij} = \begin{bmatrix}\boldsymbol{M}_{1ij}\\ \boldsymbol{M}_{2ij}\\ \boldsymbol{M}_{3ij}\\ \boldsymbol{M}_{4ij}\end{bmatrix}, \qquad \boldsymbol{N}_{ij} = \begin{bmatrix}\boldsymbol{N}_{1ij}\\ \boldsymbol{N}_{2ij}\\ \boldsymbol{N}_{3ij}\\ \boldsymbol{N}_{4ij}\end{bmatrix},
$$

$$
\boldsymbol{\Theta} = \begin{bmatrix}
\boldsymbol{\Xi}_{ij}^{11} & \boldsymbol{\Xi}_{ij}^{12} & \boldsymbol{M}_{3ij}^T - \boldsymbol{N}_{1ij} & \boldsymbol{P}\boldsymbol{B}_i + \boldsymbol{M}_{4ij}^T \\
* & \boldsymbol{\Xi}_{ij}^{22} & -\boldsymbol{M}_{3ij}^T - \boldsymbol{N}_{2ij} + \boldsymbol{N}_{3ij}^T & -\boldsymbol{M}_{4ij}^T + \boldsymbol{N}_{4ij}^T \\
* & * & -\boldsymbol{Q} - \boldsymbol{N}_{3ij} - \boldsymbol{N}_{3ij}^T & -\boldsymbol{N}_{4ij}^T \\
* & * & * & -\gamma^2\boldsymbol{I}
\end{bmatrix}.
$$

By using Schur complement, inequality (9.24) guarantees

$$
\begin{aligned}
&\boldsymbol{\Theta} + [\boldsymbol{A}_i \quad \boldsymbol{B}_i\boldsymbol{K}_j \quad 0 \quad \boldsymbol{E}_i]^T\boldsymbol{Z}[\boldsymbol{A}_i \quad \boldsymbol{B}_i\boldsymbol{K}_j \quad 0 \quad \boldsymbol{E}_i] \\
&+ d\boldsymbol{M}_{ij}\boldsymbol{Z}^{-1}\boldsymbol{M}_{ij}^T + d\boldsymbol{N}_{ij}\boldsymbol{Z}^{-1}\boldsymbol{N}_{ij}^T < 0.
\end{aligned}
$$

Thus, from Eq. (9.34), we obtain

$$
y^T(t)y(t) - \gamma^2 w^T(t)w(t) + \dot{V}(t) < 0. \tag{9.35}
$$

Therefore, from Eq. (9.35), $\|y(t)\|_2 \leq \gamma\|w(t)\|_2$ can be achieved for all nonzero $w(t) \in \boldsymbol{L}_2[0, \infty)$, and then the H_∞ performance is established. This proof is completed. □

Remark 9.2. Adequate conditions are derived to determine the asymptotical stability for the system in Eq. (9.23) by Theorem 9.1. It is noted that the free-weighting matrices are used in the theorem, which can reduce the conservativeness of the proposed delay-dependent results.

Now, based on the following theorem, the fuzzy sampled-data controller in Eq. (9.19) can be obtained.

Theorem 9.2. *Given scalars $d > 0, \gamma > 0$, the system in Eq. (9.23) is asymptotically stable, if there exist matrices $\boldsymbol{P} = \boldsymbol{P}^{\mathrm{T}} > 0, \boldsymbol{Q} = \boldsymbol{Q}^{\mathrm{T}} > 0, \boldsymbol{Z} = \boldsymbol{Z}^{\mathrm{T}} > 0$, $\boldsymbol{M}_{\mathrm{ij}} = \begin{bmatrix} \boldsymbol{M}_{1\mathrm{ij}} & \boldsymbol{M}_{2\mathrm{ij}} & \boldsymbol{M}_{3\mathrm{ij}} & \boldsymbol{M}_{4\mathrm{ij}} \end{bmatrix}^{\mathrm{T}}$ and $\boldsymbol{N}_{\mathrm{ij}} = \begin{bmatrix} \boldsymbol{N}_{1\mathrm{ij}} & \boldsymbol{N}_{2\mathrm{ij}} & \boldsymbol{N}_{3\mathrm{ij}} & \boldsymbol{N}_{4\mathrm{ij}} \end{bmatrix}^{\mathrm{T}}$, such that*

$$\begin{bmatrix} \bar{\boldsymbol{\Xi}}_{\mathrm{ij}}^{11} & \bar{\boldsymbol{\Xi}}_{\mathrm{ij}}^{12} & \bar{\boldsymbol{\Xi}}_{\mathrm{ij}}^{13} & \bar{\boldsymbol{\Xi}}_{\mathrm{ij}}^{14} & \bar{\boldsymbol{P}}\boldsymbol{A}_{\mathrm{i}}^{\mathrm{T}} & \sqrt{d}\bar{\boldsymbol{M}}_{1\mathrm{ij}} & \sqrt{d}\bar{\boldsymbol{N}}_{1\mathrm{ij}} & \bar{\boldsymbol{P}}\boldsymbol{C}_{\mathrm{i}}^{\mathrm{T}} \\ * & \bar{\boldsymbol{\Xi}}_{\mathrm{ij}}^{22} & \bar{\boldsymbol{\Xi}}_{\mathrm{ij}}^{23} & \bar{\boldsymbol{\Xi}}_{\mathrm{ij}}^{24} & \bar{\boldsymbol{K}}_{\mathrm{j}}^{\mathrm{T}}\boldsymbol{B}_{\mathrm{i}}^{\mathrm{T}} & \sqrt{d}\bar{\boldsymbol{M}}_{2\mathrm{ij}} & \sqrt{d}\bar{\boldsymbol{N}}_{2\mathrm{ij}} & \boldsymbol{0} \\ * & * & \bar{\boldsymbol{\Xi}}_{\mathrm{ij}}^{33} & -\bar{\boldsymbol{N}}_{4\mathrm{ij}}^{\mathrm{T}} & \boldsymbol{0} & \sqrt{d}\bar{\boldsymbol{M}}_{3\mathrm{ij}} & \sqrt{d}\bar{\boldsymbol{N}}_{3\mathrm{ij}} & \boldsymbol{0} \\ * & * & * & -\gamma^2\boldsymbol{I} & \boldsymbol{E}_{\mathrm{i}}^{\mathrm{T}} & \sqrt{d}\bar{\boldsymbol{M}}_{4\mathrm{ij}} & \sqrt{d}\bar{\boldsymbol{N}}_{1\mathrm{ij}} & \boldsymbol{0} \\ * & * & * & * & \bar{\boldsymbol{Z}} - 2\bar{\boldsymbol{P}} & \boldsymbol{0} & \boldsymbol{0} & \boldsymbol{0} \\ * & * & * & * & * & -\bar{\boldsymbol{Z}} & \boldsymbol{0} & \boldsymbol{0} \\ * & * & * & * & * & * & -\bar{\boldsymbol{Z}} & \boldsymbol{0} \\ * & * & * & * & * & * & * & -\boldsymbol{I} \end{bmatrix} < \boldsymbol{0}, \tag{9.36}$$

where

$$\begin{aligned} \bar{\boldsymbol{\Xi}}_{\mathrm{ij}}^{11} &= \bar{\boldsymbol{P}}\boldsymbol{A}_{\mathrm{i}} + \boldsymbol{A}_{\mathrm{i}}^{\mathrm{T}}\bar{\boldsymbol{P}} + \bar{\boldsymbol{Q}} + \bar{\boldsymbol{M}}_{1\mathrm{ij}} + \bar{\boldsymbol{M}}_{1\mathrm{ij}}^{\mathrm{T}}, \\ \bar{\boldsymbol{\Xi}}_{\mathrm{ij}}^{12} &= \boldsymbol{B}_{\mathrm{i}}\bar{\boldsymbol{K}}_{\mathrm{j}} - \bar{\boldsymbol{M}}_{1\mathrm{ij}} + \bar{\boldsymbol{M}}_{2\mathrm{ij}}^{\mathrm{T}} + \bar{\boldsymbol{N}}_{1\mathrm{ij}}, \\ \bar{\boldsymbol{\Xi}}_{\mathrm{ij}}^{13} &= \bar{\boldsymbol{M}}_{3\mathrm{ij}}^{\mathrm{T}} - \bar{\boldsymbol{N}}_{1\mathrm{ij}}, \\ \bar{\boldsymbol{\Xi}}_{\mathrm{ij}}^{14} &= \boldsymbol{E}_{\mathrm{i}} + \bar{\boldsymbol{M}}_{4\mathrm{ij}}^{\mathrm{T}}, \\ \bar{\boldsymbol{\Xi}}_{\mathrm{ij}}^{22} &= -\bar{\boldsymbol{M}}_{2\mathrm{ij}} - \bar{\boldsymbol{M}}_{2\mathrm{ij}}^{\mathrm{T}} + \bar{\boldsymbol{N}}_{2\mathrm{ij}} + \bar{\boldsymbol{N}}_{2\mathrm{ij}}^{\mathrm{T}}, \\ \bar{\boldsymbol{\Xi}}_{\mathrm{ij}}^{23} &= -\bar{\boldsymbol{M}}_{3\mathrm{ij}}^{\mathrm{T}} - \bar{\boldsymbol{N}}_{2\mathrm{ij}} + \bar{\boldsymbol{N}}_{3\mathrm{ij}}^{\mathrm{T}}, \\ \bar{\boldsymbol{\Xi}}_{\mathrm{ij}}^{24} &= -\bar{\boldsymbol{M}}_{4\mathrm{ij}}^{\mathrm{T}} + \bar{\boldsymbol{N}}_{4\mathrm{ij}}^{\mathrm{T}}, \\ \bar{\boldsymbol{\Xi}}_{\mathrm{ij}}^{33} &= -\bar{\boldsymbol{Q}} - \bar{\boldsymbol{N}}_{3\mathrm{ij}} + \bar{\boldsymbol{N}}_{3\mathrm{ij}}^{\mathrm{T}}. \end{aligned}$$

Moreover, the fuzzy sampled-data controller gains $\boldsymbol{K}_1, \boldsymbol{K}_2, \boldsymbol{K}_3$ are obtained as

$$\boldsymbol{K}_{\mathrm{j}} = \bar{\boldsymbol{K}}_{\mathrm{j}}\bar{\boldsymbol{P}}^{-1}, \quad j = 1, 2, 3. \tag{9.37}$$

Proof. Let $\eta = \mathrm{diag}\left(P^{-\mathrm{T}}, P^{-\mathrm{T}}, P^{-\mathrm{T}}, I, I, P^{-\mathrm{T}}, P^{-\mathrm{T}}\right)$. Denoting

$$\begin{gathered} \bar{\boldsymbol{P}} = \boldsymbol{P}^{-1}, \quad \bar{\boldsymbol{K}}_{\mathrm{j}} = \boldsymbol{K}_{\mathrm{j}}P^{-1}, \quad \bar{\boldsymbol{Q}} = \boldsymbol{P}^{-\mathrm{T}}\boldsymbol{Q}\boldsymbol{P}^{-1}, \\ \bar{\boldsymbol{M}}_{1\mathrm{ij}} = \boldsymbol{P}^{-\mathrm{T}}\boldsymbol{M}_{1\mathrm{ij}}\boldsymbol{P}^{-1}, \quad \bar{\boldsymbol{M}}_{4\mathrm{ij}} = \boldsymbol{M}_{4\mathrm{ij}}\boldsymbol{P}^{-1}, \quad \bar{\boldsymbol{Z}} = \boldsymbol{P}^{-\mathrm{T}}\boldsymbol{Z}\boldsymbol{P}^{-1}, \end{gathered}$$

and pre- and post-multiplying inequality (9.24) by η and η^T, respectively, one can obtain Eq. (9.36) and the following inequality:

$$\begin{bmatrix} \tilde{\Xi}_{ij}^{11} & \bar{\Xi}_{ij}^{12} & \bar{\Xi}_{ij}^{13} & \bar{\Xi}_{ij}^{14} & \bar{P}A_i^T & \sqrt{d}\bar{M}_{1ij} & \sqrt{d}\bar{N}_{1ij} \\ * & \bar{\Xi}_{ij}^{22} & \bar{\Xi}_{ij}^{23} & \bar{\Xi}_{ij}^{24} & \bar{K}_j^T B_i^T & \sqrt{d}\bar{M}_{2ij} & \sqrt{d}\bar{N}_{2ij} \\ * & * & \bar{\Xi}_{ij}^{44} & -\bar{N}_{4ij}^T & 0 & \sqrt{d}\bar{M}_{3ij} & \sqrt{d}\bar{N}_{3ij} \\ * & * & * & -\gamma^2 I & \bar{P}E_i^T & \sqrt{d}\bar{M}_{4ij} & \sqrt{d}\bar{N}_{4ij} \\ * & * & * & * & -\bar{P}Z^{-1}\bar{P} & 0 & 0 \\ * & * & * & * & * & -\bar{Z} & 0 \\ * & * & * & * & * & * & -\bar{Z} \end{bmatrix} < 0, \quad (9.38)$$

where

$$\tilde{\Xi}_{11}^{ij} = \bar{P}A_i + A_i^T\bar{P} + \bar{Q} + \bar{M}_{1ij} + \bar{M}_{1ij}^T + \bar{P}C_i^T C_i\bar{P}.$$

Noting that $-\bar{P}\bar{Z}^{-1}\bar{P} \leq \bar{Z} - 2\bar{P}$ and by Schur complement, Eq. (9.38) implies inequality (9.36). This proof is completed. □

9.3. Numerical examples

In the section, a simulation example for DP of ships is given to illustrate that the proposed methods are effective. Consider the following scaled matrices given in [8]:

$$M = \begin{bmatrix} 1.0852 & 0 & 0 \\ 0 & 2.0575 & -0.4087 \\ 0 & -0.4087 & 0.2153 \end{bmatrix},$$

$$D = \begin{bmatrix} 0.0865 & 0 & 0 \\ 0 & 0.0762 & 0.1510 \\ 0 & 0.0151 & 0.0031 \end{bmatrix}.$$

Let $\alpha = \sin 2^0$ and $\beta = \cos 88^0$. The matrices A_1, A_2, A_3, B_1, B_2, and B_3 can be calculated according to Theorem 9.2 as follows:

$$A_1 = \begin{bmatrix} 0 & 0 & 0 & 1.0000 & -0.0349 & 0 \\ 0 & 0 & 0 & 0.0349 & 1.0000 & 0 \\ 0 & 0 & 0 & 0 & 0 & 1.0000 \\ 0 & 0 & 0 & -0.0797 & 0 & 0 \\ 0 & 0 & 0 & 0 & -0.0818 & -0.1224 \\ 0 & 0 & 0 & 0 & -0.2254 & -0.2468 \end{bmatrix},$$

$$A_2 = \begin{bmatrix} 0 & 0 & 0 & 0.0349 & -1.0000 & 0 \\ 0 & 0 & 0 & 1.0000 & 0.0349 & 0 \\ 0 & 0 & 0 & 0 & 0 & 1.0000 \\ 0 & 0 & 0 & -0.0797 & 0 & 0 \\ 0 & 0 & 0 & 0 & -0.0818 & -0.1224 \\ 0 & 0 & 0 & 0 & -0.2254 & -0.2468 \end{bmatrix},$$

$$A_3 = \begin{bmatrix} 0 & 0 & 0 & 0.0349 & 1.0000 & 0 \\ 0 & 0 & 0 & -1.0000 & 0.0349 & 0 \\ 0 & 0 & 0 & 0 & 0 & 1.0000 \\ 0 & 0 & 0 & -0.0797 & 0 & 0 \\ 0 & 0 & 0 & 0 & -0.0818 & -0.1224 \\ 0 & 0 & 0 & 0 & -0.2254 & -0.2468 \end{bmatrix},$$

$$B_1 = B_2 = B_3 = \begin{bmatrix} 0 & 0 & 0 \\ 0 & 0 & 0 \\ 0 & 0 & 0 \\ 0.9215 & 0 & 0 \\ 0 & 0.7802 & 1.4811 \\ 0 & 1.4811 & 7.4562 \end{bmatrix}.$$

Assuming there is the sampling interval with $d = 0.25$ s, and the minimum value achieved by Theorem 9.2 to guarantee H_∞ performance is $\gamma_{\min} = 7.641$, then the fuzzy sampled-data controller gain matrices are obtained as:

$$K_1 = \begin{bmatrix} -0.8012 & 0.0030 & 0.0794 & -1.0835 & 0.0124 & 0.0618 \\ 0.0264 & -1.3633 & -5.7699 & -0.0088 & -2.2216 & -4.6795 \\ -0.0051 & 0.2635 & 1.0231 & 0.0017 & 0.4581 & 0.7566 \end{bmatrix},$$

$$K_2 = \begin{bmatrix} -0.5920 & -0.1130 & 2.0058 & -1.0498 & 0.5103 & 1.4343 \\ 0.7407 & -0.5576 & -6.2220 & -0.3016 & -1.9950 & -5.1017 \\ -0.1428 & 0.1078 & 1.1126 & 0.0593 & 0.4130 & 0.8411 \end{bmatrix},$$

$$K_3 = \begin{bmatrix} -0.5920 & 0.1130 & -2.0058 & -1.0498 & -0.5103 & -1.4343 \\ -0.7407 & -0.5576 & -6.2220 & 0.3016 & -1.9950 & -5.1017 \\ 0.1428 & 0.1078 & 1.1126 & -0.0593 & 0.4130 & 0.8411 \end{bmatrix}.$$

We apply the sampled-data controllers to the system in Eq. (9.23). Assume the ship's initial positions and heading are $\eta_i = [5 \quad 5 \quad 2]^{\mathrm{T}}$, and body-fixed velocity vector is $\nu_i = [0 \quad 0 \quad 0]^{\mathrm{T}}$.

Consider the external environment disturbance as follows:

$$w(t) = 0.1 \sin(t). \tag{9.39}$$

Fig. 9.2 shows the ship's x and y positions, Fig. 9.3 shows the ship's heading ψ. Figs. 9.4 and 9.5 show velocity response. From Figs. 9.2–9.5, we can see that the variables $[x \; y \; \psi \; p \; \upsilon \; r]^{\mathrm{T}}$ tend to zero in a short time. That is, the designed fuzzy sampled-data controllers can stabilize the ship and guarantee H_∞ control performance; hence, the ship can maintain desired position, heading, and velocity under the external disturbance like waves, wind, and ocean currents.

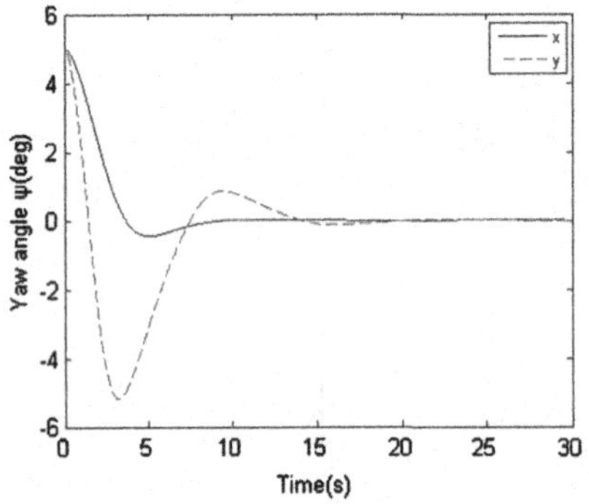

Figure 9.2 Position x and y of the ship.

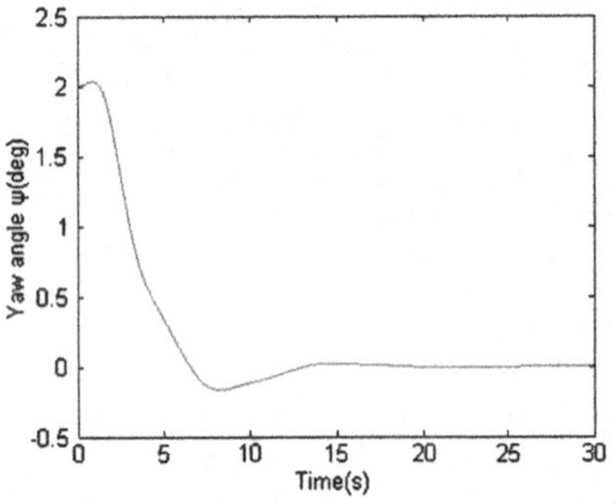

Figure 9.3 Heading ψ of the ship.

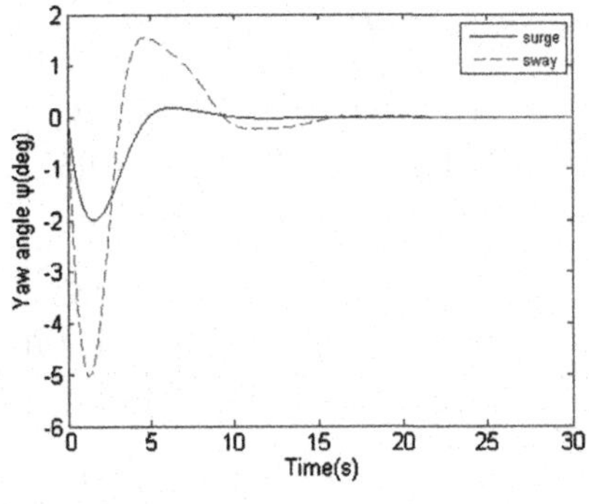

Figure 9.4 Velocity p and υ of the ship.

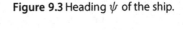

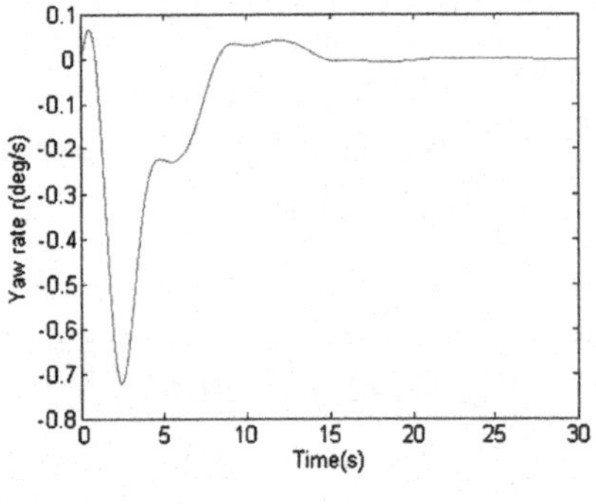

Figure 9.5 Yaw rate r of the ship.

9.4. Conclusions

This note discusses the robust H_∞ sampled-data stabilization problem for a nonlinear DP ship system. T–S fuzzy model was employed to approximate the nonlinear DP system. By using the Lyapunov approach, an adequate condition is derived to guarantee the system's asymptotical stability and obtain good H_∞ performance. The fuzzy sampled-data controller is designed by using Lyapunov stability theorems. It has been proven from a DP ship simulation that the designed method is effective so that the DP of ships can maintain the desired position, heading, and velocities under the ocean disturbance. The proposed methods can also be used for other types of ship.

References

[1] F. Liu, C. Hui, Application of moving horizon filter for dynamic positioning ship, in: International Conference on Computational Intelligence and Software Engineering (CiSE), IEEE Computer Society, United States, 2010, pp. 1–4.

[2] F. Benetao, G. Ippoliti, S. Longhi, et al., Dynamic positioning of a marine vessel using DTVSC and robust control allocation, in: 20th Mediterranean Conference on Control & Automation (MED), IEEE Computer Society, United States, 2012, pp. 1211–1216.

[3] A. Witkowska, Dynamic positioning system with vectorial backstepping controller, in: 18th International Conference on Methods and Models in Automation and Robotics (MMAR), IEEE Computer Society, United States, 2013, pp. 842–847.

[4] V. Hassani, A.M. Pascoal, A.P. Aguiar, Multiple model adaptive wave filtering for dynamic positioning of marine vessels, in: Proceedings of the American Control Conference (ACC), Institute of Electrical and Electronics Engineers, United States, 2012, pp. 6222–6228.

[5] J.L. Du, X. Hu, M. Krstic, et al., Robust dynamic positioning of ships with disturbances under input saturation, Automatica 73 (2016) 207–214.

[6] H. Li, X. Jing, H.K. Lam, et al., Fuzzy sampled-data control for uncertain vehicle suspension systems, IEEE Transactions on Cybernetics 44 (7) (2014) 1111–1126.

[7] L. Xu, Z. Liu, Design of fuzzy PID controller for ship dynamic positioning, in: Proceedings of the 28th Chinese Control and Decision Conference, Institute of Electrical and Electronics Engineers Inc, 2016, pp. 3130–3135.

[8] W.E. Ngongi, J.L. Du, R. Wang, Robust fuzzy controller design for dynamic positioning system of ships, International Journal of Control, Automation and Systems 13 (5) (2015) 1294–1305.

[9] W.H. Ho, S.H. Chen, J.H. Chou, Optimal control of Takagi–Sugeno fuzzy-model-based systems representing dynamic ship positioning systems, Applied Soft Computing 13 (7) (2013) 3197–3210.

[10] P. Shi, Filtering on sampled-data systems with parametric uncertainty, IEEE Transactions on Automatic Control 43 (7) (1998) 1022–1027.

[11] E. Fridman, A. Seuret, J.P. Richard, Robust sampled-data stabilization of linear systems: an input delay approach, Automatica 40 (8) (2004) 1441–1446.

[12] H. Gao, W. Sun, P. Shi, Robust sampled-data control for vehicle active suspension systems, IEEE Transactions on Control Systems Technology 18 (1) (2010) 238–245.

[13] Y.Y. Wang, H. Shen, H.R. Karimi, et al., Dissipativity-based fuzzy integral sliding mode control of continuous-time T-S fuzzy systems, IEEE Transactions on Fuzzy Systems 26 (3) (2018) 1164–1176.

[14] X.L. Zhu, B. Chen, D. Yue, et al., An improved input delay approach to stabilization of fuzzy systems under variable sampling, IEEE Transactions on Fuzzy Systems 20 (2) (2012) 330–341.

[15] Y.Y. Wang, Y.Q. Xia, P.F. Zhou, Fuzzy-model-based sampled-data control of chaotic systems: a fuzzy time-dependent Lyapunov–Krasovskii functional approach, IEEE Transactions on Fuzzy Systems (2016).

[16] Y.Y. Wang, P. Shi, On master–slave synchronization of chaotic Lur'e systems using sampled-data control, IEEE Transactions on Circuits and Systems II: Express Briefs (2016).

[17] W.H. Chen, Z. Wang, X. Lu, On sampled-data control for master–slave synchronization of chaotic Lur'e systems, IEEE Transactions on Circuits and Systems II: Express Briefs 59 (8) (2012) 515–519.

[18] B. Shen, Z. Wang, X. Liu, Sampled-data synchronization control of dynamical networks with stochastic sampling, IEEE Transactions on Automatic Control 57 (10) (2012) 2644–2650.

[19] C. Hua, C. Ge, X. Guan, Synchronization of chaotic Lur'e systems with time delays using sampled-data control, IEEE Transactions on Neural Networks and Learning Systems 26 (6) (2015) 1214–1221.

[20] Y.Y. Wang, H. Shen, D.P. Duan, On stabilization of quantized sampled-data neural-network-based control systems, IEEE Transactions on Cybernetics 47 (10) (2017) 3124–3135.

[21] Z.G. Wu, P. Shi, H. Su, et al., Stochastic synchronization of Markovian jump neural networks with time-varying delay using sampled data, IEEE Transactions on Cybernetics 43 (6) (2013) 1796–1806.

[22] Z.G. Wu, P. Shi, H. Su, et al., Exponential stabilization for sampled-data neural-network-based control systems, IEEE Transactions on Neural Networks and Learning Systems 25 (12) (2014) 2180–2190.

CHAPTER TEN

Finite-time control of autonomous surface vehicles

Ning Wang[a]
School of Marine Electrical Engineering, Dalian Maritime University, Dalian, China

Contents

Abstract

In this chapter, finite-time control of an autonomous surface vehicle (ASV) with complex unknowns, including unmodeled dynamics, uncertainties and/or unknown disturbances, is addressed within a proposed homogeneity-based finite-time control (HFC) framework. Major contributions are as follows: (1) In the absence of external disturbances, a nominal HFC framework is established to achieve exact trajectory tracking

[a] This work is supported by the National Natural Science Foundation of P. R. China (under Grants 51009017 and 51379002), the Fund for Liaoning Innovative Talents in Colleges and Universities (under Grant LR2017024), the Fund for Dalian Distinguished Young Scholars (under Grant 2016RJ10), the Liaoning Revitalization Talents Program (under Grant XLYC1807013), the Stable Supporting Fund of Science and Technology on Underwater Vehicle Laboratory (SXJQR2018WDKT03), and the Fundamental Research Funds for the Central Universities (under Grant 3132019344).

https://doi.org/10.1016/B978-0-12-820271-5.00015-8

control of an ASV, whereby global finite-time stability is ensured by combining homogeneous analysis and Lyapunov approach; (2) Within the HFC scheme, a finite-time disturbance observer (FDO) is further nested to rapidly and accurately reject complex disturbances, and thereby contributing to an FDO-based HFC (FDO-HFC) scheme which can realize exactness of trajectory tracking and disturbance observation; (3) Aiming to exactly deal with complicated unknowns including unmodeled dynamics and/or disturbances, a finite-time unknown observer (FUO) is deployed as a patch for the nominal HFC framework, and eventually results in an FUO-based HFC (FUO-HFC) scheme which guarantees that accurate trajectory tracking can be achieved for an ASV under harsh environments. Simulation studies and comprehensive comparisons conducted on a benchmark ship demonstrate the effectiveness and superiority of the proposed HFC schemes.

Keywords

Global finite-time stability, Accurate trajectory tracking, Finite-time disturbance observer, Finite-time unknown observer, Autonomous surface vehicle

10.1. Introduction

In recent years, autonomous surface vehicles (ASVs) have been widely deployed for various missions related with observations, military tasks, coastal and inland waters monitoring, etc. [1]. Generally, tracking control of an ASV to a prescribed trajectory/path with an acceptable accuracy plays a key role within the entire autopilot system and has thus attracted great attention from both marine engineering and control communities [2]. It is much more demanding to achieve accurate tracking of a predetermined trajectory in some applications, for example, marine surveying and mapping on the sea in the presence of complicated uncertainties and variations, including system uncertainties and external disturbances due to ocean winds, waves, and currents [3–8]. In this context, it becomes extremely challenging to achieve accurate trajectory tracking of an ASV sailing in such harsh environments.

The sliding mode control (SMC) technique has been investigated as a promising approach to achieve high accurate trajectory tracking control of an ASV [9]. However, the SMC-based approaches have to incur high-frequency chattering with conservatively large magnitude around the sliding surface to dominate unknowns and achieve the robustness. Furthermore, the SMC technique can only handle matched unknowns. By virtue of the (vectorial) backstepping technique [10], a trajectory tracking controller has been designed for an ASV in the presence of (mismatched) unknowns including time-varying disturbances and system uncertainties

[11]. It should be noted that tracking errors can only be made globally uniformly ultimately bounded. Applying the integrator backstepping [12] technique to the design of trajectory tracking control law for an underactuated ASV contributes to semiglobally exponentially stable tracking errors. Unfortunately, uncertainties and disturbances have not been addressed. In combination with neural networks (NNs) and adaptive robust control techniques [13], a saturated tracking controller that renders tracking errors semiglobally uniformly ultimately bounded has been proposed to preserve the robustness against time-varying disturbances induced by waves and ocean currents. In addition to NNs [14–16], a lot of efforts on adaptive approximation based tracking control have also been made via fuzzy systems (FS) [17–21], and fuzzy neural networks (FNN) [22–27], etc., and can roughly compensate unknown dynamics. Recently, a significant progress has been made by an innovative approximator termed self-constructing fuzzy neural network (SCFNN) [28–31] towards the dynamic-structure-approximation-based adaptive control approaches [22] with much higher accuracy of both reconstruction and trajectory tracking. It should be highlighted that accurate tracking control can still hardly be achieved by the foregoing SCFNN-based control approaches since there still exist unexpected approximation residuals. Nevertheless, the convergence rate of tracking errors is usually somewhat slow since only asymptotic or exponential closed-loop stability can be derived from previous tracking control approaches.

In order to further pursue better tracking performance, finite-time control approaches have been implemented in the literature [32,33]. In [32], the SMC technique has been employed to realize attitude tracking of a rigid spacecraft and a modified differentiator has been incorporated to compensate disturbances and inertia uncertainties, whereby finite-time convergence of tracking errors can be obtained. By virtue of the homogeneous method in [33], finite-time stability of the closed-loop control system with negative degree of homogeneous can be ensured if tracking error dynamics can be proved to be asymptotically stable. It should be noted that, compared with traditional asymptotic convergent methods, the aforementioned finite-time control approaches can achieve not only faster convergence rate within the vicinity of the origin but also stronger disturbance rejection. Meaningfully, finite-time control approaches ensure that tracking errors can reach zero within a finite time. Note that, unlike SMC-based approaches, the homogeneity-based method [33] contributes to a straightforward solution without any couplings of tracking errors. Besides, external disturbances can even excite unmodeled dynamics of the ASV, and thus require to be well estimated. Otherwise, complex disturbances pertaining to an ASV cannot be exactly observed within a short time, and make trajectory tracking inaccurate.

Recently, disturbance observer based control (DOBC) technique has also been proposed by Chen [34] to not only improve system robustness but also enhance the entire performance without sacrificing the nominal one [35,36]. In this context, the DOBC schemes have been extensively studied and widely applied to various industrial sectors, including mechatronics systems [37], and aerospace systems [38], etc. Clearly, it is innovative within the DOBC framework that all disturbances and/or unknowns are addressed as a lumped nonlinearity estimated by a nonlinear disturbance observer (NDO) which is usually involved.

In this chapter, an ambitious goal of achieving fast and accurate trajectory control of an ASV in the presence of unknowns including disturbances and unmodeled dynamics is pursued. To be specific, a homogeneity-based finite-time control (HFC) scheme is developed to achieve accurate trajectory tracking of an ASV in the absence of external disturbances. In conjunction with a finite-time disturbance observer (FDO) which can be further devised to exactly estimate external disturbances within a short time, an FDO-based HFC (FDO-HFC) scheme is thus implemented to exactly track an ASV suffering from complex disturbances. In order to further address unmodeled dynamics including uncertainties and/or excited dynamics due to disturbances, a high-order sliding mode estimator is designed to realize a finite-time unknown observer (FUO) which can exactly capture unknown dynamics. Incorporating the FUO into the HFC scheme contributes to the FUO-based HFC (FUO-HFC) scheme which can achieve fast and accurate trajectory tracking with complex unknowns including unmodeled dynamics and/or uncertainties in addition to disturbances.

The rest of this chapter is organized as follows. In Section 10.2, preliminaries together with the trajectory tracking problem associated with an ASV are addressed. The HFC, FDO-HFC, and FUO-HFC schemes together with theoretical analysis on finite-time stability are presented in Section 10.3. Simulation studies and discussions are conducted in Section 10.4. Conclusions are drawn in Section 10.5.

10.2. Preliminaries and problem statement

10.2.1 Preliminaries

For the convenience of readers, we collect the key definitions and lemmas frequently used in this chapter in the sequel.

Consider an autonomous nonlinear system as follows:

$$\dot{\boldsymbol{x}}(t) = \boldsymbol{f}(\boldsymbol{x}(t)), \quad x(0) = 0, \quad \boldsymbol{f}(0) = 0, \quad \boldsymbol{x} \in U_0 \subset \mathbb{R}^n \tag{10.1}$$

where $\boldsymbol{x} = [x_1, \ldots, x_n]^T$ and nonlinear function $\boldsymbol{f}(\cdot)$ is continuous on a open neighborhood U_0 of the origin.

Definition 10.1 (Global Asymptotic Stability [39]). The equilibrium $\boldsymbol{x}_e = 0$ of system (10.1) is globally asymptotically stable if there exits a function $V(\boldsymbol{x})$ satisfying

(a) $V(0) = 0$;
(b) $V(\boldsymbol{x}) > 0, \forall \boldsymbol{x} \neq 0$, and $V(\boldsymbol{x})$ is radically unbounded;
(c) $\dot{V}(\boldsymbol{x}) \leq 0$;
(d) $\dot{V}(\boldsymbol{x})$ does not vanish identically along any trajectory in $\mathbb{R}^n$, other than the null solution $\boldsymbol{x} = 0$.

Definition 10.2 (Homogeneity [33]). Let $V(\boldsymbol{x}) : \mathbb{R}^n \to \mathbb{R}$ be a continuous scalar function. Then $V(\boldsymbol{x})$ is said to be a homogeneous function of degree σ with respect to weights $(r_1, \ldots, r_n) \in \mathbb{R}^n$ with $r_i > 0$, $i = 1, 2, \ldots, n$, if, for any given $\varepsilon > 0$,

$$V(\varepsilon^{r_1} x_1, \ldots, \varepsilon^{x_n} x_n) = \varepsilon^{\sigma} V(\boldsymbol{x}), \quad i = 1, \ldots, n, \forall \boldsymbol{x} \in \mathbb{R}^n.$$

Let $\boldsymbol{f}(\boldsymbol{x}) = [f_1(\boldsymbol{x}), \ldots, f_n(\boldsymbol{x})]^T$ be a continuous vector field. Then $\boldsymbol{f}(\boldsymbol{x})$ is homogeneous of degree $k \in \mathbb{R}$ with respect to weights $(r_1, \ldots, r_n)$, if, for any given $\varepsilon > 0$,

$$f_i(\varepsilon^{r_1} x_1, \ldots, \varepsilon^{r_n} x_n) = \varepsilon^{k+r_i} f_i(\boldsymbol{x}), \quad i = 1, \ldots, n, \forall \boldsymbol{x} \in \mathbb{R}^n.$$

And, system (10.1) is said to be homogeneous if $\boldsymbol{f}(\boldsymbol{x})$ is homogeneous.

Using Definitions 10.1 and 10.2, a fundamental result on global finite-time stability can be obtained as follows:

Lemma 10.1 (Global Finite-Time Stability [33]). *System (10.1) is globally finite-time stable if system (10.1) is globally asymptotically stable and is homogeneous of a negative degree.*

By virtue of Lemma 10.1, we can derive a cornerstone result, whose proof is presented in details in Appendix 10.A, for finite-time observer design and analysis in this chapter.

Lemma 10.2. *The following system:*

$$\begin{aligned}\dot{z}_1 &= z_2 - l_1 \mathrm{sig}^{\alpha_2}(z_1),\\ \dot{z}_2 &= z_3 - l_2 \mathrm{sig}^{\alpha_3}(z_1),\\ &\vdots\\ \dot{z}_n &= -l_n \mathrm{sig}^{\alpha_{n+1}}(z_1),\end{aligned} \tag{10.2}$$

where $l_i > 0, i = 1, 2, \ldots, n$, are appropriate constants, and

$$\mathrm{sig}^{\alpha_i}(z_1) := |z_1|^{\alpha_i}\mathrm{sgn}(z_1) \tag{10.3}$$

with $\alpha_{i+1} = \alpha_i + \tau, i = 1, 2, \ldots, n$ and $\alpha_1 = 1$ for any $\tau < 0$, is globally finite-time stable.

10.2.2 Problem formulation

Let $\boldsymbol{\eta} = [x, y, \psi]^T$ denote the 3-DOF position (x, y) and heading angle (ψ) of the ASV in the earth-fixed inertial frame as shown in Fig. 10.1, and let $\boldsymbol{\nu} = [u, v, r]^T$ denote the corresponding linear velocities (u, v), i.e., surge and sway velocities, and angular rate (r), i.e., yaw, in the body-fixed frame. An ASV sailing in a planar space can be modeled as follows [3]:

$$\begin{aligned}\dot{\boldsymbol{\eta}} &= \mathbf{R}(\psi)\boldsymbol{\nu},\\ \mathbf{M}\dot{\boldsymbol{\nu}} &= \boldsymbol{f}(\boldsymbol{\eta}, \boldsymbol{\nu}) + \boldsymbol{\tau} + \boldsymbol{\tau}_d,\end{aligned} \tag{10.4}$$

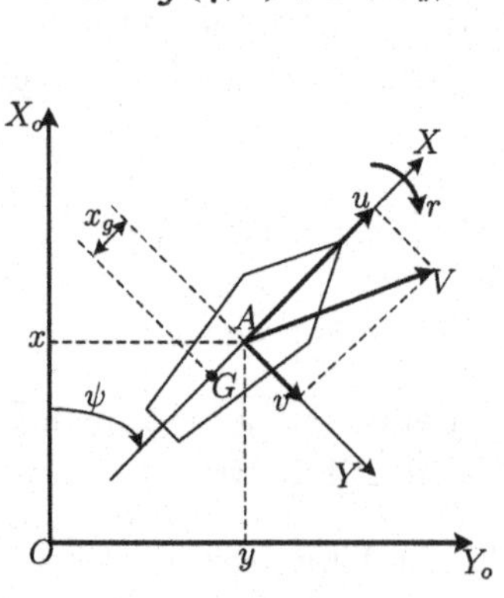

Figure 10.1 Earth-fixed OX_OY_O and body-fixed AXY coordinate frames of an ASV.

with dynamics $\boldsymbol{f}(\boldsymbol{\eta}, \boldsymbol{\nu})$ usually modeled by

$$\boldsymbol{f}(\boldsymbol{\eta}, \boldsymbol{\nu}) = -\mathbf{C}(\boldsymbol{\nu})\boldsymbol{\nu} - \mathbf{D}(\boldsymbol{\nu})\boldsymbol{\nu} - \boldsymbol{g}(\boldsymbol{\eta}, \boldsymbol{\nu}), \tag{10.5}$$

where $\boldsymbol{\tau} = [\tau_1, \tau_2, \tau_3]^T$ and $\boldsymbol{\tau}_d := \mathbf{M}\mathbf{R}^T(\psi)\boldsymbol{d}(t)$ with $\boldsymbol{d}(t) = [d_1(t), d_2(t), d_3(t)]^T$ are control input and mixed disturbances, respectively, and $\boldsymbol{g}$ denotes the restoring forces and moments due to gravitation/buoyancy. The term $\mathbf{R}(\psi)$ is a rotation matrix given by

$$\mathbf{R}(\psi) = \begin{bmatrix} \cos\psi & -\sin\psi & 0 \\ \sin\psi & \cos\psi & 0 \\ 0 & 0 & 1 \end{bmatrix} \tag{10.6}$$

with the following properties:

$$\mathbf{R}^T(\psi)\mathbf{R}(\psi) = \mathbf{I}, \text{ and } \|\mathbf{R}(\psi)\| = 1, \ \forall \psi \in [0, 2\pi], \tag{10.7a}$$

$$\dot{\mathbf{R}}(\psi) = \mathbf{R}(\psi)\mathbf{S}(r), \tag{10.7b}$$

$$\mathbf{R}^T(\psi)\mathbf{S}(r)\mathbf{R}(\psi) = \mathbf{R}(\psi)\mathbf{S}(r)\mathbf{R}^T(\psi) = \mathbf{S}(r), \tag{10.7c}$$

where $\mathbf{S}(r) = \begin{bmatrix} 0 & -r & 0 \\ r & 0 & 0 \\ 0 & 0 & 0 \end{bmatrix}$, the inertia matrix $\mathbf{M} = \mathbf{M}^T > 0$, the skew-symmetric matrix $\mathbf{C}(\boldsymbol{\nu}) = -\mathbf{C}(\boldsymbol{\nu})^T$ and the damping matrix $\mathbf{D}(\boldsymbol{\nu})$ are given by

$$\mathbf{M} = \begin{bmatrix} m_{11} & 0 & 0 \\ 0 & m_{22} & m_{23} \\ 0 & m_{32} & m_{33} \end{bmatrix}, \tag{10.8a}$$

$$\mathbf{C}(\boldsymbol{\nu}) = \begin{bmatrix} 0 & 0 & c_{13}(\boldsymbol{\nu}) \\ 0 & 0 & c_{23}(\boldsymbol{\nu}) \\ -c_{13}(\boldsymbol{\nu}) & -c_{23}(\boldsymbol{\nu}) & 0 \end{bmatrix}, \tag{10.8b}$$

$$\mathbf{D}(\boldsymbol{\nu}) = \begin{bmatrix} d_{11}(\boldsymbol{\nu}) & 0 & 0 \\ 0 & d_{22}(\boldsymbol{\nu}) & d_{23}(\boldsymbol{\nu}) \\ 0 & d_{32}(\boldsymbol{\nu}) & d_{33}(\boldsymbol{\nu}) \end{bmatrix}, \tag{10.8c}$$

where $m_{11} = m - X_{\dot{u}}$, $m_{22} = m - Y_{\dot{v}}$, $m_{23} = mx_g - Y_{\dot{r}}$, $m_{32} = mx_g - N_{\dot{v}}$, $m_{33} = I_z - N_{\dot{r}}$; $c_{13}(\boldsymbol{\nu}) = -m_{11}v - m_{23}r$, $c_{23}(\boldsymbol{\nu}) = m_{11}u$; $d_{11}(\boldsymbol{\nu}) = -X_u - X_{|u|u}|u| - X_{uuu}u^2$, $d_{22}(\boldsymbol{\nu}) = -Y_v - Y_{|v|v}|v|$, $d_{23}(\boldsymbol{\nu}) = -Y_r - Y_{|v|r}|v| - Y_{|r|r}|r|$, $d_{32}(\boldsymbol{\nu}) = -N_v - N_{|v|v}|v| - N_{|r|v}|r|$, and $d_{33}(\boldsymbol{\nu}) = -N_r - N_{|v|r}|v| - N_{|r|r}|r|$. Here, m is the mass of the vessel, I_z is the moment of inertia about the yaw rotation, $Y_{\dot{r}} = N_{\dot{v}}$, and X_*, Y_*, and N_* denote corresponding hydrodynamic derivatives which are actually difficult to be accurately obtained.

Consider the following desired trajectory:

$$\begin{aligned} \dot{\boldsymbol{\eta}}_d &= \mathbf{R}(\psi_d)\,\boldsymbol{\nu}_d, \\ \mathbf{M}\dot{\boldsymbol{\nu}}_d &= \boldsymbol{f}_0(\boldsymbol{\eta}_d, \boldsymbol{\nu}_d), \end{aligned} \tag{10.9}$$

where $\boldsymbol{f}_0(\cdot)$ is the nominal dynamics, $\boldsymbol{\eta}_d = [x_d, y_d, \psi_d]^T$, and $\boldsymbol{\nu}_d = [u_d, v_d, r_d]^T$ are the desired position and velocity vectors, respectively.

In this context, the control objective is to design a controller $\boldsymbol{\tau}$ such that the actual position and velocity vectors (i.e., $\boldsymbol{\eta}$ and $\boldsymbol{\nu}$) of the ASV in (10.4) can track exactly the desired trajectory (i.e., $\boldsymbol{\eta}_d$ and $\boldsymbol{\nu}_d$) generated by (10.9) in a finite time.

Remark 10.1. Clearly, in addition to unknown disturbances $\boldsymbol{\tau}_d$, if the dynamics $\boldsymbol{f}$ of the ASV in (10.4) cannot be sufficiently modeled due to parametric unknowns including $\mathbf{C}$, $\mathbf{D}$, and $\boldsymbol{g}$, and/or structural unmodeled dynamics, accurate trajectory tracking control of an ASV under harsh environments would become extremely challenging.

10.3. Homogeneity-based finite-time tracking control scheme

10.3.1 Nominal homogeneity-based finite-time control

In order to facilitate controller design and analysis, we introduce an auxiliary velocity vector as follows:

$$\boldsymbol{w} = \mathbf{R}\boldsymbol{\nu}, \tag{10.10a}$$

$$\boldsymbol{w}_d = \mathbf{R}(\psi_d)\boldsymbol{\nu}_d, \tag{10.10b}$$

where $\boldsymbol{w} = [w_1, w_2, w_3]^T$, $\boldsymbol{w}_d = [w_{d,1}, w_{d,2}, w_{d,3}]^T$, $\mathbf{R} = \mathbf{R}(\psi)$, and $\mathbf{R}_d = \mathbf{R}(\psi_d)$.

Together with (10.4) and (10.10a), using properties in (10.7), we have

$$\begin{aligned} \dot{\boldsymbol{\eta}} &= \boldsymbol{w}, \\ \dot{\boldsymbol{w}} &= \mathbf{R}\mathbf{M}^{-1}\boldsymbol{\tau} + \boldsymbol{h}(\boldsymbol{\eta}, \boldsymbol{w}) + \boldsymbol{d}(t), \end{aligned} \tag{10.11}$$

where

$$\boldsymbol{h}(\boldsymbol{\eta}, \boldsymbol{w}) = \mathbf{S}(w_3)\boldsymbol{w} + \mathbf{R}\mathbf{M}^{-1}\boldsymbol{f}(\boldsymbol{\eta}, \mathbf{R}^T\boldsymbol{w}). \tag{10.12}$$

Similarly, using (10.9) and (10.10b), we have

$$\dot{\boldsymbol{\eta}}_d = \boldsymbol{w}_d,$$
$$\dot{\boldsymbol{w}}_d = \mathbf{S}(w_{d,3})\boldsymbol{w}_d + \mathbf{R}_d\mathbf{M}^{-1}\boldsymbol{f}_0\left(\boldsymbol{\eta}_d, \mathbf{R}_d^T\boldsymbol{w}_d\right). \tag{10.13}$$

Combining with (10.11) and (10.13), we have

$$\dot{\boldsymbol{\eta}}_e = \boldsymbol{w}_e,$$
$$\dot{\boldsymbol{w}}_e = \mathbf{RM}^{-1}\boldsymbol{\tau} + \boldsymbol{h}_e(\boldsymbol{\eta}, \boldsymbol{w}, \boldsymbol{\eta}_d, \boldsymbol{w}_d) + \boldsymbol{d}(t), \tag{10.14}$$

where $\boldsymbol{\eta}_e = \boldsymbol{\eta} - \boldsymbol{\eta}_d := [\eta_{e,1}, \eta_{e,2}, \eta_{e,3}]^T$, $\boldsymbol{w}_e = \boldsymbol{w} - \boldsymbol{w}_d := [w_{e,1}, w_{e,2}, w_{e,3}]^T$, and

$$\begin{aligned}\boldsymbol{h}_e(\boldsymbol{\eta}, \boldsymbol{w}, \boldsymbol{\eta}_d, \boldsymbol{w}_d) = {} & \mathbf{S}\boldsymbol{w} - \mathbf{S}_d\boldsymbol{w}_d + \mathbf{RM}^{-1}\boldsymbol{f}(\boldsymbol{\eta}, \mathbf{R}^T\boldsymbol{w}) \\ & - \mathbf{R}_d\mathbf{M}^{-1}\boldsymbol{f}_0(\boldsymbol{\eta}_d, \mathbf{R}_d^T\boldsymbol{w}_d),\end{aligned} \tag{10.15}$$

with $\mathbf{S} = \mathbf{S}(w_3)$ and $\mathbf{S}_d = \mathbf{S}(w_{d,3})$.

Starting from the tracking error dynamics (10.14), we set out to design a nominal homogeneity-based finite-time control (HFC) scheme which is expected to ensure that the tracking errors $\boldsymbol{\eta}_e$ and $\boldsymbol{w}_e$ converge to zero in a finite time.

To this end, a nominal HFC scheme for the ASV in (10.4) without disturbances (i.e., $\boldsymbol{\tau}_d = 0$ or $\boldsymbol{d}(t) = 0$) is developed by employing the homogeneous theory. Moreover, we show that finite-time stability of the entire closed-loop tracking system can be ensured by using the Lyapunov synthesis.

Design the nominal HFC law $\boldsymbol{\tau}_{\text{HFC}}$ as follows:

$$\begin{aligned}\boldsymbol{\tau}_{\text{HFC}} = {} & -\mathbf{MR}^{-1}\left(K_1\text{sig}^{\beta_1}(\boldsymbol{\eta} - \boldsymbol{\eta}_d) + K_2\text{sig}^{\beta_2}(\mathbf{R}\boldsymbol{\nu} - \mathbf{R}_d\boldsymbol{\nu}_d)\right) \\ & - \mathbf{MS}\boldsymbol{\nu} + \mathbf{MR}^{-1}\mathbf{S}_d\mathbf{R}_d\boldsymbol{\nu}_d \\ & -\boldsymbol{f}(\boldsymbol{\eta}, \boldsymbol{\nu}) + \mathbf{MR}^{-1}\mathbf{R}_d\mathbf{M}^{-1}\boldsymbol{f}_0(\boldsymbol{\eta}_d, \boldsymbol{\nu}_d),\end{aligned} \tag{10.16}$$

where $\text{sig}^{\beta_i}(\boldsymbol{x}) = \left[\text{sig}^{\beta_i}(x_1), \ldots, \text{sig}^{\beta_i}(x_n)\right]^T, i = 1, 2,\ K_1 > 0,\ K_2 > 0,\ 0 < \beta_1 < 1$, and $\beta_2 = 2\beta_1/(1 + \beta_1)$.

It is essential that the proposed HFC scheme can make the ASV in (10.4) track exactly the desired trajectory generated by (10.9) in a finite time. The key result ensuring the closed-loop finite-time stability is now stated.

Theorem 10.1 (HFC). *Using the HFC scheme governed by (10.16), the ASV in (10.4) can exactly track the desired trajectory generated by (10.9) within a finite time* $0 < T < \infty$, *i.e.,* $\boldsymbol{\eta}(t) \equiv \boldsymbol{\eta}_d(t), \boldsymbol{\nu}(t) \equiv \boldsymbol{\nu}_d(t), \forall\, t \geq T$.

Proof. Substituting the HFC law (10.16) into the tracking error system (10.14) without considering disturbances yields the closed-loop tracking error dynamics as follows:

$$\begin{aligned}\dot{\eta}_{e,j} &= w_{e,j},\\ \dot{w}_{e,j} &= -K_1\text{sig}^{\beta_1}(\eta_{e,j}) - K_2\text{sig}^{\beta_2}(w_{e,j}),\end{aligned} \tag{10.17}$$

for $j = 1, 2, 3$.

In light of Lemma 10.1, global asymptotic stability and negative homogeneity of system (10.17) are expected to be guaranteed respectively in the sequel.

1) Global Asymptotic Stability. Consider the following Lyapunov function:

$$V(\boldsymbol{\eta}_e, \boldsymbol{w}_e) = \sum_{j=1}^{3}\left(K_1\int_0^{\eta_{e,j}}\text{sig}^{\beta_1}(\mu)d\mu + \frac{1}{2}w_{e,j}^2\right) \tag{10.18}$$

Differentiating $V(\boldsymbol{\eta}_e, \boldsymbol{w}_e)$ along the tracking error dynamics (10.17), we have

$$\begin{aligned}\dot{V}(\boldsymbol{\eta}_e, \boldsymbol{w}_e) &= K_1\sum_{j=1}^{3}\text{sig}^{\beta_1}(\eta_{e,j})w_{e,j}\\ &\quad - \sum_{j=1}^{3}w_{e,j}\left(K_1\text{sig}^{\beta_1}(\eta_{e,j}) + K_2\text{sig}^{\beta_2}(w_{e,j})\right)\\ &= -K_2\sum_{j=1}^{3}w_{e,j}\text{sig}^{\beta_2}(w_{e,j})\\ &= -K_2\sum_{j=1}^{3}|w_{e,j}|^{1+\beta_2}\end{aligned} \tag{10.19}$$

which yields that $V(t)$ is bounded as time t tends to infinity, i.e.,

$$\frac{1}{2}\|\boldsymbol{w}_e(t)\|^2 \leq V(t) < \infty. \tag{10.20}$$

Using $\|\boldsymbol{w}_e(t)\|^{1+\beta_2} \leq \|\boldsymbol{w}_e(t)\|^2 + 1$, we further have

$$\left\|\boldsymbol{w}_e^{(1+\beta_2)/2}(t)\right\| \leq \sqrt{2V(t)+1} < \infty. \tag{10.21}$$

Note, from (10.19), that

$$\int_0^t \left\| \boldsymbol{w}_e^{(1+\beta_2)/2}(\tau) \right\|^2 d\tau = \frac{V(0) - V(t)}{K_2} < \infty. \tag{10.22}$$

Combining with (10.21) and (10.22) and using Barbalat's lemma [40] yields

$$\lim_{t \to \infty} \boldsymbol{w}_e(t) = 0. \tag{10.23}$$

In what follows, we expect to prove that $\boldsymbol{\eta}_e(t)$ also converges to zero as time t tends to infinity. To this end, a proof by contradiction is employed by assuming that $\boldsymbol{\eta}_e(t)$ converges to a nonzero constant $\boldsymbol{\eta}_0 \neq 0$.

Together with (10.17) and (10.23), we have, as $t \to \infty$,

$$\begin{aligned} \dot{\eta}_{e,j} &= 0, \\ \dot{w}_{e,j} &= -K_1 \mathrm{sig}^{\beta_1}(\eta_{e,j}), \end{aligned} \tag{10.24}$$

which, combining with the hypothesis, implies that $\eta_{e,j}$ converges to a nonzero constant $\eta_{0,j} \neq 0$, and thereby $\dot{w}_{e,j} = -K_1 \mathrm{sig}^{\beta_1}(\eta_{e,j}) \neq 0$. In this context, $w_{e,j}$ deviates from the origin and makes $\dot{\eta}_{e,j} \neq 0$, and thereby resulting in a new convergent constant $\bar{\eta}_{0,j}$ different from the assumed one, i.e., $\bar{\eta}_{0,j} \neq \eta_{0,j}$. This leads to a contradiction and thus yields

$$\lim_{t \to \infty} \boldsymbol{\eta}_e(t) = 0. \tag{10.25}$$

It follows from (10.23) and (10.25) that using the HFC law in (10.16), system (10.14) without disturbances (i.e., $\boldsymbol{d}(t) = 0$) is globally asymptotically stable.

2) Negative Homogeneity. For system (10.17), selecting a dilation as follows:

$$(r_1, r_2) = (1, \frac{1+\beta_1}{2}), \tag{10.26}$$

for any given $\varepsilon > 0$, yields

$$\begin{aligned} f_1(\varepsilon^{r_1}\eta_{e,j}, \varepsilon^{r_2} w_{e,j}) &= \varepsilon^{\sigma + r_1} f_1(\eta_{e,j}, w_{e,j}), \\ f_2(\varepsilon^{r_1}\eta_{e,j}, \varepsilon^{r_2} w_{e,j}) &= \varepsilon^{\sigma + r_2} f_2(\eta_{e,j}, w_{e,j}), \end{aligned} \tag{10.27}$$

with $f_1(\cdot) = w_{e,j}$, $f_2 = -K_1 \mathrm{sig}^{\beta_1}(\eta_{e,j}) - K_2 \mathrm{sig}^{\beta_2}(w_{e,j})$, and a negative degree of homogeneous with respect to the dilation in (10.26), i.e., $\sigma = (\beta_1 - 1)/2 < 0$.

By Lemma 10.1, we can conclude that the tracking error system (10.14) without disturbances controlled by the HFC scheme (10.16) is globally finite-time stable, i.e., there exists a finite time $0 < T < \infty$ such that

$$\boldsymbol{\eta}_e(t) \equiv 0, \ \boldsymbol{w}_e(t) \equiv 0, \ \forall\, t \geq T. \tag{10.28}$$

Together with (10.10), we further have

$$\boldsymbol{\eta}_e(t) \equiv 0, \ \boldsymbol{\nu}_e(t) \equiv 0, \ \forall\, t \geq T. \tag{10.29}$$

This concludes the proof. ■

Remark 10.2. If the powers β_1 and β_2 of the HFC scheme in (10.16) are set as $\beta_1 = \beta_2 = 1$, the closed-loop system composed of the tracking error system (10.14) without disturbances and the HFC law (10.16) degrades to be globally asymptotically stable, i.e.,

$$\dot{\eta}_{e,j} = w_{e,j}, \tag{10.30}$$

$$\dot{w}_{e,j} = -K_1\eta_{e,j} - K_2 w_{e,j}, \tag{10.31}$$

which can be derived easily from the conventional backstepping technique.

Remark 10.3. Note that the external disturbance $\boldsymbol{d}(t)$ in (10.14) is not addressed in the nominal HFC scheme. In this context, within the HFC framework, the disturbance observer is expected to be developed for enhancing the robustness and even achieving exact disturbance rejection.

10.3.2 Finite-time disturbance observer based HFC

In this subsection, the finite-time disturbance observer based HFC (FDO-HFC) scheme is proposed. To this end, a generic assumption on the disturbance $\boldsymbol{d}(t)$ is required as follows:

Assumption 10.1. The external time-varying disturbance $\boldsymbol{d}(t)$ in (10.11) satisfies

$$\boldsymbol{d}^{(n)}(t) = \sum_{i=0}^{n-1} \mathbf{H}_{n-i}\boldsymbol{d}^{(i)}(t), \ \ i = 0, 1, \ldots, n-1, \tag{10.32}$$

where n is a positive integer and $\mathbf{H}_i = \mathrm{diag}(h_{i,1}, h_{i,2}, h_{i,3})$ with any constants $h_{i,j} \in \mathbb{R}, j = 1, 2, 3.$ □

The key result pertaining to the FDO-HFC scheme is now summarized as follows:

Theorem 10.2 (FDO-HFC). *Consider the ASV in (10.11) with unknown external disturbances $\boldsymbol{d}(t)$ satisfying Assumption 1, an FDO-HFC scheme designed as follows:*

$$\begin{aligned}\boldsymbol{\tau}_{\mathrm{FDO}} =& -\mathbf{M}\mathbf{R}^{-1}\big(K_1\mathrm{sig}^{\beta_1}(\boldsymbol{\eta}-\boldsymbol{\eta}_d)+K_2\mathrm{sig}^{\beta_2}(\mathbf{R}\boldsymbol{\nu}-\mathbf{R}_d\boldsymbol{\nu}_d)\big)\\ &-\mathbf{M}\mathbf{S}\boldsymbol{\nu}+\mathbf{M}\mathbf{R}^{-1}\mathbf{S}_d\mathbf{R}_d\boldsymbol{\nu}_d\\ &-\boldsymbol{f}(\boldsymbol{\eta},\boldsymbol{\nu})+\mathbf{M}\mathbf{R}^{-1}\mathbf{R}_d\mathbf{M}^{-1}\boldsymbol{f}_0(\boldsymbol{\eta}_d,\boldsymbol{\nu}_d)-\mathbf{M}\mathbf{R}^{-1}\widehat{\boldsymbol{d}},\end{aligned} \tag{10.33}$$

with the FDO governed by

$$\widehat{\boldsymbol{d}} = \widehat{\boldsymbol{p}}_1 + \mathbf{H}_1\widehat{\boldsymbol{p}}_0, \tag{10.34}$$

where $\widehat{\boldsymbol{p}}_1$ and $\widehat{\boldsymbol{p}}_0$ are derived by

$$\begin{aligned}\dot{\widehat{\boldsymbol{p}}}_0 &= \widehat{\boldsymbol{p}}_1 + \mathbf{H}_1\boldsymbol{p}_0 + \boldsymbol{u} + \mathbf{L}_0\mathrm{sig}^{\alpha_1}(\boldsymbol{p}_0-\widehat{\boldsymbol{p}}_0),\\ &\vdots\\ \dot{\widehat{\boldsymbol{p}}}_{n-1} &= \widehat{\boldsymbol{p}}_n + \mathbf{H}_n\boldsymbol{p}_0 - \mathbf{H}_{n-1}\boldsymbol{u} + \mathbf{L}_{n-1}\mathrm{sig}^{\alpha_n}(\boldsymbol{p}_0-\widehat{\boldsymbol{p}}_0),\\ \dot{\widehat{\boldsymbol{p}}}_n &= -\mathbf{H}_n\boldsymbol{u} + \mathbf{L}_n\mathrm{sig}^{\alpha_{n+1}}(\boldsymbol{p}_0-\widehat{\boldsymbol{p}}_0),\end{aligned} \tag{10.35}$$

with

$$\boldsymbol{p}_0 = \boldsymbol{w}_e,\ \boldsymbol{u} = \mathbf{R}\mathbf{M}^{-1}\boldsymbol{\tau} + \boldsymbol{h}_e \tag{10.36}$$

and $\mathbf{L}_i = \mathrm{diag}(l_{i,1}, l_{i,2}, l_{i,3})$, $i = 0, 1, \ldots, n$, $\alpha_i = 1 + i\vartheta$ with $-1/(n+1) < \vartheta < 0$, and $\vartheta = -q_1/q_2$, with q_1 and q_2 being positive even and odd integers, can render the ASV in (10.4) exactly track the desired trajectory generated by (10.9) within a short time $0 < T < \infty$, i.e., $\boldsymbol{\eta}(t) \equiv \boldsymbol{\eta}_d(t), \boldsymbol{\nu}(t) \equiv \boldsymbol{\nu}_d(t), \forall\, t \geq T$.

Proof. In order to examine finite-time stability of the closed-loop system (10.14) and (10.33) including an FDO (10.35), we need to obtain the disturbance observation error dynamics. To this end, we define auxiliary variables as follows:

$$\boldsymbol{\varepsilon}_0 = \boldsymbol{w}_e,\ \boldsymbol{\varepsilon}_1 = \boldsymbol{d}(t),\ \boldsymbol{\varepsilon}_2 = \dot{\boldsymbol{d}}(t),\ \ldots,\ \boldsymbol{\varepsilon}_n = \boldsymbol{d}^{(n-1)}(t). \tag{10.37}$$

Together with (10.14) and (10.32), we thus have

$$\dot{\boldsymbol{\varepsilon}}_0 = \boldsymbol{\varepsilon}_1 + \boldsymbol{u},$$

$$\dot{\boldsymbol{\varepsilon}}_i = \boldsymbol{\varepsilon}_{i+1}, \quad i = 1, 2, \ldots, n-1,$$
$$\dot{\boldsymbol{\varepsilon}}_n = \mathbf{H}_n \boldsymbol{\varepsilon}_1 + \mathbf{H}_{n-1} \boldsymbol{\varepsilon}_2 + \cdots + \mathbf{H}_1 \boldsymbol{\varepsilon}_n. \tag{10.38}$$

Considering a coordinate transformation given by

$$\begin{aligned}
\boldsymbol{p}_0 &= \boldsymbol{\varepsilon}_0, \\
\boldsymbol{p}_1 &= \boldsymbol{\varepsilon}_1 - \mathbf{H}_1 \boldsymbol{\varepsilon}_0, \\
&\vdots \\
\boldsymbol{p}_n &= \boldsymbol{\varepsilon}_n - \mathbf{H}_1 \boldsymbol{\varepsilon}_{n-1} - \cdots - \mathbf{H}_n \boldsymbol{\varepsilon}_0,
\end{aligned} \tag{10.39}$$

we further have

$$\begin{aligned}
\dot{\boldsymbol{p}}_0 &= \boldsymbol{p}_1 + \mathbf{H}_1 \boldsymbol{p}_0 + \boldsymbol{u}, \\
&\vdots \\
\dot{\boldsymbol{p}}_{n-1} &= \boldsymbol{p}_n + \mathbf{H}_n \boldsymbol{p}_0 - \mathbf{H}_{n-1} \boldsymbol{u}, \\
\dot{\boldsymbol{p}}_n &= -\mathbf{H}_n \boldsymbol{u}.
\end{aligned} \tag{10.40}$$

Together with (10.35) and (10.40), the disturbance observation error dynamics can be derived as follows:

$$\begin{aligned}
\dot{\boldsymbol{p}}_{e,0} &= \boldsymbol{p}_{e,1} - \mathbf{L}_0 \mathrm{sig}^{\alpha_1}(\boldsymbol{p}_{e,0}), \\
&\vdots \\
\dot{\boldsymbol{p}}_{e,n-1} &= \boldsymbol{p}_{e,n} - \mathbf{L}_{n-1} \mathrm{sig}^{\alpha_n}(\boldsymbol{p}_{e,0}), \\
\dot{\boldsymbol{p}}_{e,n} &= -\mathbf{L}_n \mathrm{sig}^{\alpha_{n+1}}(\boldsymbol{p}_{e,0}),
\end{aligned} \tag{10.41}$$

where $\boldsymbol{p}_{e,i} = \boldsymbol{p}_i - \widehat{\boldsymbol{p}}_i := [p_{e,i}^1, p_{e,i}^2, p_{e,i}^3]^T$, $i = 0, 1, \ldots, n$, and

$$\begin{aligned}
\dot{p}_{e,0}^j &= p_{e,1}^j - l_{0,j} \mathrm{sig}^{\alpha_1}(p_{e,0}^j), \\
&\vdots \\
\dot{p}_{e,n-1}^j &= p_{e,n}^j - l_{n-1,j} \mathrm{sig}^{\alpha_n}(p_{e,0}^j), \\
\dot{p}_{e,n}^j &= -l_{n,j} \mathrm{sig}^{\alpha_{n+1}}(p_{e,0}^j), \quad j = 1, 2, 3.
\end{aligned} \tag{10.42}$$

Applying Lemma 10.2 to system (10.42), we can conclude that the disturbance observation errors $\boldsymbol{p}_{e,i}$, $i = 0, 1, \ldots, n$ are globally finite-time stable, i.e., the FDO in (10.35) can exactly observe the dynamics in (10.40) within a finite time $0 < T < \infty$. Together with (10.34), (10.37), and (10.39), we can immediately obtain that $\widehat{\boldsymbol{p}}_0$ and $\widehat{\boldsymbol{p}}_1$ can exactly estimate $\boldsymbol{w}_e$ and $\boldsymbol{d}-\mathbf{H}_1\boldsymbol{w}_e$, respectively, and $\widehat{\boldsymbol{d}}$ governed by (10.34) can thus exactly observe the disturbance $\boldsymbol{d}$ in a finite time. Actually, the derivatives $\boldsymbol{d}^{(i)}(t), i=1,2,\ldots,n-1$ can also be exactly observed within a finite time by $\widehat{\boldsymbol{p}}_{i+1}+\mathbf{H}_1\widehat{\boldsymbol{p}}_i+\cdots+\mathbf{H}_{i+1}\widehat{\boldsymbol{p}}_0$.

In this context, we eventually have

$$\widehat{\boldsymbol{d}}^{(i)}(t)\equiv\boldsymbol{d}^{(i)}(t),\ \forall\, t>T,\quad i=0,1,\ldots,n-1, \tag{10.43}$$

with $\widehat{\boldsymbol{d}}^{(i)}=\widehat{\boldsymbol{p}}_{i+1}+\mathbf{H}_1\widehat{\boldsymbol{p}}_i+\cdots+\mathbf{H}_{i+1}\widehat{\boldsymbol{p}}_0$.

Substituting (10.33) together with (10.34) and (10.43) into (10.14) yields

$$\begin{aligned}\dot{\eta}_{e,j}&=w_{e,j},\\ \dot{w}_{e,j}&=-K_1\mathrm{sig}^{\beta_1}(\eta_{e,j})-K_2\mathrm{sig}^{\beta_2}(w_{e,j})+\tilde{d}_j,\end{aligned} \tag{10.44}$$

where $\tilde{d}_j=d_j-\widehat{d}_j$.

Together with (10.43) and (10.44), we further have

$$\begin{aligned}\dot{\eta}_{e,j}(t)&=w_{e,j}(t),\\ \dot{w}_{e,j}(t)&=-K_1\mathrm{sig}^{\beta_1}(\eta_{e,j}(t))-K_2\mathrm{sig}^{\beta_2}(w_{e,j}(t)),\end{aligned} \tag{10.45}$$

for any $t>T$ with a finite time $0<T<\infty$.

In what follows, similar to the proof of Theorem 10.1, global finite-time stability of the closed-loop system (10.45) can be ensured. As a consequence, in the presence of complex disturbances, using the FDO-HFC in (10.33), the ASV in (10.4) can exactly track the desired trajectory $(\boldsymbol{\eta}_d,\boldsymbol{\nu}_d)$ generated by (10.9) in a finite time. This concludes the proof. ■

Remark 10.4. In addition to disturbances $\boldsymbol{\tau}_d$, the dynamics $\boldsymbol{h}_e$ given by (10.15) might be at least partially unknown due to nonlinearities $\boldsymbol{f}(\cdot)$ and $\boldsymbol{f}_0(\cdot)$, and will make the foregoing HFC in (10.16) and FDO-HFC in (10.33) unavailable. In this context, the observer for accurate estimate on mixed unknowns including not only unmodeled dynamics but also external disturbances is required to be a patch within the HFC framework.

10.3.3 Finite-time unknown observer based HFC

Rewrite the tracking error dynamics in (10.14) as follows:

$$\dot{\boldsymbol{\eta}}_e = \boldsymbol{w}_e,$$
$$\dot{\boldsymbol{w}}_e = \mathbf{R}\mathbf{M}^{-1}\boldsymbol{\tau} + \mathbf{S}\boldsymbol{w} - \mathbf{S}_d\boldsymbol{w}_d + \boldsymbol{f}_u(\boldsymbol{\eta}, \boldsymbol{w}, \boldsymbol{\eta}_d, \boldsymbol{w}_d, t), \tag{10.46}$$

with the lumped unknowns $\boldsymbol{f}_u$ including unmodeled dynamics $\boldsymbol{f}$, desired dynamics $\boldsymbol{f}_0$, and disturbances $\boldsymbol{d}$, i.e.,

$$\boldsymbol{f}_u(\boldsymbol{\eta}, \boldsymbol{w}, \boldsymbol{\eta}_d, \boldsymbol{w}_d, t) = \mathbf{R}\mathbf{M}^{-1}\boldsymbol{f}(\boldsymbol{\eta}, \mathbf{R}^T\boldsymbol{w}) - \mathbf{R}_d\mathbf{M}^{-1}\boldsymbol{f}_0(\boldsymbol{\eta}_d, \mathbf{R}_d^T\boldsymbol{w}_d) + \boldsymbol{d}(t). \tag{10.47}$$

From (10.47), it reasonably requires to assume that dynamics $\boldsymbol{f}, \boldsymbol{f}_0$, and $\boldsymbol{d}$ are twice differentiable, and thereby contributing to the following hypothesis:

$$\|\ddot{\boldsymbol{f}}_u\| \leq L_u \tag{10.48}$$

for a bounded constant $L_u < \infty$.

In this context, a finite-time unknown observer based HFC (FUO-HFC) scheme will be proposed to accurately track the ASV in (10.4) with complex unknowns including both unmodeled dynamics and disturbances to a desired trajectory with completely unknown dynamics. This challenging problem will be solved by the proposed FUO-HFC scheme with global finite-time stability presented as follows.

Theorem 10.3 (FUO-HFC). *Considering the ASV in (10.4) with unmodeled dynamics $\boldsymbol{f}$ and unknown external disturbances $\boldsymbol{d}(t)$, an FUO-HFC scheme designed as follows:*

$$\boldsymbol{\tau}_{\text{FUO}} = -\mathbf{M}\mathbf{R}^{-1}\Big(K_1\text{sig}^{\beta_1}(\boldsymbol{\eta} - \boldsymbol{\eta}_d) + K_2\text{sig}^{\beta_2}(\mathbf{R}\boldsymbol{\nu} - \mathbf{R}_d\boldsymbol{\nu}_d) + \boldsymbol{z}_1\Big), \tag{10.49}$$

with $\boldsymbol{z}_1$ estimated by the following FUO:

$$\dot{\boldsymbol{z}}_0 = \boldsymbol{\zeta}_0 + \mathbf{R}\mathbf{M}^{-1}\boldsymbol{\tau} + \mathbf{S}\boldsymbol{w} - \mathbf{S}_d\boldsymbol{w}_d,$$
$$\boldsymbol{\zeta}_0 = -\lambda_1\boldsymbol{\mathcal{L}}^{1/3}\text{sig}^{2/3}(\boldsymbol{z}_0 - \boldsymbol{w}_e) + \boldsymbol{z}_1,$$
$$\dot{\boldsymbol{z}}_1 = \boldsymbol{\zeta}_1,$$
$$\boldsymbol{\zeta}_1 = -\lambda_2\boldsymbol{\mathcal{L}}^{1/2}\text{sig}^{1/2}(\boldsymbol{z}_1 - \boldsymbol{\zeta}_0) + \boldsymbol{z}_2,$$
$$\dot{\boldsymbol{z}}_2 = -\lambda_3\boldsymbol{\mathcal{L}}\text{sgn}(\boldsymbol{z}_2 - \boldsymbol{\zeta}_1), \tag{10.50}$$

where $\boldsymbol{z}_j := [z_{j,1}, z_{j,2}, z_{j,3}]^T, j = 0, 1, 2,$ $\boldsymbol{\zeta}_k := [\zeta_{k,1}, \zeta_{k,2}, \zeta_{k,3}]^T, k = 0, 1,$ $\lambda_i > 0, i = 1, 2, 3,$ and $\boldsymbol{\mathcal{L}} = \text{diag}(\ell_1, \ell_2, \ell_3)$, can render the ASV exactly track the *desired trajectory generated by (10.9) with completely unknown dynamics* $\boldsymbol{f}_0$ *in a short time* $0 < T < \infty$*, i.e.,* $\boldsymbol{\eta}(t) \equiv \boldsymbol{\eta}_d(t), \boldsymbol{\nu}(t) \equiv \boldsymbol{\nu}_d(t), \forall t \geq T$.

Proof. Define the FUO observation errors as follows:

$$\boldsymbol{e}_1 = \boldsymbol{z}_0 - \boldsymbol{w}_e,\ \boldsymbol{e}_2 = \boldsymbol{z}_1 - \boldsymbol{f}_u,\ \boldsymbol{e}_3 = \boldsymbol{z}_2 - \dot{\boldsymbol{f}}_u. \tag{10.51}$$

Combining with (10.46) and (10.50), we have the FUO observation error dynamics as follows:

$$\begin{aligned}
\dot{\boldsymbol{e}}_1 &= -\lambda_1 \boldsymbol{\mathcal{L}}^{1/3} \mathrm{sig}^{2/3}(\boldsymbol{e}_1) + \boldsymbol{e}_2, \\
\dot{\boldsymbol{e}}_2 &= -\lambda_2 \boldsymbol{\mathcal{L}}^{1/2} \mathrm{sig}^{1/2}(\boldsymbol{e}_2 - \dot{\boldsymbol{e}}_1) + \boldsymbol{e}_3, \\
\dot{\boldsymbol{e}}_3 &= -\lambda_3 \boldsymbol{\mathcal{L}} \mathrm{sgn}(\boldsymbol{e}_3 - \dot{\boldsymbol{e}}_2) - \ddot{\boldsymbol{f}}_u,
\end{aligned} \tag{10.52}$$

i.e.,

$$\begin{aligned}
\dot{e}_{1,j} &= -\lambda_1 \ell_j^{1/3} \mathrm{sig}^{2/3}(e_{1,j}) + e_{2,j}, \\
\dot{e}_{2,j} &= -\lambda_2 \ell_j^{1/2} \mathrm{sig}^{1/2}(e_{2,j} - \dot{e}_{1,j}) + e_{3,j}, \\
\dot{e}_{3,j} &\in -\lambda_3 \ell_j \mathrm{sgn}(e_{3,j} - \dot{e}_{2,j}) + [-L_u, L_u],
\end{aligned} \tag{10.53}$$

where "$\in$" denotes the differential inclusion understood in the Filippov sense [41].

According to [42, Lemma 2], and from its detailed proof, we can immediately conclude that the tracking error dynamics in (10.53) is globally finite-time stable, i.e., there exists a finite time $0 < T < \infty$ such that

$$\boldsymbol{z}_0(t) \equiv \boldsymbol{w}_e(t), \boldsymbol{z}_1(t) \equiv \boldsymbol{f}_u(t), \boldsymbol{z}_2(t) \equiv \dot{\boldsymbol{f}}_u(t),\ \forall t > T. \tag{10.54}$$

Substituting (10.49) into (10.46) and using (10.54) yields

$$\begin{aligned}
\dot{\boldsymbol{\eta}}_e(t) &= \boldsymbol{w}_e(t), \\
\dot{\boldsymbol{w}}_e(t) &= -K_1 \mathrm{sig}^{\beta_1}(\boldsymbol{\eta}_e(t)) - K_2 \mathrm{sig}^{\beta_2}(\boldsymbol{w}_e(t)),\ \forall t > T,
\end{aligned} \tag{10.55}$$

which has been proven to be globally finite-time stable in Theorem 10.1. In this context, the proposed FUO-HFC in (10.49) renders the ASV in (10.4) with lumped unmodeled dynamics and disturbances can exactly track the desired trajectory $(\boldsymbol{\eta}_d, \boldsymbol{\nu}_d)$ generated by (10.9) with completely unknown dynamics in a finite time. This concludes the proof. ∎

Table 10.1 Main parameters of CyberShip II.

m	23.8000	Y_v	−0.8612	$X_{\dot{u}}$	−2.0
I_z	1.7600	$Y_{\lvert v\rvert v}$	−36.2823	$Y_{\dot{v}}$	−10.0
x_g	0.0460	Y_r	0.1079	$Y_{\dot{r}}$	−0.0
X_u	−0.7225	N_v	0.1052	$N_{\dot{v}}$	−0.0
$X_{\lvert u\rvert u}$	−1.3274	$N_{\lvert v\rvert v}$	5.0437	$N_{\dot{r}}$	−1.0
X_{uuu}	−5.8664				

Remark 10.5. It should be noted that in addition to unmodeled dynamics and external disturbances pertaining to the ASV, the desired dynamics of the trajectory to be tracked are also not necessarily known for the FUO-HFC scheme. It implies that the proposed FUO-HFC approach is completely independent on dynamics of the ASV and the desired trajectory, and thereby contributing to a both task-free and model-free methodology.

10.4. Simulation studies and discussions

In order to demonstrate the effectiveness and superiority of the proposed control schemes for trajectory tracking control of an ASV, simulation studies and comprehensive comparisons are conducted on a well-known surface vehicle CyberShip II [43] of which the main parameters are listed in Table 10.1.

Our objective is to track exactly the desired trajectory $(\boldsymbol{\eta}_d, \boldsymbol{\nu}_d)$ governed by (10.9) with assumed dynamics $\boldsymbol{f}_0(\boldsymbol{\eta}_d, \boldsymbol{\nu}_d) = -\mathbf{C}(\boldsymbol{\nu}_d)\boldsymbol{\nu}_d - \mathbf{D}(\boldsymbol{\nu}_d)\boldsymbol{\nu}_d - \mathbf{g}(\boldsymbol{\eta}_d, \boldsymbol{\nu}_d) + \boldsymbol{\tau}_0$, where $\boldsymbol{\tau}_0 = [6, 3\cos^2(0.2\pi t), \sin^2(0.2\pi t)]^T$, and the initial conditions are set as $\boldsymbol{\eta}_d(0) = [15.6, 6.8, \pi/4]^T$, $\boldsymbol{\nu}_d(0) = [1, 0, 0]^T$, $\boldsymbol{\eta}(0) = [15.5, 5.5, \pi/2]^T$, and $\boldsymbol{\nu}(0) = [0, 0, 0]^T$.

In what follows, 3 cases will be deployed to evaluate the performance of the proposed HFC, FDO-HFC, and FUO-HFC approaches, respectively.

10.4.1 Performance evaluation on the HFC

In this subsection, a nominal case where the ASV is sufficiently modeled (i.e., $\boldsymbol{f}$ is known) and external disturbances are not considered (i.e., $\boldsymbol{d} = 0$) is employed to demonstrate the effectiveness and superiority of the proposed HFC scheme. User-defined parameters are chosen as follows: $K_1 = 0.3$, $K_2 = 0.3$, $\beta_1 = 1/3$, and $\beta_2 = 1/2$.

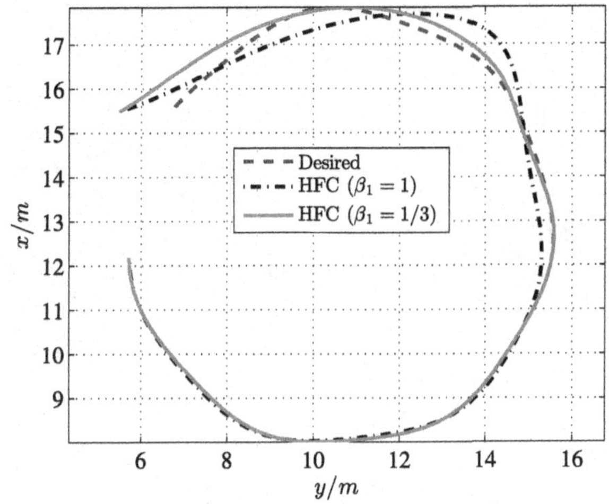

Figure 10.2 Desired and actual trajectories in the *xy*-plane.

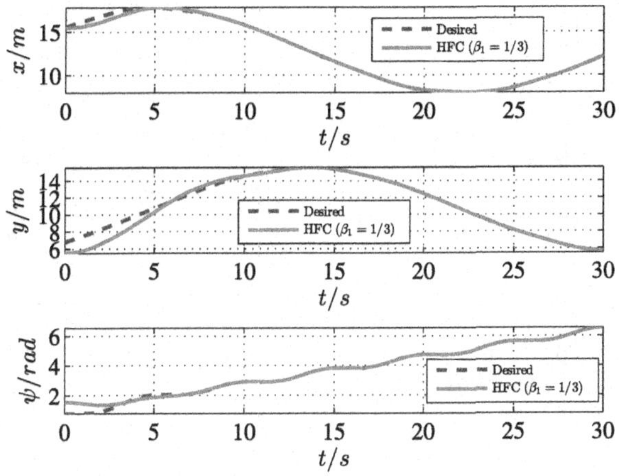

Figure 10.3 Desired and actual states *x*, *y*, and ψ.

Simulation results are shown in Figs. 10.2–10.7. Comparing with the traditional asymptotic approach, i.e., $\beta_1 = \beta_2 = 1$ within the HFC scheme, we can see from Fig. 10.2 that the HFC with $\beta_1 = 1/3$ can achieve much faster convergence. In addition, as shown in Figs. 10.3–10.6, the ASV can exactly track the desired trajectory within a finite time by virtue of the HFC laws shown in Fig. 10.7. In comparison with the asymptotic approach (i.e.,

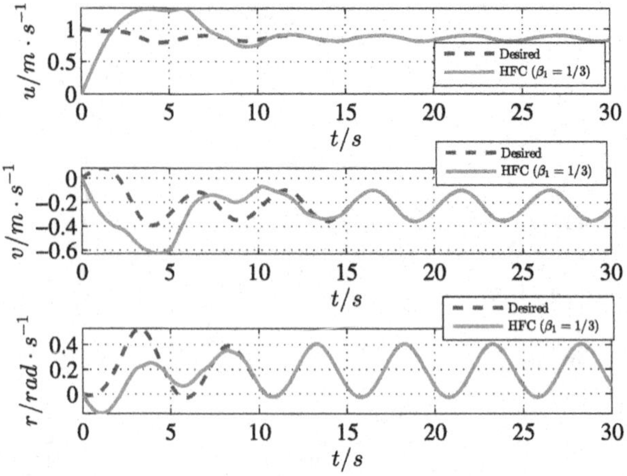

Figure 10.4 Desired and actual states *u*, *v*, and *r*.

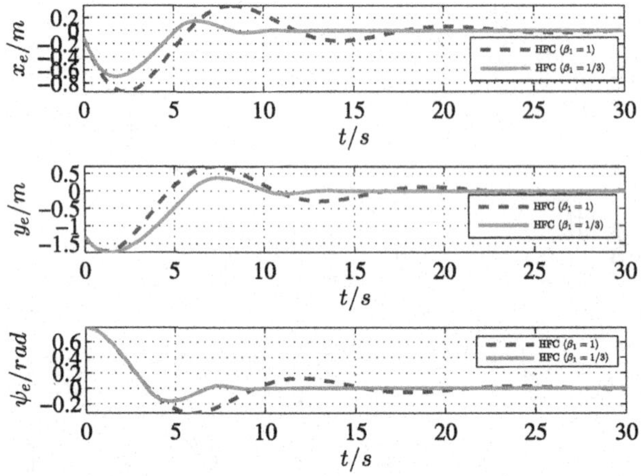

Figure 10.5 Tracking errors x_e, y_e, and ψ_e.

$\beta_1 = 1$), the HFC scheme with $\beta_1 = 1/3$ is able to render tracking errors converge to the origin in a very short time.

10.4.2 Performance evaluation on the FDO-HFC

In this subsection, a much more practical case with unknown disturbances is deployed to demonstrate the performance evaluation and comparisons. In order to facilitate simulation studies, the external disturbances are assumed

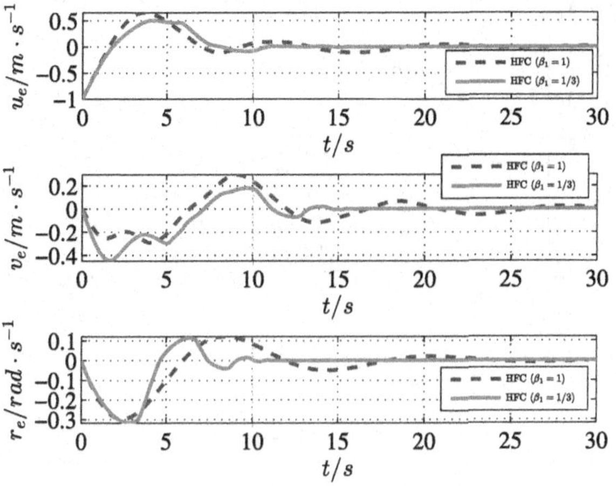

Figure 10.6 Tracking errors u_e, v_e, and r_e.

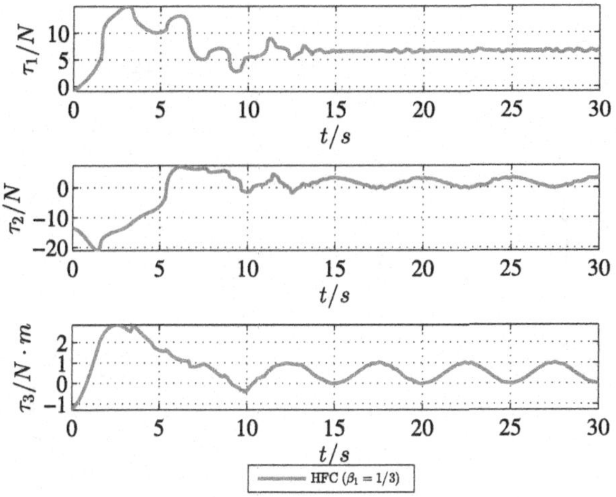

Figure 10.7 Control forces τ_1, τ_2, and torque τ_3.

to be governed by

$$\boldsymbol{d}(t) = \begin{bmatrix} 10\cos(0.1\pi t - \pi/5) \\ 8\cos(0.3\pi t + \pi/6) \\ 6\cos(0.2\pi t + \pi/3) \end{bmatrix}, \tag{10.56}$$

with $\mathbf{H}_1 = \text{diag}(0, 0, 0)$ and $\mathbf{H}_2 = \text{diag}(-0.01\pi^2, -0.09\pi^2, -0.04\pi^2)$.

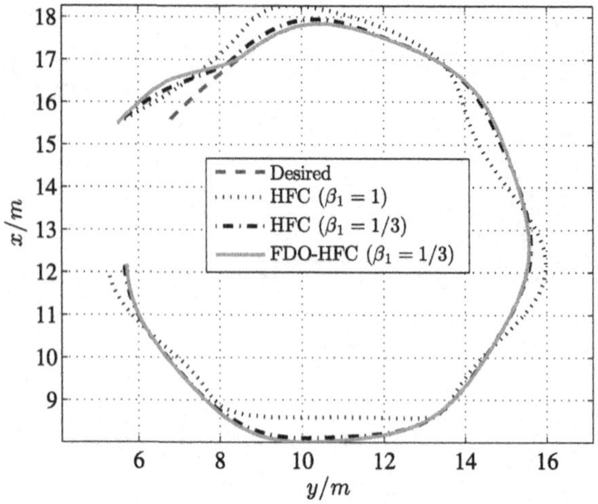

Figure 10.8 Desired and actual trajectories in the *xy*-plane.

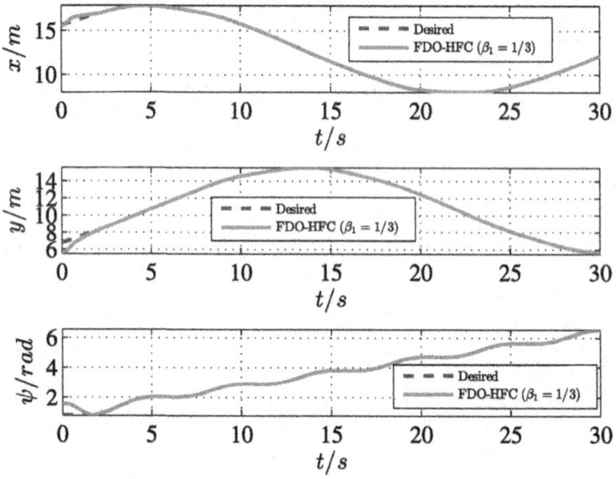

Figure 10.9 Desired and actual states *x*, *y*, and ψ.

Accordingly, user-defined parameters of the HFC and FDO-HFC schemes are commonly chosen as $K_1 = 2.6$, $K_2 = 2.6$, $\beta_1 = 1/3$ and $\beta_2 = 1/2$. The other parameters of the FDO are selected as follows: $\mathbf{L}_0 = \text{diag}(10, 10, 10)$, $\mathbf{L}_1 = \text{diag}(32, 32, 32)$, $\mathbf{L}_2 = \text{diag}(20, 20, 20)$, $\alpha_1 = 7/9$, $\alpha_2 = 5/9$, and $\alpha_3 = 1/3$.

The actual and desired trajectories in the planar space are shown in Fig. 10.8–10.12, from which we can see that, in comparison with the

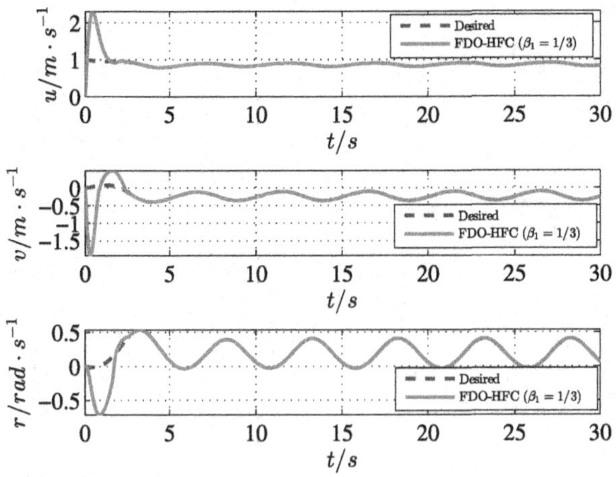

Figure 10.10 Desired and actual states u, v, and r.

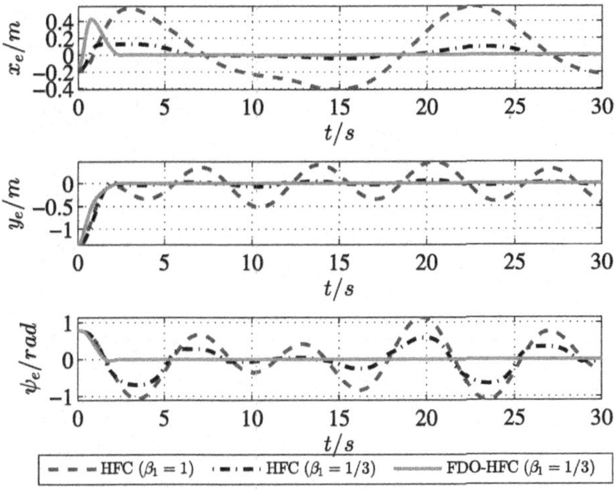

Figure 10.11 Tracking errors x_e, y_e, and ψ_e.

conventional asymptotic control scheme (i.e., $\beta_1 = 1$), the proposed HFC scheme (i.e., $\beta_1 = 1/3$) achieves faster convergence and stronger disturbance rejection simultaneously, and thereby resulting in higher tracking accuracy, and the FDO-HFC approach is able to realize exact trajectory tracking within a short time since unknown disturbances can be finite-time observed exactly as shown in Fig. 10.13.

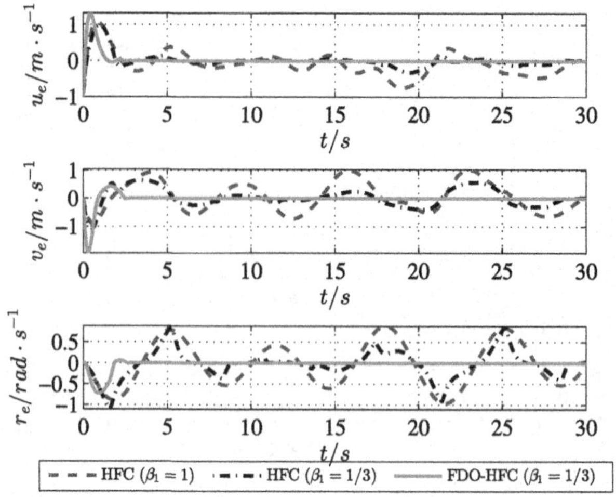

Figure 10.12 Tracking errors u_e, v_e, and r_e.

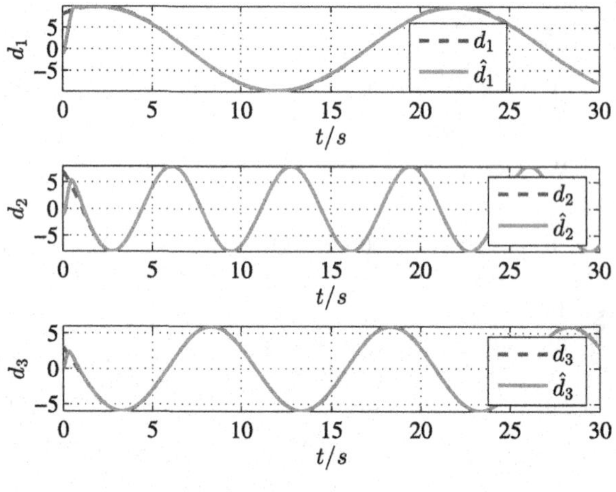

Figure 10.13 Disturbances and their finite-time observation.

10.4.3 Performance evaluation on the FUO-HFC

In order to demonstrate the superiority of the proposed FUO-HFC scheme, we employ a much more complex case where both unmodeled dynamics of the ASV and desired trajectory and unknown disturbances are included. The control parameters are selected as follows: $K_1 = 2.6$, $K_2 = 2.6$, $\beta_1 = 1/3$, $\beta_2 = 1/2$, $\lambda_1 = 2$, $\lambda_2 = 1.5$, $\lambda_3 = 1.1$, and

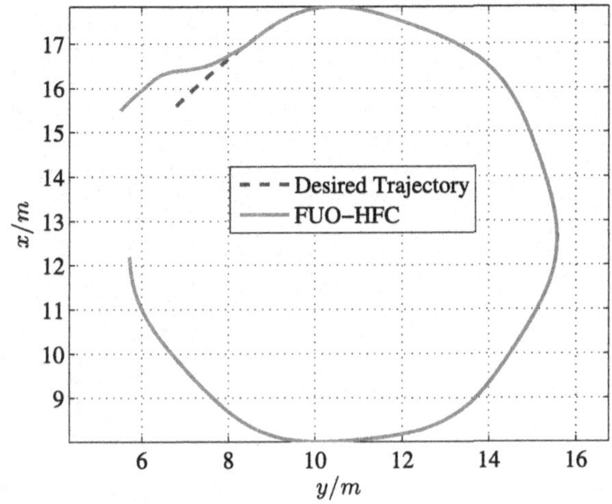

Figure 10.14 Desired and actual trajectories in the *xy*-plane.

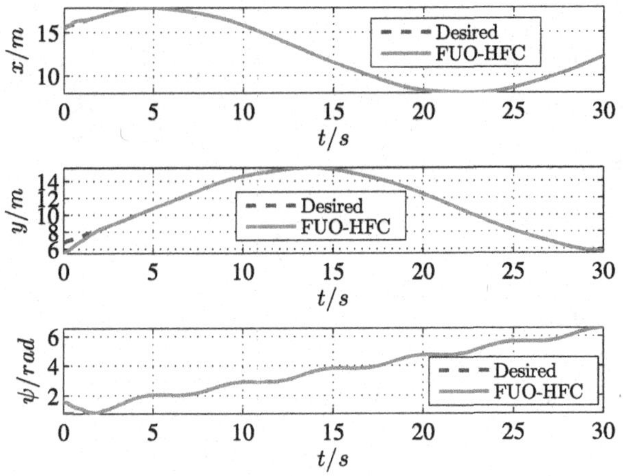

Figure 10.15 Desired and actual states *x*, *y*, and ψ.

$\mathcal{L} = \text{diag}(30, 30, 30)$. Corresponding simulation results and comparisons are shown in Figs. 10.14–10.19. The actual and reference trajectories in the planar space are shown in Fig. 10.14, which indicates that the actual trajectory can exactly track the desired one in a very short time although the ASV suffers from unmodeled dynamics and unknown disturbances. Actually, the previous HFC and FDO-HFC approaches become unavailable due

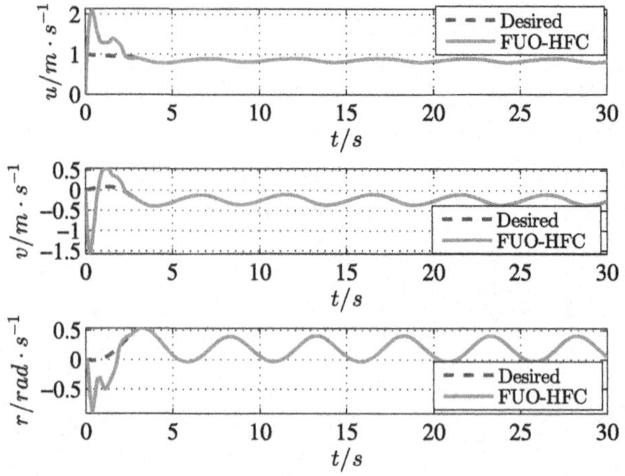

Figure 10.16 Desired and actual states u, v, and r.

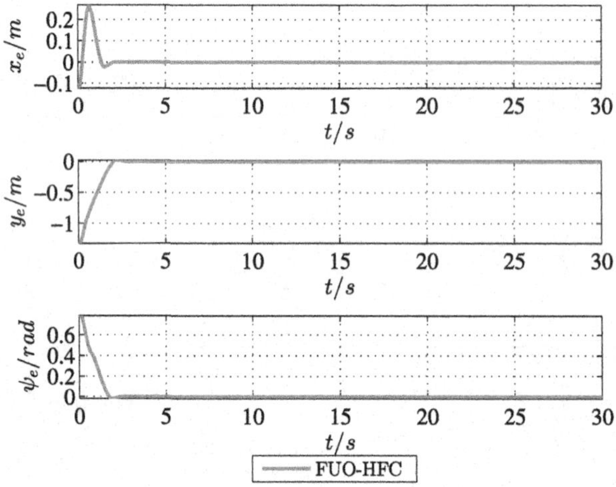

Figure 10.17 Tracking errors x_e, y_e, and ψ_e.

to the unexpected unmodeled dynamics of the ASV and the desired trajectory. From the tracking performance on position and velocity illustrated in Figs. 10.15–10.18, we can see that trajectory tracking errors converge to zero in a very short time in spite of complex unknowns including unmodeled dynamics and unknown disturbances. In essence, the remarkable performance of the proposed FUO-HFC scheme on exact trajectory track-

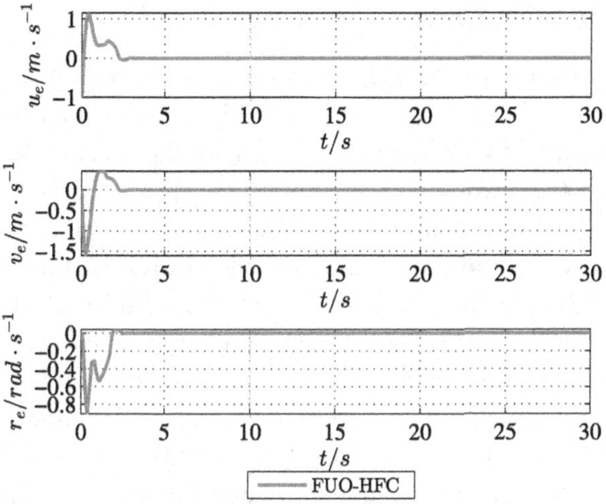

Figure 10.18 Tracking errors u_e, v_e, and r_e.

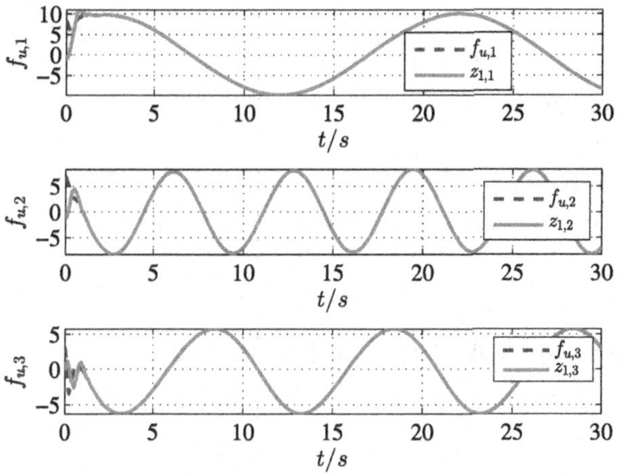

Figure 10.19 Lumped unknowns and their finite-time observation.

ing relies on the accurate observation on the lumped unknowns via an FUO, whereby the finite-time observation results are shown in Fig. 10.19. In this context, the FUO-HFC methodology can achieve fast and exact trajectory tracking together with accurate reconstruction on complex unknowns, including unmodeled dynamics, uncertainties, and unknown disturbances.

10.5. Conclusions

In this chapter, in order to achieve trajectory fast and accurate tracking control of an autonomous surface vehicle (ASV) subject to unmodeled dynamics and unknown disturbances, a homogeneity-based finite-time control (HFC) framework has been innovatively proposed. For exactly dealing with external disturbances, a finite-time disturbance observer (FDO) has been developed and has been incorporated into the HFC framework, thereby contributing the FDO-based HFC (termed FDO-HFC) scheme which can realize exact trajectory tracking control of an ASV in the presence of complex disturbances. To further accurately handle complicated unknowns, including both unmodeled dynamics and unknown disturbances, a finite-time unknown observer based HFC (FUO-HFC) scheme has been proposed to enhance the entire performance, including both trajectory tracking and unknowns identification, whereby high accuracy and fast convergence can be ensured simultaneously. Simulation studies and comprehensive comparisons have been conducted on a benchmark ship, i.e., CyberShip II, and have demonstrated the effectiveness and superiority of the proposed HFC schemes in term of exact trajectory tracking and unknowns rejection.

Acknowledgments

The authors would like to thank the Editor-in-Chief, Associate Editor, and anonymous referees for their invaluable comments and suggestions.

Appendix 10.A. Proof of Lemma 10.2

Proof. In the light of Lemma 10.1, the entire proof can be divided into 2 phases, i.e., proofs of global asymptotic stability and negative homogeneity.

Phase I: Global Asymptotic Stability

In order to facilitate an inductive proof, applying a set of coordinate transformations as follows:

$$z_i = l_{i-1} x_i, \quad i = 1, 2, \ldots, n, \tag{A.1}$$

with $l_0 = 1$, to system (10.2), we have

$$\dot{x}_1 = c_1 \left(x_2 - x_1^{\alpha_2} \right),$$

$$\dot{x}_2 = c_2\left(x_3 - x_1^{\alpha_3}\right),$$
$$\vdots$$
$$\dot{x}_n = -c_n x_1^{\alpha_{n+1}}, \tag{A.2}$$

where $x_1^{\alpha_i} := \mathrm{sig}^{\alpha_i}(x_1)$ and $c_i = l_i/l_{i-1}, i = 1, 2, \ldots, n$.

Using (A.2), a backward recursive procedure will be established in the sequel.

Initial step: We first consider the following system:

$$\dot{x}_n = -c_n x_n^{\alpha_{n+1}/\alpha_n}. \tag{A.3}$$

Choosing a Lyapunov function as follows:

$$V_n(x_n) = \frac{\alpha_n}{2}|x_n|^{2/\alpha_n}, \tag{A.4}$$

we have

$$\dot{V}_n(x_n)|_{(A.3)} \leq -k_n|x_n|^{(2+\tau)/\alpha_n}, \tag{A.5}$$

with $k_n = c_n$.

Inductive step: Assume there exists a Lyapunov function as follows:

$$V_{i+1}(x_{i+1}, x_{i+2}, \ldots, x_n) = \sum_{j=i+1}^{n} \int_{x_{j+1}^{\alpha_j/\alpha_{j+1}}}^{x_j} \left(s^{(2-\alpha_j)/\alpha_j} - x_{j+1}^{(2-\alpha_j)/\alpha_{j+1}}\right) ds \tag{A.6}$$

such that $\dot{V}_{i+1}$ along the following system:

$$\dot{x}_{i+1} = c_{i+1}\left(x_{i+2} - x_{i+1}^{\alpha_{i+2}/\alpha_{i+1}}\right),$$
$$\dot{x}_{i+2} = c_{i+2}\left(x_{i+3} - x_{i+1}^{\alpha_{i+3}/\alpha_{i+1}}\right),$$
$$\vdots$$
$$\dot{x}_n = -c_n x_{i+1}^{\alpha_{n+1}/\alpha_{i+1}} \tag{A.7}$$

satisfies

$$\dot{V}_{i+1}|_{(A.7)} \leq -k_{i+1} \sum_{j=i+1}^{n} \left|x_j^{\alpha_{j+1}/\alpha_j} - x_{j+1}\right|^{(2+\tau)/\alpha_{j+1}}, \tag{A.8}$$

with $k_{i+1} > 0$. It is easily verified that (A.8) holds for system (A.3) at the initial step, i.e., $i = n - 1$.

In this context, we are expected to prove that for the system as follows:

$$\begin{aligned} \dot{x}_i &= c_i\left(x_{i+1} - x_i^{\alpha_{i+1}/\alpha_i}\right), \\ \dot{x}_{i+1} &= c_{i+1}\left(x_{i+2} - x_i^{\alpha_{i+2}/\alpha_i}\right), \\ &\vdots \\ \dot{x}_n &= -c_n x_i^{\alpha_{n+1}/\alpha_i}, \end{aligned} \tag{A.9}$$

there exists the following Lyapunov function:

$$\begin{aligned} V_i(x_i, x_{i+1}, \ldots, x_n) &= V_{i+1}(x_{i+1}, x_{i+2}, \ldots, x_n) \\ &+ \int_{x_{i+1}^{\alpha_i/\alpha_{i+1}}}^{x_i} \left(s^{(2-\alpha_i)/\alpha_i} - x_{i+1}^{(2-\alpha_i)/\alpha_{i+1}}\right) ds \end{aligned} \tag{A.10}$$

such that $\dot{V}_i$ along (A.9) satisfies an inequality like (A.8).

To this end, together with (A.8), we have

$$\begin{aligned} \dot{V}_i|_{(A.9)} &= \dot{V}_{i+1}|_{(A.7)} - \sum_{j=1+1}^{n} c_j \frac{\partial V_{i+1}}{\partial x_j}\left(x_i^{\alpha_{j+1}/\alpha_i} - x_{i+1}^{\alpha_{j+1}/\alpha_{i+1}}\right) \\ &\quad + \frac{d}{dt}\int_{x_{i+1}^{\alpha_i/\alpha_{i+1}}}^{x_i} \left(s^{(2-\alpha_i)/\alpha_i} - x_{i+1}^{(2-\alpha_i)/\alpha_{i+1}}\right) ds\Big|_{(A.9)} \\ &\le -k_{i+1}\sum_{j=i+1}^{n}\left|x_j^{\alpha_{j+1}/\alpha_j} - x_{j+1}\right|^{(2+\tau)/\alpha_{j+1}} \\ &\quad - c_i\left(x_i^{(2-\alpha_i)/\alpha_i} - x_{i+1}^{(2-\alpha_i)/\alpha_{i+1}}\right)\left(x_i^{\alpha_{i+1}/\alpha_i} - x_{i+1}\right) \\ &\quad - \frac{c_{i+1}(2-\alpha_i)}{\alpha_{i+1}} x_{i+1}^{(2-\alpha_i-\alpha_{i+1})/\alpha_{i+1}} \\ &\quad \times \left(x_{i+2} - x_i^{\alpha_{i+2}/\alpha_i}\right)\left(x_i - x_{i+1}^{\alpha_i/\alpha_{i+1}}\right) \\ &\quad - \sum_{j=1+1}^{n} c_j \frac{\partial V_{i+1}}{\partial x_j}\left(x_i^{\alpha_{j+1}/\alpha_i} - x_{i+1}^{\alpha_{j+1}/\alpha_{i+1}}\right). \end{aligned} \tag{A.11}$$

Note that

$$\begin{aligned} &- c_i\left(x_i^{(2-\alpha_i)/\alpha_i} - x_{i+1}^{(2-\alpha_i)/\alpha_{i+1}}\right)\left(x_i^{\alpha_{i+1}/\alpha_i} - x_{i+1}\right) \\ &\le -c_i k_{i+1}\left|x_i^{\alpha_{i+1}/\alpha_i} - x_{i+1}\right|^{(2+\tau)/\alpha_{i+1}} \end{aligned} \tag{A.12}$$

$$
\begin{aligned}
&- \frac{c_{i+1}(2-\alpha_i)}{\alpha_{i+1}} x_{i+1}^{(2-\alpha_i-\alpha_{i+1})/\alpha_{i+1}} \\
&\quad \times \left(x_{i+2} - x_i^{\alpha_{i+2}/\alpha_i}\right)\left(x_i - x_{i+1}^{\alpha_i/\alpha_{i+1}}\right) \\
&\le \mu_1 \sum_{j=i+1}^{n} \left|x_j^{\alpha_{j+1}/\alpha_j} - x_{j+1}\right|^{(2+\tau)/\alpha_{j+1}} \\
&\quad + \rho_1(\mu_1)\left|x_i^{\alpha_{i+1}/\alpha_i} - x_{i+1}\right|^{(2+\tau)/\alpha_{i+1}} \qquad \text{(A.13)}
\end{aligned}
$$

$$
\begin{aligned}
&- \sum_{j=1+1}^{n} c_j \frac{\partial V_{i+1}}{\partial x_j}\left(x_i^{\alpha_{j+1}/\alpha_i} - x_{i+1}^{\alpha_{j+1}/\alpha_{i+1}}\right) \\
&\le \mu_2 \sum_{j=i+1}^{n} \left|x_j^{\alpha_{j+1}/\alpha_j} - x_{j+1}\right|^{(2+\tau)/\alpha_{j+1}} \\
&\quad + \rho_2(\mu_2)\left|x_i^{\alpha_{i+1}/\alpha_i} - x_{i+1}\right|^{(2+\tau)/\alpha_{i+1}}, \qquad \text{(A.14)}
\end{aligned}
$$

with positive continuous functions $\rho_1(\cdot)$ and $\rho_2(\cdot)$ with respect to any positive constants μ_1 and μ_2, respectively.

Substituting (A.12)–(A.14) into (A.11) yields

$$
\begin{aligned}
\dot{V}_i|_{(A.9)} &\le -(k_{i+1} - \mu_1 - \mu_2) \sum_{j=i+1}^{n} \left|x_j^{\alpha_{j+1}/\alpha_j} - x_{j+1}\right|^{(2+\tau)/\alpha_{j+1}} \\
&\quad - (c_i k_{i+1} - \rho_1 - \rho_2)\left|x_i^{\alpha_{i+1}/\alpha_i} - x_{i+1}\right|^{(2+\tau)/\alpha_{i+1}}. \qquad \text{(A.15)}
\end{aligned}
$$

Selecting parameters μ_1, μ_2, k_{i+1}, and c_i such that

$$
c_i \ge \frac{k_i + \rho_1(\mu_1) + \rho_2(\mu_2)}{k_{i+1}}, \quad k_i \le k_{i+1} - \mu_1 - \mu_2 \qquad \text{(A.16)}
$$

yields

$$
\dot{V}_i|_{(A.9)} \le -k_i \sum_{j=i}^{n} \left|x_j^{\alpha_{j+1}/\alpha_j} - x_{j+1}\right|^{(2+\tau)/\alpha_{j+1}}, \qquad \text{(A.17)}
$$

which implies that (A.8) also holds for the ith inductive step.

Recursively, we can eventually construct a Lyapunov function as follows:

$$
\begin{aligned}
&V_1(x_1, x_2, \ldots, x_n) \\
&\quad = \sum_{j=1}^{n} \int_{x_{j+1}^{\alpha_j/\alpha_{j+1}}}^{x_j} \left(s^{(2-\alpha_j)/\alpha_j} - x_{j+1}^{(2-\alpha_j)/\alpha_{j+1}}\right) ds, \qquad \text{(A.18)}
\end{aligned}
$$

with $x_{n+1} = 0$ and $\alpha_1 = 1$, such that

$$\dot{V}_1|_{(A.2)} \leq -k_1 \sum_{j=1}^{n} \left|x_j^{\alpha_{j+1}/\alpha_j} - x_{j+1}\right|^{(2+\tau)/\alpha_{j+1}}, \tag{A.19}$$

with $k_1 > 0$, recursively determined by (A.16).

It follows from (A.19) that system (A.2) is globally asymptotically stable. Together with the global diffeomorphism (A.1), we have that system (10.2) is globally asymptotically stable.

Phase II: Negative Homogeneity

Applying a dilation

$$(r_1, r_2, \ldots, r_n) = (\alpha_1, \alpha_2, \ldots, \alpha_n), \tag{A.20}$$

with $\alpha_1 = 1$, to system (10.2) yields a negative homogeneous of degree, i.e., $\tau = \alpha_{i+1} - \alpha_i < 0, i = 1, 2, \ldots, n$.

In this context, combining Phases I and II, and using Lemma 10.1, we immediately have that system (10.2) is globally asymptotically stable. This concludes the proof. ∎

References

[1] L. Moreira, T.I. Fossen, C.G. Soares, Path following control system for a tanker ship model, Ocean Engineering 34 (14) (Oct. 2007) 2074–2085.

[2] N. Wang, H.R. Karimi, H. Li, S.-F. Su, Accurate trajectory tracking of disturbed surface vehicles: a finite-time control approach, IEEE/ASME Transactions on Mechatronics 24 (3) (Mar. 2019) 1064–1074.

[3] T.I. Fossen, Marine Control Systems: Guidance, Navigation and Control of Ships, Rigs and Underwater Vehicles, Marine Cybernetics, Trondheim, Norway, 2002.

[4] N. Wang, Z. Deng, Finite-time fault estimator based fault-tolerance control for a surface vehicle with input saturations, IEEE Transactions on Industrial Informatics (2019), https://doi.org/10.1109/TII.2019.2930471.

[5] N. Wang, G. Xie, X. Pan, S.-F. Su, Full-state regulation control of asymmetric underactuated surface vehicles, IEEE Transactions on Industrial Electronics 66 (11) (Nov. 2019) 8741–8750.

[6] H. Qin, H. Chen, Y. Sun, L. Chen, Distributed finite-time fault-tolerant containment control for multiple ocean Bottom Flying node systems with error constraints, Ocean Engineering (2019), https://doi.org/10.1016/j.oceaneng.2019.106341.

[7] H. Qin, H. Chen, Y. Sun, Distributed finite-time fault-tolerant containment control for multiple Ocean Bottom Flying Nodes, Journal of the Franklin Institute (2019), https://doi.org/10.1016/j.jfranklin.2019.05.034.

[8] X. Xiang, L. Lapierre, B. Jouvencel, Smooth transition of AUV motion control: from fully-actuated to under-actuated configuration, Robotics and Autonomous Systems 67 (5) (Mar. 2015) 14–22.

[9] S. Serdar, J.B. Bradley, P.P. Ron, A chattering-free sliding-mode controller for underwater vehicles with fault-tolerant infinity-norm thrust allocation, Ocean Engineering 35 (16) (Nov. 2008) 1647–1659.
[10] Y. Yang, J.L. Du, H.B. Liu, C. Guo, A. Abraham, A trajectory tracking robust controller of surface vessels with disturbance uncertainties, IEEE Transactions on Control Systems Technology 22 (4) (Jul. 2014) 1511–1518.
[11] N. Wang, S.-F. Su, X. Pan, X. Yu, G. Xie, Yaw-guided trajectory tracking control of an asymmetric underactuated surface vehicle, IEEE Transactions on Industrial Informatics 15 (6) (Jun. 2019) 3502–3513.
[12] K.Y. Pettersen, H. Nijmeijer, Tracking control of an underactuated surface vessel, in: Proceedings of the 37th IEEE Conference on Decision and Control, 1998, pp. 4561–4566.
[13] K. Shojaei, Neural adaptive robust control of underactuated marine surface vehicles with input saturation, Applied Ocean Research 53 (Sep. 2015) 267–278.
[14] G.Q. Zhang, X.K. Zhang, Concise robust adaptive path-following control of underactuated ships using DSC and MLP, IEEE Journal of Oceanic Engineering 39 (4) (Oct. 2014) 685–694.
[15] Z. Zhao, W. He, S.S. Ge, Adaptive neural network control of a fully actuated marine surface vessel with multiple output constraints, IEEE Transactions on Control Systems Technology 22 (4) (Jul. 2014) 1536–1543.
[16] Z. Yan, J. Wang, Model predictive control for tracking of underactuated vessels based on recurrent neural networks, IEEE Journal of Oceanic Engineering 37 (4) (Oct. 2012) 717–726.
[17] H. Li, P. Shi, D. Yao, Adaptive sliding mode control of Markov jump nonlinear systems with actuator faults, IEEE Transactions on Automatic Control 62 (4) (Apr. 2017) 1933–1939.
[18] H. Li, C. Wu, X. Jing, L. Wu, Fuzzy tracking control for nonlinear networked systems, IEEE Transactions on Cybernetics 47 (8) (Aug. 2017) 2020–2031.
[19] Y.M. Li, S.C. Tong, T.S. Li, Adaptive fuzzy output feedback dynamic surface control of interconnected nonlinear pure-feedback systems, IEEE Transactions (Cybernetics) 45 (1) (Jan. 2015) 138–149.
[20] Y.M. Li, S.C. Tong, T.S. Li, Composite adaptive fuzzy output feedback control design for uncertain nonlinear strict-feedback systems with input saturation, IEEE Transactions (Cybernetics) 45 (10) (Oct. 2015) 2299–2308.
[21] B. Xu, F.C. Sun, Y.P. Pan, B.D. Chen, Disturbance observer based composite learning fuzzy control of nonlinear systems with unknown dead zone, IEEE Transactions on Systems, Man, and Cybernetics: Systems 47 (8) (Aug. 2017) 1854–1862.
[22] C.S. Chen, Dynamic structure neural-fuzzy networks for robust adaptive control of robot manipulators, IEEE Transactions on Industrial Electronics 55 (9) (Sep. 2008) 3402–3414.
[23] N. Wang, M.J. Er, Self-constructing adaptive robust fuzzy neural tracking control of surface vehicles with uncertainties and unknown disturbances, IEEE Transactions on Control Systems Technology 23 (3) (May 2015) 991–1002.
[24] A. Wu, P.K. Tam, Stable fuzzy neural tracking control of a class of unknown nonlinear systems based on fuzzy hierarchy error approach, IEEE Transactions on Fuzzy Systems 10 (6) (Dec. 2002) 779–789.
[25] W. He, Y.H. Chen, Z. Yin, Adaptive neural network control of an uncertain robot with full-state constraints, IEEE Transactions (Cybernetics) 46 (3) (Mar. 2016) 620–629.

[26] W. He, Y. Dong, C. Sun, Adaptive neural impedance control of a robotic manipulator with input saturation, IEEE Transactions on Systems, Man and Cybernetics: Systems 46 (3) (Mar. 2016) 334–344.

[27] S.I. Han, J.M. Lee, Recurrent fuzzy neural network backstepping control for the prescribed output tracking performance of nonlinear dynamic systems, ISA Transactions 53 (1) (Jan. 2014) 33–43.

[28] N. Wang, M.J. Er, J.C. Sun, Y.C. Liu, Adaptive robust online constructive fuzzy control of a complex surface vehicle system, IEEE Transactions (Cybernetics) 46 (7) (Jul. 2016) 1511–1523.

[29] N. Wang, J.C. Sun, M.J. Er, Y.C. Liu, A novel extreme learning control framework of unmanned surface vehicles, IEEE Transactions (Cybernetics) 46 (5) (May 2016) 1106–1117.

[30] N. Wang, M.J. Er, Direct adaptive fuzzy tracking control of marine vehicles with fully unknown parametric dynamics and uncertainties, IEEE Transactions on Control Systems Technology 24 (5) (Sep. 2016) 1845–1852.

[31] N. Wang, M.J. Er, M. Han, Dynamic tanker steering control using generalized ellipsoidal-basis-function-based fuzzy neural networks, IEEE Transactions on Fuzzy Systems 23 (5) (Oct. 2015) 1414–1427.

[32] K.F. Lu, Y.Q. Xia, Z. Zhu, M.V. Basin, Sliding mode attitude tracking of rigid spacecraft with disturbances, Journal of the Franklin Institute 349 (2) (Mar. 2012) 423–440.

[33] Y.G. Hong, Y.S. Xu, H. Jie, Finite-time control for robot manipulators, Systems & Control Letters 46 (Jan. 2002) 243–253.

[34] W.-H. Chen, D.J. Ballance, P.J. Gawthrop, J. O'Reilly, A nonlinear disturbance observer for robotic manipulators, IEEE Transactions on Industrial Electronics 47 (4) (Aug. 2000) 932–938.

[35] W.-H. Chen, K. Ohnishi, L. Guo, Advances in disturbance/uncertainty estimation and attenuation, IEEE Transactions on Industrial Electronics 62 (9) (Sep. 2015) 5758–5762.

[36] W.-H. Chen, J. Yang, L. Guo, S.H. Li, Disturbance-observer-based control and related methods-an overview, IEEE Transactions on Industrial Electronics 63 (2) (Sep. 2015) 1083–1095.

[37] T. Umeno, Y. Hori, Robust speed control of DC servomotors using modern two degrees-of-freedom controller design, IEEE Transactions on Industrial Electronics 38 (5) (Oct. 1991) 363–368.

[38] W.-H. Chen, Nonlinear disturbance observer-enhanced dynamic inversion control of missiles, Journal of Guidance, Control, and Dynamics 26 (1) (Jan. 2003) 161–166.

[39] H.J. Marquez, Nonlinear Control Systems: Analysis and Design, John Wiley & Sons, Inc, Hoboken, New Jersey, 2003.

[40] H. Khalil, Nonlinear Systems, 2nd ed., Prentice-Hall, Upper Saddle River, NJ, 1996.

[41] A. Levant, Homogeneity approach to high-order sliding mode design, Automatica 41 (5) (May 2005) 823–830.

[42] Y.B. Shtessel, I.A. Shkolnikov, A. Levant, Smooth second-order sliding modes: missile guidance application, Automatica 43 (8) (Aug. 2007) 1470–1476.

[43] R. Skjetne, T.I. Fossen, P.V. Kokotović, Adaptive maneuvering, with experiments, for a model ship in a marine control laboratory, Automatica 41 (2) (Feb. 2005) 289–298.

CHAPTER ELEVEN

Way-point tracking control of underactuated USV based on GPC path planning

Yueying Wang, Tao Jiang
School of Mechatronic Engineering and Automation, Shanghai University, Shanghai, China

Contents

Abstract

In order to solve the problems of large tracking error and slow error convergence in the tracking process of underactuated asymmetric unmanned surface vehicles (USVs) in sharp turns and other extreme paths, an adaptive sliding mode control method based on generalized predictive control (GPC) algorithm (LOS-GPC-SMC) is proposed. Firstly, the path generation part takes into account the mobility constraints of a USV, and realizes path generation by GPC combined with Line-of-Sight (LOS). Secondly, due to the asymmetry of the dynamic model of a USV, the global homeomorphic differential transformation is used to transform and decouple the state variables of the system. After that, in order to effectively compensate the model uncertainty and external disturbance in the tracking process of a USV, the adaptive sliding mode control method is used to design the actual control law to ensure error stabilization and realize path tracking. Finally, the effectiveness of the proposed control strategy is verified by a large number of simulation experiments.

Fundamental Design and Automation Technologies in Offshore Robotics
https://doi.org/10.1016/B978-0-12-820271-5.00016-X

Keywords

Underactuated Unmanned Surface Vehicle, Way-point Tracking, Path Generation

11.1. Introduction

Way-point tracking control for underactuated USV includes two aspects: path generation and path tracking for the generated path. The expected trajectory of an underactuated USV is usually obtained by giving the expected position points at each moment, but the usually given trajectory often does not contain the ship's mobility constraints. At the same time, the path tracking control of an underactuated USV is also a long-standing control problem, which has occupied the control community for many years [1]. Since the USV has only longitudinal driving force and yaw moment generated by the tail rudder and no lateral driving force, it cannot directly eliminate lateral drift and can only indirectly eliminate lateral error by adjusting yaw angle through yaw moment. Moreover, USV is a typical system with large inertia, nonlinearity, and uncertainty of motion model. It is also affected by external environment interference such as wind, wave, and current when sailing on the sea surface, which makes accurate path tracking of USV very difficult [2]. Gao et al. presented a predictive optimization-based model reference adaptive control (MRAC) approach for dynamic positioning (DP) of a fully actuated underwater vehicle subject to dynamic uncertainties and actuator saturation [3]. Wu et al. combined GPC with PID control, and designed GPC-PID cascade controller to control the heading motion and steering motion of USV, indirectly realizing the path tracking control of USV [4]. Yu et al. proposed a sliding mode control law based on surge velocity tracking error and lateral motion tracking error for the robust tracking control of underactuated USV with parameter uncertainty, thus realizing trajectory tracking [5]. Soryeok et al. designed a model predictive control scheme with LOS path generation capability, and used the three-degree-of-freedom model of USV in the controller design to realize the path following of underactuated surface vehicle with input constraints [1]. Liu et al. proposed a new path following guidance law for underactuated marine surface vessels, and formed a cascade system of tracking error system and prediction error system to realize path tracking [6]. McNinch et al. used a nonlinear discrete model predictive controller to solve the propulsion and heading control inputs of USV with input saturation and input constraints, and to accomplish specific targets such as minimum error or minimum time, thus realizing trajectory tracking [7]. Wang et al. described ship dynamics using Serreret framework and designed an adaptive path following control law to enable underactuated ships to travel along a predetermined path at a constant forward speed with uncertain parameters [8]. Zhu et al. put forward a sliding mode control method based on upper and lower bounds to solve the problem of model parameter uncertainty and disturbance of external wind waves and currents encountered by underactuated USV in realizing track tracking control of horizontal plane [9]. Zhang et al. combined Kalman filter, disturbance observer, and robust constrained model predictive control to propose a new rolling constrained ship path following control method [10]. Soltan et al. proposed a method based on nonlinear sliding mode control, which combines trajectory planning, tracking, and coordinated control of USV [11]. Bibuli et al. proposed a linear following guidance scheme for underactuated marine systems, the difference of which is in the definition of the error variables to be stabilized to zero [12]. Tribou et al. proposed a new path following controller that mediates visual control stability [13]. Xu et al. proposed a backstepping adaptive dynamic sliding mode control method for the path following control system of underactuated surface vessel [14]. Pettersen et al. designed the full-state feedback control law by cascade method, and realized the tracking control of ship's way points by yaw torque control [15].

Way-point tracking control usually includes two aspects, path generation and path tracking. The virtual path of most of the above articles is given directly through mathematical expressions, and does not include the maneuvering characteristics constraints of ships. If the given trajectory does not conform to the sailing limit that the hull can reach, especially at the inflection point of the path, and if the path includes the physical limitations of the hull, can the ship bend in a more reasonable way? Therefore, how to include the physical properties of the hull in the generated way-points is the focus of this paper. In addition, most of the models used in the above articles are based on symmetric models to design controllers, but the model of "jinghai 8-b" USV has the characteristics of asymmetry. How to realize the tracking control of the virtual trajectory under the interference of the model is also the focus of this article.

11.2. Problem description and model identification

The dynamic motion of a USV is very complicated. Analyzing its motion state involves six degrees of freedom. However, in the actual control analysis, it is often simplified as a three-degree-of-freedom motion analysis, that is, it contains only surge, sway, and yaw motion. The research object of this paper is "jinghai 8-b" USV. When analyzing its motion, only the surge, sway, and yaw motions are discussed, that is, only the surge velocity u, sway velocity v, and yaw velocity r are investigated [16].

The motion System in the Fixed Geodetic Coordinate System is:

$$\begin{cases} \dot{x} = v_x \cos\psi - v_y \sin\psi \\ \dot{y} = v_y \cos\psi + v_x \sin\psi \\ \dot{\psi} = r \end{cases} \tag{11.1}$$

$$\begin{cases} m_{11}\dot{u} - m_{22}vr + d_{11}u = \tau_x + \tau_1 \\ m_{22}\dot{v} + m_{11}ur + d_{22}v = \tau_2 \\ m_{33}\dot{r} + (m_{22} - m_{11})\,ur + d_{33}r = \tau_r + \tau_3 \end{cases} \tag{11.2}$$

where x, y, ψ are the position and heading angle in the fixed coordinate system of the earth; v_x, v_y are the velocity components of the USV in the hull coordinate system, and r is the yaw rate of the USV; $d_{11}, d_{22}, d_{33}, m_{11}, m_{22}, m_{33}$ are the hydrodynamic damping coefficients and inertia parameters including additional mass of the ship system, respectively. Control inputs τ_u and τ_r are the longitudinal thrust and steering torque of the ship, respectively; τ_1, τ_2, τ_3 are the interference force and moment generated by the external environment.

Since there is no lateral propulsion device and no control input is available in the lateral direction, the studied ship path tracking is an underactuated control problem. The problem of way-point tracking control for underactuated USV can be divided into virtual path generation and path tracking control (as shown in Fig. 11.1). Path generation uses GPC combined with LOS to generate the desired path. The virtual path tracking is realized by designing an adaptive sliding mode controller. How to make the control target drive the USV with asymmetric model to sail along the virtual trajectory by designing feedback control law is also a research focus of this chapter.

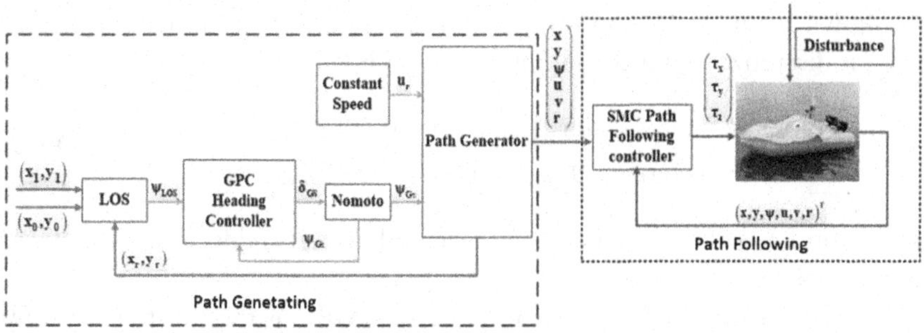

Figure 11.1 "Jinghai 8-b" control block diagram.

11.3. Path generation and path tracking

LOS-GPC is used to generate the desired path. The desired heading angle AA is generated by LOS, and the heading controller is designed by GPC to track AA. The control object is the Nomoto model of a USV, to generate a path considering the constraint of ship maneuvering characteristics. After the virtual path is generated, at the next moment an adaptive sliding mode control method is adopted to track the virtual path. Because the dynamic model of "jinghai 8-b" USV is asymmetric, the model needs to be preprocessed and decoupled.

11.3.1 Virtual path generation

Design of virtual path generation controller. The path generation design block diagram is shown in Fig. 11.1. There (x_1, y_1) is the expected way-point in LOS, (x_0, y_0) is the initial point in LOS, δ_{Gr} is the rudder angle value calculated by GPC heading controller at the next moment, ψ_{Gs} is the heading angle that the ship can reach under the action of δ_{Gr}, which also serves as the expected heading angle in the path tracking process; u_r is the desired speed. Through integration and coordinate transformation, the desired position (x_r, y_r) at the next moment can be calculated, thus generating the desired path.

The relevant coefficients $K = 0.1555$ and $T_0 = 0.2214$ in Nomoto model of "jinghai 8-b" are identified through experiments. Therefore, the Nomoto model of "jinghai 8-b" USV is obtained as follows:

$$G(s)_{\delta\psi} = \frac{0.1555}{0.2214s^2 + s}. \tag{11.3}$$

As shown in the following formula, the identified wild model of "jinghai 8-b" is discretized, and the sampling period is 0.2 s:

$$\begin{aligned}&\psi_{GS}(t)-1.405\psi_{GS}(t-1)+0.4052\psi_{GS}(t-2)\\&=0.01062\delta_{GS}(t-1)+0.007875\delta_{GS}(t-2),\end{aligned}\tag{11.4}$$

$\psi_{GS}(t)$ and $\delta_{GS}(t)$ are the heading angle and virtual rudder angle at time t, respectively.

In order to find a local optimal solution, GPC is the most commonly used method. We start by constructing an evaluation function and two Diophantine equations for recursive solution [17][18]:

$$\begin{cases}J=\varepsilon\{\sum\limits_{j=N_0}^{N_1}\left(\psi_{GS}(t+j)-\psi_{LOS}(t+j)\right)^2+\sum\limits_{j=1}^{N_u}\lambda(j)\left(\Delta\delta_{GS}(t+j-1)\right)^2\}\\1=E_j(z^{-1})A_1(z^{-1})\Delta+z^{-j}F_j(z^{-1})\\E_j(z^{-1})B_1(z^{-1})=G_j(z^{-1})+z^{-j}H_j(z^{-1})\end{cases}\tag{11.5}$$

where N_0 is the minimum prediction time domain, N_1 is the maximum prediction time domain, N_u is the control time domain, $\lambda(j)$ is the control weighting sequence; A_j, B_j, E_j, F_j, G_j, H_j are the polynomials of the difference operator $\Delta=1-z^{-1}$. At the same time, rudder angle saturation should also be taken as an important factor in solving for the optimal deflection angle, namely $-30°\leq\delta_{GS}\leq 30°$.

In order to minimize the evaluation function, J provides the control law that minimizes the evaluation function as follows:

$$\Delta\delta_{GS}(t)=P^T[\psi_{LOS}-H\Delta\delta_{GS}(t-1)-F\psi_{GS}(t)]\tag{11.6}$$

where P^T takes the first row of $\left(G^TG+\lambda I\right)^{-1}G^T$, and the optimal control law at the next moment is:

$$\delta_{GS}(t)=\delta_{GS}(t-1)+\Delta\delta_{GS}(t).\tag{11.7}$$

After the design of the heading controller of the virtual ship is completed, the expected heading angle of the actual ship can be generated by combining the identified Nomoto model, and the path is generated at a fixed speed. Through the expected heading angle and speed at each moment, the expected position point (x_r, y_r) at the next moment can be generated. The next problem to be considered is how to track the expected position point (x_r, y_r) at each moment.

11.3.2 Path tracking

Because there are off-diagonal elements m_{23} and m_{32} in the inertia matrix **M** of the "jinghai 8-b" model, there will be coupling terms after the formula is expanded, so the system model needs to be preprocessed before the controller design.

Model preprocessing. In order to eliminate the coupling term and facilitate the design of the controller, a homeomorphic differential transformation is carried out on the dynamics model of the USV. We design the converted position and velocity state variables z_i $(i = 1, \ldots, 6)$ as follows:

$$\begin{cases} \dot{z}_{1e} &= z_4 - v_p \cos(z_{3e}) + z_6 z_{2e} \\ z_1 &= x + \dfrac{m_{23}}{m_{22}}(\cos(\psi) - 1) \\ z_2 &= y + \dfrac{m_{23}}{m_{22}}\sin(\psi) \\ z_3 &= \psi \\ z_4 &= u \\ z_5 &= v + \frac{m_{23}}{m_{22}} r \\ z_6 &= r \end{cases} \tag{11.8}$$

Next we carry out the following transformation design on the control input thrust and torque:

$$\begin{cases} f_1 = \dfrac{1}{m_{11}}(\tau_u + m_{22}vr + \dfrac{(m_{23} + m_{32})}{2} r^2 - d_{11}u) \\ f_2 = \dfrac{m_{22}}{m_{22}m_{33} - m_{32}m_{23}}\{\tau_r + (m_{11} - m_{22})uv + \\ [\dfrac{m_{11}m_{32}}{m_{22}} - \dfrac{(m_{23} + m_{32})}{2}]ur + \\ (m_{32}d_{22} - m_{22}d_{32})v/m_{22} - (m_{22}d_{33} - m_{32}d_{23})r/m_{22}\} \end{cases} \tag{11.9}$$

Similarly, the following transformation design is carried out on the external interference force and moment:

$$\begin{cases} f_{e1} = \dfrac{1}{m_{11}}\tau_1 \\ f_{e2} = \dfrac{m_{23}}{m_{22}m_{33} - m_{32}m_{23}}(\dfrac{m_{33}}{m_{23}}\tau_2 - \dfrac{m_{32}}{m_{22}}\tau_3) \\ f_{e3} = -\dfrac{m_{32}}{m_{22}m_{33} - m_{32}m_{23}}\tau_3 \end{cases} \tag{11.10}$$

The transformed system can be obtained by derivation as shown in the following formula:

$$\begin{cases} \dot{z}_1 = z_4\cos(z_3) - z_5\sin(z_3) \\ \dot{z}_2 = z_4\sin(z_3) + z_5\cos(z_3) \\ \dot{z}_3 = z_6 \\ \dot{z}_4 = f_1 + f_{e1} \\ \dot{z}_5 = -\alpha z_4 z_6 - \beta z_5 + \gamma z_6 + f_{e2} \\ \dot{z}_6 = f_2 + f_{e3} \end{cases} \tag{11.11}$$

where $\alpha = m_{11}/m_{22}$, $\beta = d_{22}/m_{22}$, and $\gamma = d_{22}m_{23}/m_{22}^2 - d_{23}/m_{22}$.

The actual position and actual speed of the system are $\mathbf{z} = [z_1, z_2, z_3, z_4, z_5, z_6]^{\mathrm{T}}$. The expected path and expected speed $(x_r, y_r, \psi_r, u_r, v_r, r_r)$ generated by LOS-GPC are transformed according to Eq. (11.10) to obtain the transformed expected pose and expected speed $\mathbf{z}_d = [z_{1d}, z_{2d}, z_{3d}, z_{4d}, z_{5d}, z_{6d}]^{\mathrm{T}}$, and the tracking error is defined as $\mathbf{z}_e = [z_{1e}, z_{2e}, z_{3e}, z_{4e}, z_{5e}, z_{6e}]^{\mathrm{T}}$. The control objective is to design the control inputs τ_u and τ_r so that the tracking error can converge to the initial state field with any small error. At the same time, in order to deal with the uncertainty of the model and external interference, an adaptive term is added to reduce the impact.

Design of path tracking controller. Referring to the design method of adaptive sliding mode control law in [19], the path tracking error is converted first. Then, longitudinal sliding mode adaptive law, transverse sliding mode adaptive law, steering sliding mode flow, and adaptive law are designed, respectively.

Transition from track tracking to tracking error. LOS-GPC generates the desired way-point at the next moment. After transformation as in (11.8), the X- and Y-axis can be expressed as z_{1d} and z_{2d} in the fixed coordinate system, and the heading angle is denoted as z_{3d}. Define the position and heading tracking errors under the hull coordinate system as follows:

$$\begin{bmatrix} z_{1e} \\ z_{2e} \\ z_{3e} \end{bmatrix} = \begin{bmatrix} \cos(z_3) & \sin(z_3) & 0 \\ -\sin(z_3) & \cos(z_3) & 0 \\ 0 & 0 & 1 \end{bmatrix} \begin{bmatrix} z_1 - z_{1d} \\ z_2 - z_{2d} \\ z_3 - z_{3d} \end{bmatrix}; \tag{11.12}$$

$\eta_e = [z_{1e}, z_{2e}, z_{3e}]^{\mathrm{T}}$ is defined as the tracking pose error in the hull coordinate system, and $\eta_d = [z_{1d}, z_{2d}, z_{3d}]^{\mathrm{T}}$ is the expected pose in the geodetic coordinate system.

Define $v_p = \sqrt{\dot{z}_{1d}^2 + \dot{z}_{2d}^2}$.

In order to analyze the position and pose tracking error, the derivative of Eq. (11.11) is computed as

$$\begin{cases} \dot{z}_{1e} = z_4 - v_p \cos(z_{3e}) + z_6 z_{2e} \\ \dot{z}_{2e} = z_5 + v_p \sin(z_{3e}) - z_6 z_{1e} \end{cases} \tag{11.13}$$

In order to stabilize the tracking error of the system, the Lyapunov function of the designed system is considered as

$$V_1 = \frac{1}{2}(z_{1e}^2 + z_{2e}^2). \tag{11.14}$$

The derivative of Lyapunov function V_1 is given as

$$\dot{V}_1 = z_{1e}[z_4 - v_p \cos(z_{3e})] + z_{2e}[z_5 + v_p \sin(z_{3e})]. \tag{11.15}$$

Then, using the idea of backstepping, in order to make the system satisfy Lyapunov stability condition, the derivative of V_1 needs to be negative. Here z_4 and z_3 are selected as virtual control inputs to stabilize tracking errors z_{1e} and z_{2e}, respectively. In order to avoid the Euler angle singularity of $z_3 = \psi$, the following virtual velocity variable ϖ_v is selected instead of z_{3e}:

$$\varpi_v = v_p \sin(z_{3e}); \tag{11.16}$$

z_4 and ϖ_v are taken as virtual control variables to stabilize the tracking error, and their expected values are respectively selected as follows:

$$\begin{cases} z_{4d} = v_p \cos(z_{3e}) - k_1 z_{1e}/E \\ \varpi_d = -z_5 - k_2 z_{2e}/E \end{cases} \tag{11.17}$$

where $E = \sqrt{1 + z_{1e}^2 + z_{2e}^2}$ while k_1 and k_2 are real numbers.

Based on the expected value of the virtual control quantity, the tracking errors of the virtual control quantities z_4 and ϖ_v are further defined as follows:

$$\begin{cases} z_{4e} = z_4 - z_{4d} \\ \varpi_{ve} = \varpi_v - \varpi_{vd} \end{cases} \tag{11.18}$$

Then $\dot{V}_1$ can be rewritten as:

$$\dot{V}_1 = -(k_1 z_{1e}^2 + k_2 z_{2e}^2)/E + z_{1e} z_{4e} + \varpi_{ve} z_{2e}. \tag{11.19}$$

When the tracking error of the virtual control input tends to zero, $\dot{V}_1$ can reach a negative value. So far, the tracking problem has been transformed into the problem of stabilizing the tracking error of the system. When the tracking error of the virtual control input approaches zero, the system will reach the equilibrium point.

Design of longitudinal sliding mode flow and adaptive law. After the trajectory tracking problem is transformed into the tracking error stabilization problem, it is first stabilized according to Eq. (11.15) as follows:

$$\dot{z}_{4e} = \frac{1}{m_{11}}\left(\tau_u + \tau_1 + m_{22}vr + \frac{(m_{23}+m_{32})}{2}r^2 - d_{11}u - m_{11}\dot{z}_{4d}\right). \tag{11.20}$$

In Eq. (11.20), take the uncertainty of the system as

$$U_1 = \tau_1 + m_{22}vr + \frac{(m_{23}+m_{32})}{2}r^2 - d_{11}u - m_{11}\dot{z}_{4d}. \tag{11.21}$$

The estimated value of the system uncertainty U_1 is defined as $\hat{U}_1$, and the difference between the true uncertainty and its estimated value is $\tilde{U}_1$. Select the following Lyapunov function:

$$V_2 = V_1 + \frac{1}{2}m_{11}z_{4e}^2 + \frac{1}{2}\tilde{U}_1^2. \tag{11.22}$$

The sliding mode flow of the designed control input z_4 is

$$S_1 = \lambda_1 z_{4e} + \dot{z}_{4e} + \frac{z_{1e}}{m_{11}} - \frac{\tilde{U}_1}{m_{11}}. \tag{11.23}$$

The Lyapunov function of the selected control system is

$$V_3 = V_2 + \frac{1}{2}m_{11}S_1^2. \tag{11.24}$$

The derivative of V_3 is

$$\begin{aligned}\dot{V}_3 = &-(k_1 z_{1e}^2 + k_2 z_{2e}^2)/E + \varpi_{ve} z_{2e} + m_{11}z_{4e}S_1 - \lambda_1 m_{11} z_{4e}^2 \\ &+ (z_{4e} - \dot{\hat{U}}_1)\tilde{U}_1 + S_1[\lambda_1(U_1 + \tau_u) + \dot{\tau}_u + \dot{\hat{U}}_1 + \dot{z}_{1e}].\end{aligned} \tag{11.25}$$

The designed $\dot{\tau}_u$ becomes

$$\dot{\tau}_u = -\lambda_1(\hat{U}_1 + \tau_u) - \dot{\hat{U}}_1 - \dot{z}_{1e} - m_{11}z_{4e} - k_{s1}\mathrm{sgn}(S_1) - w_{s1}S_1; \tag{11.26}$$

k_{s1} and w_{s1} are constants of reaching rate of synovial membrane controller. The adaptive law is designed as follows:

$$\dot{\hat{U}}_1 = z_{4e} + \lambda_1 S_1. \tag{11.27}$$

Design of transverse adaptive law. Stabilization of virtual control input error ϖ_{ve} is

$$\dot{\varpi}_{ve} = \dot{v}_p \sin(z_{3e}) + v_p \cos(z_{3e})(z_6 - \dot{z}_{3d}) + \frac{U_2}{m_{22}} + Q_1 \tag{11.28}$$

where U_2 is an uncertain term and Q_1 is an abbreviated term. The virtual control is selected to stabilize ϖ_{ve}; $\hat{U}_2$ is defined as the estimated value of the system uncertainty U_2, and the difference between U_2 and $\hat{U}_2$ is defined as $\tilde{U}_2$.

The influence of system uncertainty is eliminated by an adaptive law. The expected input to define z_6 is

$$z_{6d} = \dot{z}_{3d} + \frac{-\dot{v}_p \sin(z_{3e}) - \hat{U}_2/m_{22} - Q_1 - k_3 \varpi_{ve} - z_{2e}/m_{22}}{v_p \cos(z_{3e})} \tag{11.29}$$

where k_3 is a positive number. Define z_{6e} as

$$z_{6e} = z_6 - z_{6d}. \tag{11.30}$$

Select Lyapunov function as

$$V_4 = V_3 + \frac{1}{2} m_{22} \varpi_{ve}^2 + \frac{1}{2} \tilde{U}_2^2. \tag{11.31}$$

The adaptive law defining the system uncertainty estimate $\hat{U}_2$ is

$$\dot{\hat{U}}_2 = \varpi_{ve}. \tag{11.32}$$

Deriving V_4 gives

$$\begin{aligned} \dot{V}_4 = &-(k_1 z_{1e}^2 + k_2 z_{2e}^2)/E - \lambda_1 m_{11} z_{4e}^2 - k_{s1}|S_1| - w_{s1} S_1^2 - k_3 m_{22} \varpi_{ve}^2 \\ &+ m_{22} \varpi_{ve} z_{6e} v_p \cos(z_{3e}). \end{aligned} \tag{11.33}$$

Design of steering sliding mode flow and adaptive law. The auxiliary virtual control quantity z_6 is introduced above, and then the error of z_6 should be stabilized. Deriving z_{6e} gives

$$\dot{z}_{6e} = \frac{1}{\zeta}(\tau_r + U_3) \tag{11.34}$$

where $\zeta = (m_{22}m_{33} - m_{32}m_{23})/m_{22}$, U_3 is the uncertainty of the system; $\hat{U}_3$ is defined as the estimated value of the system uncertainty U_3, and the difference between the uncertainty and the estimated value is defined as $\tilde{U}_3$. The stability of r_e is proved by using the following Lyapunov function:

$$V_5 = V_4 + \frac{1}{2}\zeta z_{6e}^2 + \frac{1}{2}\tilde{U}_3^2. \tag{11.35}$$

The sliding mode flow of designed z_6 is

$$S_2 = \lambda_2 z_{6e} + \dot{z}_{6e} + \frac{m_{22}\varpi_{ve}v_p\cos(z_{3e}) - \tilde{U}_3}{\zeta} \tag{11.36}$$

where λ_2 is a constant, and $Q_2 = m_{22}\varpi_{ve}v_p\cos(z_{3e})/\zeta$.

In order to verify the stability of the sliding mode controller, the following Lyapunov function is taken for the system:

$$V_6 = V_5 + \frac{1}{2}\zeta S_2^2. \tag{11.37}$$

The designed actual control input τ_r is

$$\tau_r = -\lambda_2(\tau_r + U_3) - \dot{\hat{U}}_3 - \zeta\dot{Q}_2 - \zeta z_{6e} - k_{s2}\operatorname{sgn}(S_2) - w_{s2}S_2^2. \tag{11.38}$$

The k_{s2} and w_{s2} here are constants of the reaching rate of the sliding mode controller and are used to adjust the time when the system approaches the sliding mode flow. An adaptive law is designed as

$$\dot{\hat{U}}_3 = z_{6e} + \lambda_2 S_2. \tag{11.39}$$

We also have

$$\begin{aligned}\dot{V}_6 = &-(k_1 z_{1e}^2 + k_2 z_{2e}^2)/E - k_3\varpi_{ve}^2 - \lambda_1 z_{4e}^2 - \lambda_2 z_{6e}^2 \\ &- k_{s1}\left|S_1\right| - w_{s1}S_1^2 - k_{s2}\left|S_2\right| - w_{s2}S_2^2 \le 0.\end{aligned} \tag{11.40}$$

System stability analysis. When the USV uses the adaptive sliding mode controller designed as above, the tracking error of the control system is assumed to be $\boldsymbol{\varepsilon} = [z_{1e}, z_{2e}, z_{3e}, z_{4e}, \varpi_e, z_{6e}, \tilde{U}_1, \tilde{U}_2, \tilde{U}_3]$. After the system changes from the initial state, the tracking error will tend to zero over a period of time. When $\boldsymbol{\varepsilon}$ tends to zero, the tracking error of the system before transformation will also tend to zero, and the system is globally uniformly asymptotically stable.

11.4. Simulation experiment

11.4.1 Tracking effect

The simulation object is "jinghai 8-b" USV. The values $K = 0.1555$ and $T_0 = 0.2214$ were identified through experiments, therefore Nomoto model of "jinghai 8-b" USV was obtained. In LOS-GPC-SMC system, two kinds of simulation experiment are carried out:

1) The tracking effect of two way-points. Starting from (0 m, 0 m) to (100 m, 100 m) at a constant speed of 1 m/s. The initial heading angle of the ship is 0 rad, and the simulation is carried out in the presence of constant disturbance. The blue line (light gray line in the print version) indicates the trajectory points generated by LOS-GPC at each moment, and the red line (medium gray line in the print version) indicates the tracking situation under the action of the controller. The tracking effect is shown in Fig. 11.2.

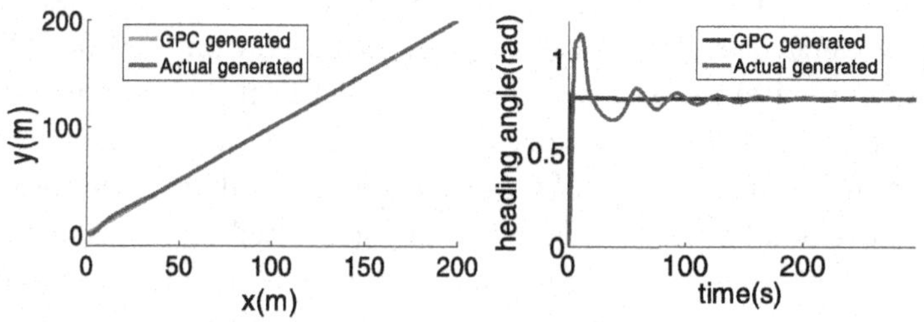

Figure 11.2 Path tracking and heading tracking diagram.

2) The tracking effect of multiple way-points. Starting from (0 m, 0 m)–(100 m, 100 m)–(300 m, 100 m)–(450 m, 0 m), task is realized at a constant speed of 1 m/s. The initial heading angle of the ship is 0 rad, and the simulation is carried out in the presence of constant disturbance. As shown in Fig. 11.3, the blue line (light gray line in the print version) represents the trajectory points generated by LOS-GPC at each time, and the red line (medium gray line in the print version) represents the tracking path under the action of the adaptive sliding mode controller. It can be seen from the figure that the USV has reached all the designated way-points, and the trajectory is relatively smooth. Only at (100 m, 100 m) did the trajectory show large amplitude jitter, but it also stabilized the trajectory quickly.

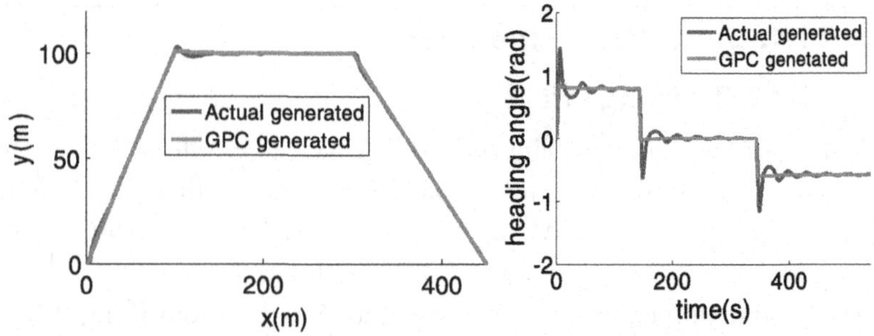

Figure 11.3 Path tracking and heading tracking diagram.

11.4.2 Comparative analysis of simulation

If LOS-GPC is not used to generate the desired trajectory at each moment and LOS guidance is directly used, the actual trajectory will deflect greatly during tracking because the desired trajectory does not consider the physical limitations of its driver. As shown in Fig. 11.4, the desired trajectory generated by the hull from (0 m, 0 m)–(100 m, 100 m)–(200 m, 0 m)–(300 m, 100 m) blue line (light gray line in the print version). The black line is the desired trajectory generated by LOS, and the path tracking (hereinafter referred to as LOS-SMC system) is directly performed by the adaptive sliding mode. The trajectory generated by the actual ship is illustrated when the red line (medium gray line in the print version) is in LOS-GPC-SMC state. Comparing the LOS-SMC path and heading tracking results with the LOS-GPC-SMC tracking result, it can be seen that LOS-GPC-SMC is superior to LOS-SMC in the recovery performance of tracking the desired trajectory. After passing through the inflection point, LOS-GPC-SMC can return to the desired trajectory faster.

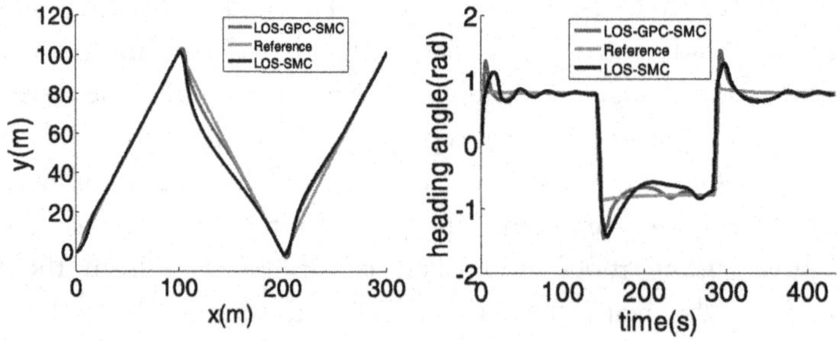

Figure 11.4 Path tracking and heading tracking.

11.5. Conclusions

In order to solve the problem of how to realize way-point tracking control of an underactuated USV with asymmetric model under the condition of interference, this chapter proposes using LOS-GPC to generate the desired trajectory, decouple the model through homeomorphic differential transformation, and design the hull controller with sliding mode control to design the whole system. In order to simplify the control, the transformation relationship between the virtual rudder angle of "jinghai 8-b" and the double propellers is identified through experiments. In addition, in order to include the physical properties of the hull in the generated virtual path, the Nomoto model of the hull is identified through experiments, and the model is applied to LOS-GPC to generate the desired trajectory. The simulation results show that the whole system has good robustness in the presence of external interference.

References

[1] S. Oh, J. Sun, Path following of underactuated marine surface vessels using line-of-sight based model predictive control, Ocean Engineering 37 (2) (2010) 289–295.

[2] L. Changxi, F. Yuan, Y. Li, Design of a linear track controller for incomplete driving ships based on back stepping method, Ship Engineering 30 (4) (2008) 64–67.

[3] J. Gao, P. Wu, T. Li, et al., Optimization-based model reference adaptive control for dynamic positioning of a fully actuated underwater vehicle, Nonlinear Dynamics 87 (4) (2016) 1–13.

[4] P. Yan, W. Weiqing, L. Mei, UAV track tracking GPC–PID cascade control, Control Engineering 21 (2) (2014) 245–248.

[5] R. Yu, Q. Zhu, G. Xia, Sliding mode tracking control of an underactuated surface vessel, IET Control Theory & Applications 6 (3) (2012) 461–466.

[6] L. Liu, W. Dan, P. Zhouhua, Predictor-based line-of-sight guidance law for path following of underactuated marine surface vessels, in: Sixth International Conference on Intelligent Control & Information Processing, 2016.

[7] L.C. McNinch, K.R. Muske, H. Ashrafiuon, Model-based predictive control of an unmanned surface vessel, in: Proceedings of the 11th IASTED International Conference on Intelligent Systems and Control, 2008, pp. 385–390.

[8] W. Xiaofei, Z. Zaojian, L. Tianshan, Adaptive path following controller of underactuated ships using Serret–Frenet frame, Journal of Shanghai Jiaotong University (Science) 15 (3) (2010) 334–339.

[9] Z. Qidan, Y. Ruiting, X. Guihua, Sliding mode robust control for track tracking of underactuated ships with wind wave and current disturbance and parameter uncertainty, Control Theory and Application 29 (7) (2012) 000959.

[10] Z. Jun, S. Tairen, L. Zhilin, Robust model predictive control for path-following of underactuated surface vessels with roll constraints, Ocean Engineering 143 (2017) 125–132.

[11] R.A. Soltan, H. Ashrafiuon, K.R. Muske, State-dependent trajectory planning and tracking control of unmanned surface vessels, in: American Control Conference, IEEE, 2009.
[12] M. Bibuli, G. Bruzzone, M. Caccia, Line following guidance control: application to the Charlie USV, in: IEEE/RSJ International Conference on Intelligent Robots & Systems, DBLP, 2008.
[13] J.M. Daly, M.J. Tribou, S.L. Waslander, A nonlinear path following controller for an underactuated unmanned surface vessel, in: 2012 IEEE/RSJ International Conference on Intelligent Robots and Systems, IEEE, 2012, pp. 82–87.
[14] D. Xu, Y. Liao, Y. Pang, Backstepping control method for the path following for the underactuated surface vehicles, Procedia Engineering 15 (7) (2011) 256–263.
[15] K.Y. Pettersen, E. Lefeber, Way-point tracking control of ships, in: IEEE Conference on Decision & Control, 2001.
[16] T.I. Fossen, Handbook of Marine Craft Hydrodynamics and Motion Control, John Wiley & Sons, 2011.
[17] W. Wei, Generalized Predictive Control Theory and Its Application, Science Press, 1998.
[18] D.W. Clarke, C. Mohtadi, P.S. Tuffs, Generalized predictive control—Part I. The basic algorithm, Automatica 23 (87) (1987) 137–148.
[19] W. Jian, M. Wang, L. Qiao, Dynamical sliding mode control for the trajectory tracking of underactuated unmanned underwater vehicles, Ocean Engineering 105 (2015) 54–63.

CHAPTER TWELVE

ESO-based guidance law for distributed path maneuvering of multiple autonomous surface vehicles with a time-varying formation

Zhouhua Peng, Nan Gu, Lu Liu, Dan Wang, Tieshan Li

Contents

Abstract

This chapter is concerned with the distributed guidance law design for cooperative path maneuvering of multiple autonomous surface vehicles guided by one parameterized path. At first, distributed guidance laws based on neighboring information are proposed for fully-actuated autonomous surface vehicles over a directed network where an observer is used to estimate the unknown velocity information of neighboring surface vehicles and the derivatives of relative deviations. Then, this work is extended to an underactuated case. For both cases, it is proven that the error signals in the guidance loop are input-to-state stable via cascade stability analysis. Moreover, the prescribed guidance signals are bounded in spite of any initial positions, thereby facilitating practical implementations. Simulation results illustrate the effectiveness of

Fundamental Design and Automation Technologies in Offshore Robotics
https://doi.org/10.1016/B978-0-12-820271-5.00017-1

the proposed distributed guidance law for cooperative path maneuvering of fully-actuated and underactuated autonomous surface vehicles.

Keywords

Autonomous Surface Vehicles, Distributed Guidance Law, Cooperative Path Maneuvering

During the past two decades, formation control of multiple autonomous surface vehicles (ASVs) has drawn a wide of research interest [1–5]. In order to achieve various formation patterns, many formation control methods have been proposed such as leader–follower approach [6–12], virtual architecture, behavioral mechanism, and graph-based method [13–40]. By virtue of mission scenario, distributed cooperative schemes can be roughly classified into cooperative trajectory tracking, cooperative target tracking, and cooperative path following [20–23,26–29,31–38]. In particular, cooperative path maneuvering of ASVs guided by one or multiple parameterized paths has been widely studied due to its salient feature of spatial and temporal decoupling [20–23,26–29,31–38]. Cooperative path maneuvering guided by multiple parameterized paths is considered in [20–26,30,33,34,38] where the path variables are synchronized. In these studies, each surface vehicle should be assigned a parameterized path known as a priori. This means that the path information is global. Without using the global knowledge of the path information, distributed path maneuvering of ASVs guided by one or multiple paths is considered in [27–29,31,32,35–37].

This chapter is concerned with the guidance law design for distributed path maneuvering of a swarm of fully-actuated and under-actuated ASVs guided by one parameterized path. At first, robust guidance laws based on relative position and yaw information are proposed for multiple fully-actuated ASVs where extended state observers (ESOs) are used to recover the unavailable linear velocities and yaw rates of neighbors. Next, ESO-based guidance laws based on an auxiliary variable design are proposed for multiple under-actuated ASVs where ESOs are used to estimate the unknown velocity information of neighbors and the derivatives of relative deviations. For both cases, it is proven that the distributed path maneuvering errors in the guidance loop are input-to-state stable based on based on cascade stability analysis. Besides, the prescribed guidance signals are bounded in spite of any initial positions, facilitating practical implementations. Simulation results are given to illustrate the effectiveness of the proposed ESO-based guidance law for distributed path maneuvering of fully-actuated and underactuated ASVs without using the linear velocity and yaw rate information of neighbors.

This chapter is organized as follows. Section 12.1 gives the problem formulation of guidance laws for distributed path maneuvering of ASVs. Sections 12.2 and 12.3 present the distributed guidance law design and analysis for fully-actuated and underactuated ASVs, respectively. Section 12.4 concludes this chapter.

12.1. Problem formulation

As shown in Fig. 12.1, consider a swarm of N ASVs moving along a parameterized path. The kinematic equation of the ith ASV is described by [20]

$$\dot{\eta}_i = R_i(\psi_i)\nu_i, \tag{12.1}$$

where $\eta_i = [p_i, \psi_i]^T \in \mathbb{R}^3$ with $p_i = [x_i, y_i]^T \in \mathbb{R}^2$ being the position in the earth-fixed frame, and $\psi_i \in (-\pi, \pi]$ being the yaw angle; $\nu_i = [u_i, v_i, r_i]^T \in \mathbb{R}^3$ denotes the surge, sway, and yaw velocities in the body-fixed reference frame; $R_i(\psi_i) = \text{diag}\{\bar{R}_i(\psi_i), 0\} \in \mathbb{R}^{3\times3}$ is a rotation matrix with

$$\bar{R}_i(\psi_i) = \begin{bmatrix} \cos(\psi_i) & -\sin(\psi_i) \\ \sin(\psi_i) & \cos(\psi_i) \end{bmatrix}.$$

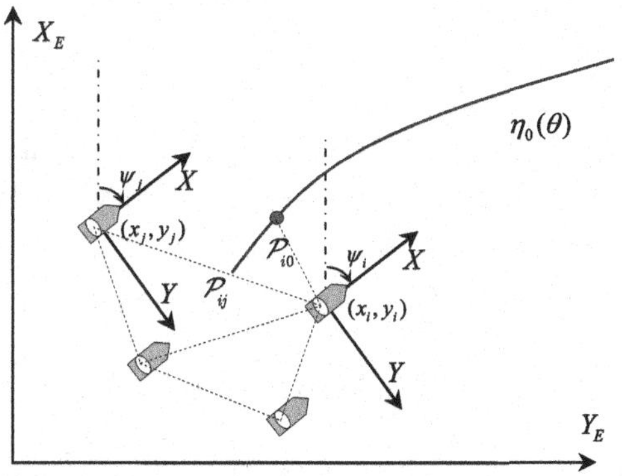

Figure 12.1 Illustration of distributed cooperative path maneuvering.

The vehicle fleet is guided by a parameterized path information known as a priori as follows

$$\begin{cases} \eta_0(\theta) = [p_0^T(\theta), \psi_0(\theta)]^T, \quad \text{with } p_0(\theta) = [x_0(\theta), y_0(\theta)]^T, \\ \psi_0(\theta) = 2k\pi + \text{atan2}(y_0^\theta(\theta), x_0^\theta(\theta)), \end{cases} \tag{12.2}$$

where $\theta \in \mathbb{R}$ is a path variable, and $k \in \mathbb{R}$ is an integer which is used to assure that ψ_0 is continuous. Let $\eta_0^\theta(\theta) = \partial\eta_0(\theta)/\partial\theta$ and $p_0^\theta(\theta) = \partial p_0(\theta)/\partial\theta$ be the partial derivatives of $\eta_0(\theta)$ and $p_0(\theta)$, respectively. It is assumed that they are bounded.

The communication graph among the N ASVs and the path can be described by an graph $\overline{\mathcal{G}} = \{\overline{\mathcal{V}}, \overline{\mathcal{E}}\}$ where $\overline{\mathcal{V}} = \{0, \mathcal{V}\}$ with $\mathcal{V} = \{1, \ldots, N\}$, $\overline{\mathcal{E}} = \{(i,j) \in \overline{\mathcal{V}} \times \overline{\mathcal{V}}\}$. For the graph $\overline{\mathcal{G}}$, we make the following assumption.

Assumption 12.1. The graph $\overline{\mathcal{G}}$ contains a spanning tree with the root node being the node 0.

The guidance law design for distributed path maneuvering of multiple ASVs at the kinematic level is to achieve the following two objectives:

- *Geometric objective* – force the ith vehicle to converge to a parameterized path with a prescribed formation pattern

$$\text{For a fully-actuated vehicle,} \quad \lim_{t\to\infty} \|\eta_i(t) - \mathcal{P}_{id}(t) - \eta_0(\theta(t))\| \le l_1, \tag{12.3}$$

$$\text{For an underactuated vehicle,} \quad \lim_{t\to\infty} \|p_i(t) - p_{id}(t) - p_0(\theta(t))\| \le l_2, \tag{12.4}$$

  where $\mathcal{P}_{id}(t) \in \mathbb{R}^3$ and $p_{id}(t) \in \mathbb{R}^2$ denote the relative deviation between the ith vehicle and the reference path, and they can be time-varying; $l_1 \in \mathbb{R}$ and $l_2 \in \mathbb{R}$ are small positive constant.

- *Dynamic objective* – drive $\dot{\theta}$ to satisfy

$$\lim_{t\to\infty} \|\dot{\theta}(t) - v_s\| \le l_3, \tag{12.5}$$

  where v_s is a desired path update speed and $l_3 \in \mathbb{R}$ is a small positive constant.

12.2. ESO-based distributed guidance law for distributed path maneuvering of fully-actuated ASVs

In this section, ESO-based guidance laws based on relative position and yaw information of neighbors are proposed for distributed path maneuvering of fully-actuated ASVs. Then, the stability of the closed-loop system is presented via cascade analysis. Finally, simulation results are provided to verify the effectiveness of the proposed ESO-based distributed guidance law.

12.2.1 ESO-based distributed guidance law design

First, a distributed path maneuvering error based on the positions of neighbors is defined as follows:

$$e_i = R_i^T\Big\{\sum_{j\in\mathcal{N}_i} a_{ij}(\eta_i-\eta_j-\mathcal{P}_{ijd}) + a_{i0}[\eta_i-\eta_0(\theta)-\mathcal{P}_{i0d}]\Big\}, \tag{12.6}$$

where R_i denotes $R_i(\psi_i)$; $\mathcal{P}_{ijd}=\mathcal{P}_{id}-\mathcal{P}_{jd}\in\mathbb{R}^3$; for $j=0,\ldots,N$, $a_{ij}=1$ if the ith node accesses the information of the jth node, otherwise, $a_{ij}=0$; $\mathcal{N}_i=\{j\in\mathcal{V}|(i,j)\in\overline{\mathcal{E}}\}$ denotes the neighbors set of the ith vehicle.

Using the vehicle kinematics (12.1), the error dynamics of e_i can be expressed by

$$\dot{e}_i = -r_i S e_i + \sum_{j=0}^{N} a_{ij}\nu_i - \sum_{j\in\mathcal{N}_i} a_{ij}R_i^T\dot{\eta}_j - a_{i0}R_i^T\eta_0^\theta(\theta)\dot{\theta} - \sum_{j\in\overline{\mathcal{N}}_i} a_{ij}R_i^T\dot{\mathcal{P}}_{ijd}, \tag{12.7}$$

where $\overline{\mathcal{N}}_i=\{j\in\overline{\mathcal{V}}|(i,j)\in\overline{\mathcal{E}}\}$; $S=\mathrm{diag}\{\bar{S},0\}$ with

$$\bar{S}=\begin{bmatrix}0 & -1\\ 1 & 0\end{bmatrix}.$$

Define

$$\dot{\theta}=\nu_s-\zeta,\quad e_{i\nu}=\nu_i-\nu_{ic}, \tag{12.8}$$

where $\zeta\in\mathbb{R}$ is a variable to be designed later; $\nu_{ic}=[u_{ic},v_{ic},r_{ic}]^T\in\mathbb{R}^3$ is a guidance signal vector.

Substituting (12.8) into (12.7) yields

$$\dot{e}_i = -r_i S e_i + \sum_{j=0}^{N} a_{ij}(\nu_{ic} + e_{i\nu}) - \sum_{j\in\mathcal{N}_i} a_{ij} R_i^T \dot{\eta}_j - a_{i0} R_i^T \eta_0^\theta(\theta)(\nu_s - \zeta)$$
$$- \sum_{j\in\overline{\mathcal{N}}_i} a_{ij} R_i^T \dot{\mathcal{P}}_{ijd}. \tag{12.9}$$

Note that the velocity information of neighboring surface vehicles, $\dot{\eta}_j$, and the derivative of the relative deviations, $\dot{\mathcal{P}}_{ijd}$, could be unknown. To address it, an ESO is established to estimate them as follows [41–43]:

$$\begin{cases} \dot{\hat{e}}_i = -K_{i1}^o(\hat{e}_i - e_i) + \sum_{j=0}^N a_{ij}(e_{i\nu} + \nu_{ic}) + \hat{\sigma}_i - a_{i0} R_i^T \eta_0^\theta(\theta)(\nu_s - \zeta), \\ \dot{\hat{\sigma}}_i = -K_{i2}^o(\hat{e}_i - e_i), \end{cases} \tag{12.10}$$

where K_{i1}^o and $K_{i2}^o \in \mathbb{R}^{3\times 3}$ are diagonal observer gains; $\hat{e}_i$ and $\hat{\sigma}_i$ are the estimates of e_i and σ_i, respectively, with $\sigma_i = -r_i S e_i - \sum_{j\in\mathcal{N}_i} a_{ij} R_i^T \dot{\eta}_j - \sum_{j\in\overline{\mathcal{N}}_i} a_{ij} R_i^T \dot{\mathcal{P}}_{ijd}$.

The following assumption is made to facilitate the stability analysis.

Assumption 12.2 ([44]). The unknown signal σ_i and its derivative $\dot{\sigma}_i$ are bounded. There exists a positive constant σ_i^* such that $\|\sigma_i\| + \|\dot{\sigma}_i\| \leq \sigma_i^*$.

Based on (12.10), the guidance signal ν_{ic} is proposed as follows:

$$\nu_{ic} = \Big(-\frac{K_i \hat{e}_i}{\Pi_i} - \hat{\sigma}_i + a_{i0} R_i^T \eta_0^\theta(\theta)\nu_s\Big) / \sum_{j=0}^{N} a_{ij}, \tag{12.11}$$

where $K_i = \mathrm{diag}\{k_{ix}, k_{iy}, k_{i\psi}\}$ with $k_{ix} \in \mathbb{R}$, $k_{iy} \in \mathbb{R}$, and $k_{i\psi} \in \mathbb{R}$ being positive constants; $\Pi_i = \sqrt{\|e_i\|^2 + \varepsilon_i^2}$ where $\varepsilon_i \in \mathbb{R}$ is used to regulate the transient performance.

Defining the estimation errors

$$\tilde{e}_i = \hat{e}_i - e_i, \quad \tilde{\sigma}_i = \hat{\sigma}_i - \sigma_i, \tag{12.12}$$

one has

$$\dot{\tilde{e}}_i = -K_{i1}^o \tilde{e}_i + \tilde{\sigma}_i, \quad \dot{\tilde{\sigma}}_i = -K_{i2}^o \tilde{e}_i - \dot{\sigma}_i. \tag{12.13}$$

Letting $E_{i1} = [\tilde{e}_i^T, \tilde{\sigma}_i^T]^T$, the estimation error dynamics can be expressed by

$$\dot{E}_{i1} = A_i E_{i1} + B_i \xi_i, \tag{12.14}$$

where

$$A_i = \begin{bmatrix} -K_{i1}^o & I_3 \\ -K_{i2}^o & 0_3 \end{bmatrix}, \quad B_i = \begin{bmatrix} 0_3 & 0_3 \\ 0_3 & I_3 \end{bmatrix}, \quad \xi_i = [0, 0, -\dot{\sigma}_i^T].$$

Noting that A_i is a Hurwitz matrix, there exists a positive definite matrix P_i satisfying $A_i^T P_i + P_i A_i = -I_6$.

Substituting (12.11) into (12.10), the error dynamics is expressed as

$$\dot{\hat{e}}_i = -K_i \hat{e}_i / \Pi_i - K_{i1}^o \tilde{e}_i + \sum_{j=0}^{N} a_{ij} e_{iv} + a_{i0} R_i^T \eta_0^\theta(\theta) \zeta. \tag{12.15}$$

To design an update law for $\dot{\theta}$, the evolution of ζ is chosen as

$$\dot{\zeta} = -\lambda[\zeta + \mu \sum_{i \in \mathcal{N}_0} a_{0i} e_i^T R_i^T \eta_0^\theta(\theta)], \tag{12.16}$$

where $\mu \in \mathbb{R}$ and $\lambda \in \mathbb{R}$ are positive constants.

In light of (12.15) and (12.16), the error dynamics for distributed path maneuvering is given by

$$\begin{cases} \dot{\hat{e}}_i = -K_i \hat{e}_i / \Pi_i - K_{i1}^o \tilde{e}_i + \sum_{j=0}^{N} a_{ij} e_{iv} + a_{i0} R_i^T \eta_0^\theta(\theta) \zeta, \\ \dot{\zeta} = -\lambda[\zeta + \mu \sum_{i \in \mathcal{N}_0} a_{0i} e_i^T R_i^T \eta_0^\theta(\theta)]. \end{cases} \tag{12.17}$$

12.2.2 Stability analysis

The following lemma presents the stability of (12.14).

Lemma 12.1. *Under Assumption 12.2, the estimation error dynamics in (12.14) is input-to-state stable.*

Proof. Choose the Lyapunov function

$$V_{i1} = \frac{1}{2} E_{i1}^T P_i E_{i1}. \tag{12.18}$$

Taking the time derivative of V_{i1} along (12.14), it follows that

$$\dot{V}_{i1} = E_{i1}^T P_i (A_i E_{i1} + B_i \xi_i)$$

$$
\begin{aligned}
&= -\frac{1}{2}E_{i1}^T E_{i1} + E_{i1}^T P_i B_i \xi_i \\
&\le -\frac{1}{2}\|E_{i1}\|^2 + \|E_{i1}\|\|P_i B_i\|\|\xi_i\|.
\end{aligned} \tag{12.19}
$$

Under Assumption 12.2, the input vector ξ_i is bounded. When $\|E_{i1}\| \ge 4\|P_i B_i\|\|\xi_i\|$, one has $\dot{V}_{i1} \le -1/4\|E_{i1}\|^2$. Using Theorem 4.6 in [45], the estimation error dynamics in (12.14) is input-to-state stable. □

Lemma 12.2. *The kinematic error dynamics in (12.17) is input-to-state stable provided that* $e_{iv} = 0$.

Proof. Consider the following Lyapunov function

$$
V_2 = \sum_{i=1}^{N} \frac{\hat{e}_i^T \hat{e}_i}{2} + \frac{\zeta^2}{2\lambda\mu}, \tag{12.20}
$$

then its time derivative along (12.17) can be expressed as

$$
\begin{aligned}
\dot{V}_2 &= \sum_{i=1}^{N}\{-\frac{\hat{e}_i^T K_i \hat{e}_i}{\Pi_i} - \hat{e}_i^T K_{i1}^o \tilde{e}_i + \hat{e}_i^T \sum_{j=0}^{N} a_{ij} e_{iv} + a_{i0}\hat{e}_i^T \eta_0^\theta(\theta)\zeta\} + \frac{\zeta\dot{\zeta}}{\lambda\mu}, \\
&= -\sum_{i=1}^{N} \frac{\hat{e}_i^T K_i \hat{e}_i}{\Pi_i} - \sum_{i=1}^{N} \hat{e}_i^T K_{i1}^o \tilde{e}_i + \sum_{i=1}^{N} \hat{e}_i^T \sum_{j=0}^{N} a_{ij} e_{iv} - \frac{\zeta^2}{\mu}.
\end{aligned} \tag{12.21}
$$

By assuming a perfect velocity tracking, it follows that

$$
\begin{aligned}
\dot{V} &= -\sum_{i=1}^{N} \frac{\hat{e}_i^T K_i \hat{e}_i}{\Pi_i} - \sum_{i=1}^{N} \hat{e}_i^T K_{i1}^o \tilde{e}_i - \frac{\zeta^2}{\mu}, \\
&\le -\sum_{i=1}^{N} \frac{\lambda_{\min}(K_i)}{\Pi_i}\|\hat{e}_i\|^2 - \frac{\zeta^2}{\mu} + \sum_{i=1}^{N} \lambda_{\max}(K_{i1}^o)\|\tilde{e}_i\|\|\hat{e}_i\|, \\
&\le -c_1\|E_2\|^2 + c_2(\sum_{i=1}^{N}\|E_{i1}\|)\|E_2\|,
\end{aligned} \tag{12.22}
$$

where $E_2 = [\hat{e}_1^T, \ldots, \hat{e}_N^T, \zeta]^T$, $c_1 = \min\{\lambda_{\min}(K_1)/\Pi_1, \ldots, \lambda_{\min}(K_N)/\Pi_N, 1/\mu\}$, and $c_2 = \max\{\lambda_{\max}(K_{11}^o), \ldots, \lambda_{\max}(K_{N1}^o)\}$. Since $\|E_{i1}\|$ is bounded, according to Lemma 12.1, it can be concluded that system in (12.17) is input-to-state stable. This proof is complete. □

Theorem 12.1. *Consider a swarm of fully-actuated ASVs with kinematics in (12.1), the guidance laws in (12.11) and (12.16), and the ESO in (12.10). Under Assumptions 12.1 and 12.2, the closed-loop system cascaded by (12.14) and (12.17) is input-to-state stable provided that* $e_{iv} = 0$.

Proof. The closed-loop system can be seen as a cascade system consisting of (12.14) and (12.17). By using Theorem 4.7 in [45], the closed-loop system is ISS. This proof is complete. □

12.2.3 Simulation results

In this subsection, an example is provided to illustrate the effectiveness of the proposed distributed guidance law in (12.11) and (12.16). Consider a networked system consisting of five vehicles, labeled 1 to 5, with the model parameters given in [46], and a parameterized path given by

$$\eta_0(\theta) = \begin{cases} [0.1\theta + 20; 0; 0], \quad \theta < 400, \\ [60 + 60\sin(0.003(\theta - 400)); 60(1 - \cos(0.003(\theta - 400))); \\ \qquad 0.003(\theta - 400)], \quad 400 \leq \theta < 400 + \pi/0.003, \\ [-0.1(\theta - 400 - \pi/0.003) + 60; 120; \pi], 400 + \pi/0.003 \leq \theta, \end{cases}$$

with θ being the path variable. The communication graph among the ASVs and the path is given in Fig. 12.2. The formation pattern is set to $\mathcal{P}_1 = [0, 0, 0]^T$, $\mathcal{P}_2 = [-4, 0, 0]^T$, $\mathcal{P}_3 = [-8, 0, 0]^T$, $\mathcal{P}_4 = [-12, 0, 0]^T$, and $\mathcal{P}_5 = [-16, 0, 0]^T$. The control parameters are chosen as $K_i = \text{diag}\{0.2, 0.2, 0.2\}$, $\Delta_i = 2$, $\mu = 10$, $\lambda = 10$, $K_{i1}^o = \text{diag}\{20, 20, 20\}$, $K_{i2}^o = \text{diag}\{100, 100, 100\}$.

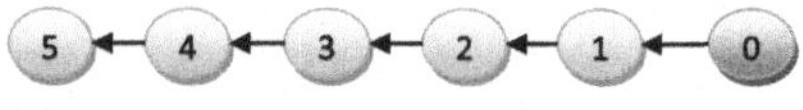

Figure 12.2 Communication graph.

The simulation results are illustrated in Figs. 12.3–12.8. Fig. 12.3 reveals that a desired formation is well maintained. Fig. 12.4 depicts the distributed path maneuvering errors e, and it reveals that the distributed path maneuvering errors converge to zero. Fig. 12.5 shows that the velocity guidance signals are all bounded with the bound known as a priori. Fig. 12.6 demonstrates the evolution of the path variable.

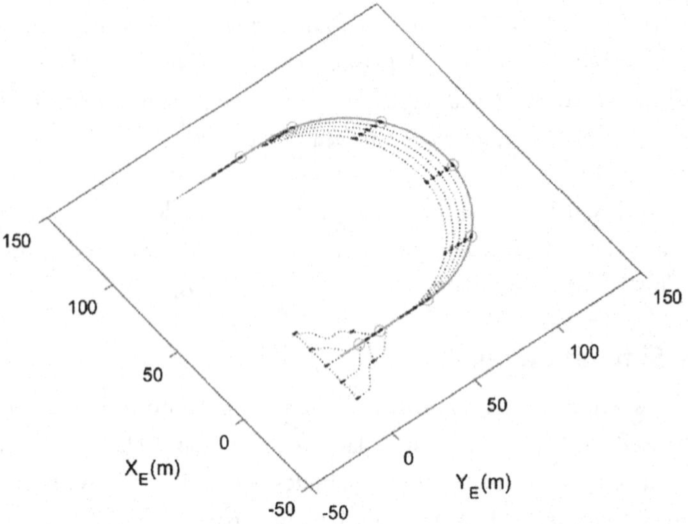

Figure 12.3 Formation pattern using the proposed distributed guidance law.

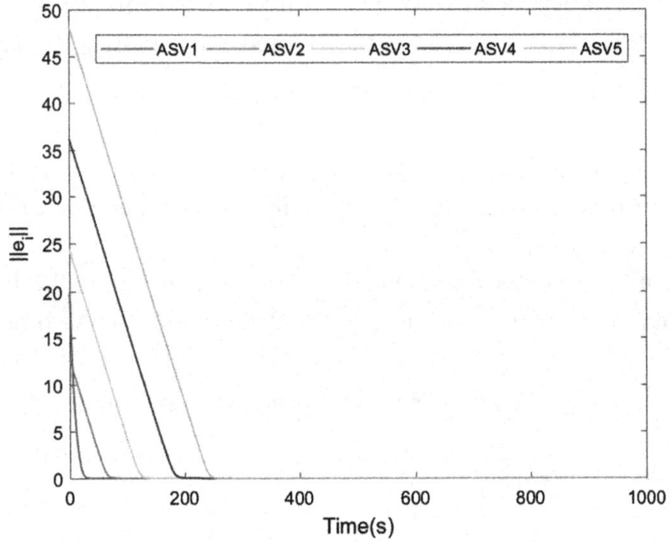

Figure 12.4 Profile of the distributed path maneuvering errors of five fully-actuated ASVs.

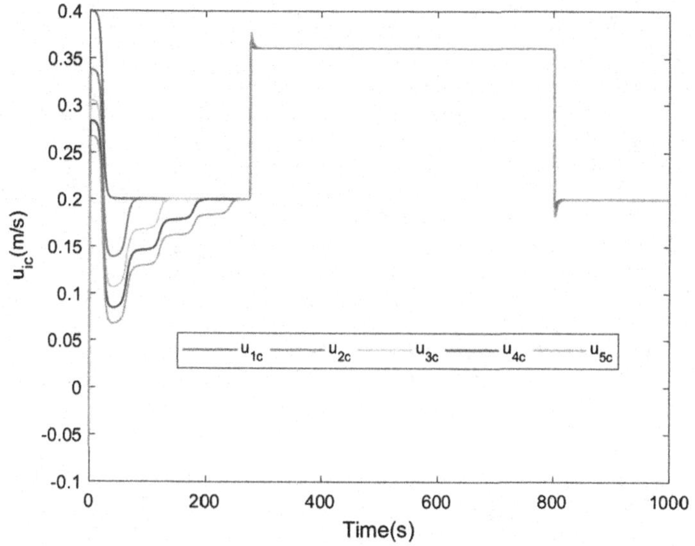

Figure 12.5 Profile of the commanded surge velocities for five fully-actuated ASVs.

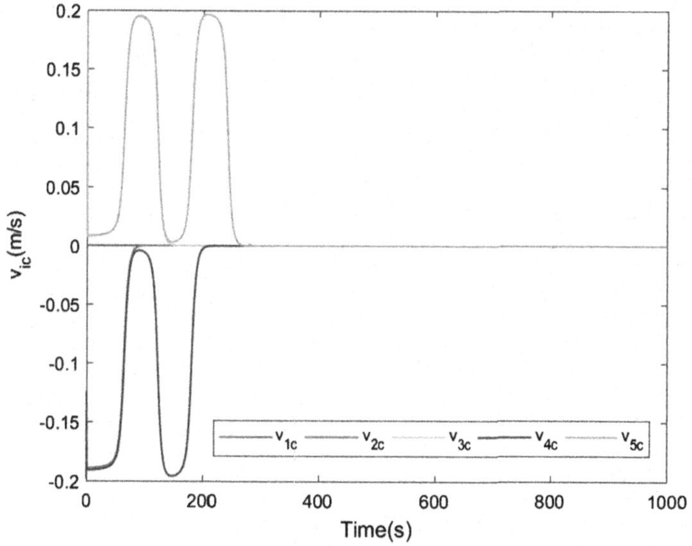

Figure 12.6 Profile of the commanded sway velocities for five fully-actuated ASVs.

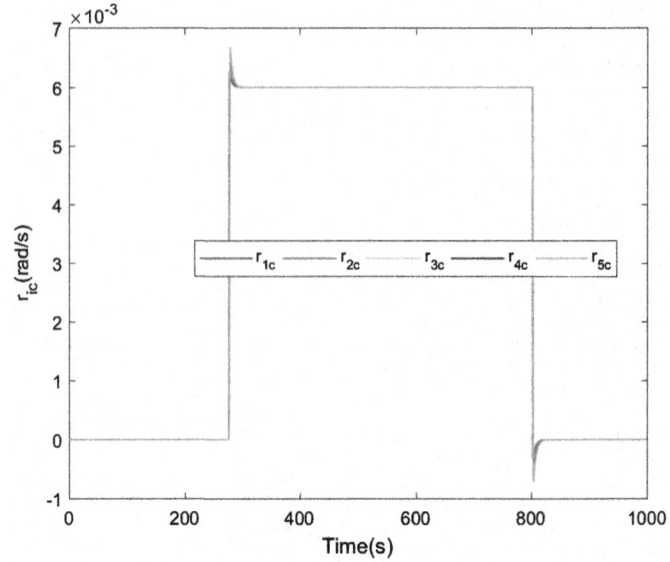

Figure 12.7 Profile of the commanded yaw rates for five fully-actuated ASVs.

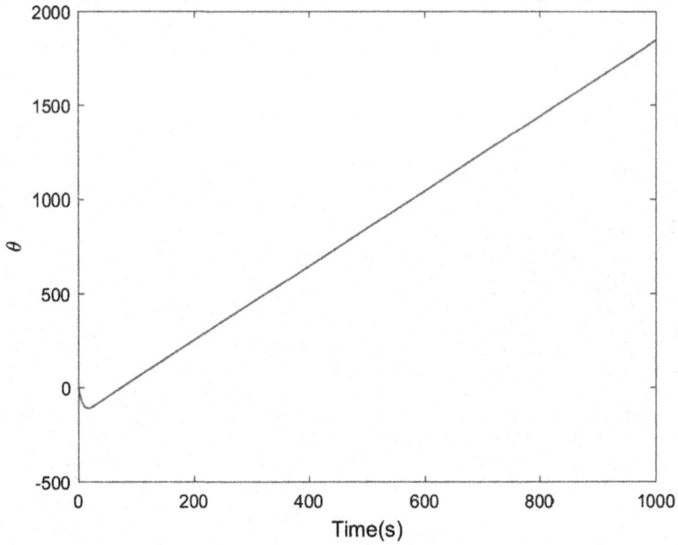

Figure 12.8 Profile of the path variable.

12.3. ESO-based distributed guidance law for distributed path maneuvering of underactuated ASVs

In the previous section, the ESO-based guidance laws are designed for distributed path maneuvering of fully-actuated ASVs. In this section, ESO-based guidance laws are designed for distributed path maneuvering of under-actuated ASVs. In order to address the underactuation, an auxiliary variable method is incorporated into the ESO-based guidance laws design. Then, the stability of the closed-loop guidance system is presented via cascade analysis. A simulation example is given to illustrate the effectiveness of the proposed ESO-based guidance laws for under-actuated ASVs.

12.3.1 ESO-based distributed guidance law design

First, a distributed path maneuvering error is defined as

$$e_i = \bar{R}_i^T\{\sum_{j\in\mathcal{N}_i} a_{ij}(p_i - p_j - p_{ijd}) + a_{i0}[p_i - p_0(\theta) - p_{i0d}]\}, \tag{12.23}$$

where $p_{ijd} = p_{id} - p_{jd}$ is a relative deviation which can be time-varying.

Define

$$\dot{\theta} = v_s - \zeta, \quad \bar{e}_{iv} = \bar{v}_i - \bar{v}_{ic}, \tag{12.24}$$

where $\zeta \in \mathbb{R}$ is a variable to be designed later; $\bar{v}_{ic} = [u_{ic}, r_{ic}]^T \in \mathbb{R}^2$ is a guidance signal to be designed for underactuated autonomous surface vehicle.

Taking the time derivative of (12.23), it follows that

$$\begin{aligned}\dot{e}_i = & -r_i\bar{S}e_i + \sum_{j=0}^{N} a_{ij}\begin{bmatrix} u_i \\ v_i \end{bmatrix} - \sum_{j\in\mathcal{N}_i} a_{ij}\bar{R}_i^T\dot{p}_j - a_{i0}\bar{R}_i^T p_0^\theta(\theta)(v_s - \zeta) \\ & - \sum_{j\in\overline{\mathcal{N}}_i} a_{ij}\bar{R}_i^T\dot{p}_{ijd}.\end{aligned} \tag{12.25}$$

In order to address the underactuation, an auxiliary variable $\bar{\delta}_0$ is introduced such that

$$\dot{\bar{e}}_i = -r_i\bar{S}\bar{e}_i + \sum_{j=0}^{N} a_{ij}\begin{bmatrix} u_i \\ v_i \end{bmatrix} - r_i\bar{S}\bar{\delta}_0 - \sum_{j\in\mathcal{N}_i} a_{ij}\bar{R}_i^T\dot{p}_j - a_{i0}\bar{R}_i^T p_0^\theta(\theta)(v_s-\zeta) - \sum_{j\in\overline{\mathcal{N}}_i} a_{ij}\bar{R}_i^T\dot{p}_{ijd}, \tag{12.26}$$

where $\bar{e}_i = e_i - \bar{\delta}_0$ and $\bar{\delta}_0 = [-\delta_0, 0]^T \in \mathbb{R}^2$ with $\delta_0 \in \mathbb{R}$ being a positive constant.

Transforming (12.26) into a compact form, one has

$$\dot{\bar{e}}_i = -r_i\bar{S}\bar{e}_i + B_i(\bar{e}_{iv} + \bar{v}_{ic}) - \sum_{j\in\mathcal{N}_i} a_{ij}\bar{R}_i^T\dot{p}_j - a_{i0}\bar{R}_i^T p_0^\theta(\theta)(v_s-\zeta) - \sum_{j\in\overline{\mathcal{N}}_i} a_{ij}\bar{R}_i^T\dot{p}_{ijd} + \sum_{j=0}^{N} a_{ij}\begin{bmatrix} 0 \\ v_i \end{bmatrix}, \tag{12.27}$$

where $B_i = \text{diag}\{\sum_{j=0}^N a_{ij}, \delta_0\}$.

In (12.27), the velocity information of neighboring surface vehicles $\dot{p}_j$ and the derivative of the relative deviations $\dot{p}_{ijd}$ could be unknown. An ESO is used to estimate them as follows:

$$\begin{cases} \dot{\hat{\bar{e}}}_i = -K_{i1}^o(\hat{\bar{e}}_i - \bar{e}_i) + B_i(\bar{e}_{iv} + \bar{v}_{ic}) + \hat{\sigma}_i - a_{i0}\bar{R}_i^T p_0^\theta(\theta)(v_s - \zeta), \\ \dot{\hat{\sigma}}_i = -K_{i2}^o(\hat{\bar{e}}_i - \bar{e}_i), \end{cases} \tag{12.28}$$

where $K_{i1}^o, K_{i2}^o \in \mathbb{R}^{2\times 2}$ are diagonal observer gains; $\hat{\bar{e}}_i$ and $\hat{\sigma}_i$ are the estimates of $\bar{e}_i$ and σ_i, respectively, with $\sigma_i = -r_i\bar{S}\bar{e}_i - \sum_{j\in\mathcal{N}_i} a_{ij}\bar{R}_i^T\dot{p}_j - \sum_{j\in\overline{\mathcal{N}}_i} a_{ij}\bar{R}_i^T\dot{p}_{ijd} + [0, \sum_{j=0}^N a_{ij}v_i]^T$.

Based on (12.28), a distributed guidance law is designed as follows:

$$\bar{v}_{ic} = B_i^{-1}\left(-\frac{K_i\hat{\bar{e}}_i}{\Pi_i} - \hat{\sigma}_i + a_{i0}v_s\bar{R}_i^T p_0^\theta(\theta)\right), \tag{12.29}$$

where $K_i = \text{diag}\{k_{ix}, k_{iy}\}$ with $k_{ix} \in \mathbb{R}$ and $k_{iy} \in \mathbb{R}$ being positive constants; $\Pi_i = \sqrt{\|\hat{\bar{e}}_i\|^2 + \varepsilon_i^2}$ and $\varepsilon_i \in \mathbb{R}$ is a positive constant; ζ is updated as follows:

$$\dot{\zeta} = -\lambda\left(\zeta + \mu\sum_{i\in\mathcal{N}_0} a_{0i}\hat{\bar{e}}_i^T\bar{R}_i^T p_0^\theta(\theta)\right), \tag{12.30}$$

where $\lambda \in \mathbb{R}$ and $\mu \in \mathbb{R}$ are positive constants.

Defining the estimation errors

$$\tilde{\bar{e}}_i = \hat{\bar{e}}_i - \bar{e}_i, \quad \tilde{\sigma}_i = \hat{\sigma}_i - \sigma_i, \tag{12.31}$$

one has

$$\dot{\tilde{\bar{e}}}_i = -K_{i1}^o \tilde{\bar{e}}_i + \tilde{\sigma}_i, \quad \dot{\tilde{\sigma}}_i = -K_{i2}^o \tilde{\bar{e}}_i - \dot{\sigma}_i. \tag{12.32}$$

Letting $E_{i1} = [\tilde{\bar{e}}_i^T, \tilde{\sigma}_i^T]^T$, the estimation error dynamics can be expressed by

$$\dot{E}_{i1} = A_i E_{i1} + B_i \xi_i, \tag{12.33}$$

where

$$A_i = \begin{bmatrix} -K_{i1}^o & I_2 \\ -K_{i2}^o & 0_2 \end{bmatrix}, \quad B_i = \begin{bmatrix} 0_2 & 0_2 \\ 0_2 & I_2 \end{bmatrix}, \quad \xi_i = [0, 0, -\dot{\sigma}_i^T].$$

Noting that A_i is a Hurwitz matrix, there exists a positive definite matrix P_i satisfying $A_i^T P_i + P_i A_i = -I_4$.

Substituting (12.29) into (12.28) and using (12.30), the guidance loop error dynamics is expressed by

$$\begin{cases} \dot{\hat{\bar{e}}}_i = -K_i \hat{\bar{e}}_i / \Pi_i - K_{i1}^o \tilde{\bar{e}}_i + B_i \bar{e}_{iv} + a_{i0} \bar{R}_i^T p_0^\theta(\theta)\zeta, \\ \dot{\zeta} = -\lambda(\zeta + \mu \sum_{i \in \mathcal{N}_0} a_{0i} \hat{\bar{e}}_i^T \bar{R}_i^T p_0^\theta(\theta)). \end{cases} \tag{12.34}$$

12.3.2 Stability analysis

We first present the stability result of (12.33).

Lemma 12.3. *Under Assumption 12.2, the estimation error dynamics in (12.33) is input-to-state stable.*

Proof. This proof is the same as the proof of Lemma 12.1 and is omitted here. □

Lemma 12.4. *The kinematic error dynamics in (12.34) is input-to-state stable provided that $\bar{e}_{iv} = 0$.*

Proof. Consider the following Lyapunov function

$$V_2 = \sum_{i=1}^{N} \frac{\hat{\bar{e}}_i^T \hat{\bar{e}}_i}{2} + \frac{\zeta^2}{2\lambda\mu}, \tag{12.35}$$

then its time derivative along (12.34) can be expressed as

$$
\begin{aligned}
\dot{V}_2 &= -\sum_{i=1}^{N}\{\frac{\hat{e}_i^T K_i \hat{e}_i}{\Pi_i} - \hat{e}_i^T K_{i1}^o \tilde{\bar{e}}_i + \hat{e}_i^T B_i \bar{e}_{iv} + a_{i0}\hat{e}_i^T p_0^\theta(\theta)\zeta\} + \frac{\zeta\dot{\zeta}}{\lambda\mu}, \\
&= -\sum_{i=1}^{N}\frac{\hat{e}_i K_i \hat{e}_i}{\Pi_i} - \sum_{i=1}^{N}\hat{e}_i^T K_{i1}^o \tilde{\bar{e}}_i + \sum_{i=1}^{N}\hat{e}_i^T B_i \bar{e}_{iv} - \frac{\zeta^2}{\mu}.
\end{aligned} \tag{12.36}
$$

Assuming a perfect velocity tracking, it follows that

$$
\begin{aligned}
\dot{V}_2 &= -\sum_{i=1}^{N}\frac{\hat{e}_i K_i \hat{e}_i}{\Pi_i} - \sum_{i=1}^{N}\hat{e}_i^T K_{i1}^o \tilde{\bar{e}}_{i1} - \frac{\zeta^2}{\mu}, \\
&\leq -\sum_{i=1}^{N}\frac{\lambda_{\min}(K_i)}{\Pi_i}\|\hat{e}_i\|^2 - \frac{\zeta^2}{\mu} + \sum_{i=1}^{N}\lambda_{\max}(K_{i1}^o)\|\tilde{\bar{e}}_i\|\,\|\hat{e}_i\|, \\
&\leq -c_1\|E_2\|^2 + c_2(\sum_{i=1}^{N}\|E_{i1}\|)\|E_2\|,
\end{aligned} \tag{12.37}
$$

where $E_2 = [\hat{e}_1^T, \ldots, \hat{e}_N^T, \zeta]^T$, $c_1 = \min\{\lambda_{\min}(K_1)/\Pi_1, \ldots, \lambda_{\min}(K_N)/\Pi_N, 1/\mu\}$ and $c_2 = \max\{\lambda_{\max}(K_{11}^o), \ldots, \lambda_{\max}(K_{N1}^o)\}$. Since $\|E_{i1}\|$ is bounded, according to Lemma 12.1, it can be concluded that the system in (12.34) is input-to-state stable. This proof is complete. □

Theorem 12.2. *Consider a swarm of under-actuated ASVs with kinematics in (12.1), guidance laws in (12.29) and (12.30), and the ESO in (12.28). Under Assumptions 12.1 and 12.2, the system cascaded by (12.33) and (12.34) is input-to-state stable provided that $\bar{e}_{iv} = 0$.*

Proof. The closed-loop system can be seen as a cascade system consisting of (12.29) and (12.30). By using Theorem 4.7 in [45], the closed-loop system is ISS. This proof is complete. □

12.3.3 Simulation results

In this subsection, simulation results are provided to illustrate the efficacy of the proposed distributed guidance law for underactuated ASVs. Consider a swarm of five ASVs and one path with $p_0(\theta) = [0.06\theta + 2, 0.06\theta + 2]^T$. The initial states of five ASVs are given as $p_1 = [1.8, 1.8]^T$, $p_2 = [2, 6]^T$, $p_3 = [2, -2]^T$, $p_4 = [2, 9]^T$, $p_5 = [2, -5]^T$. The desired formation pattern are set as $p_{1d} = [0, 0]^T$, $p_{2d} = [5\sin(t/16), 5\cos(t/16)]^T$,

$p_{3d} = [-5\sin(t/16), -5\cos(t/16)]^T$, $p_{4d} = [10\sin(t/16), 10\cos(t/16)]^T$, and $p_{5d} = [-10\sin(t/16), -10\cos(t/16)]^T$. The communication graph among the ASVs and the path is given in Fig. 12.9. The parameters for the proposed distributed guidance law are set to $K_i = \text{diag}\{0.1, 0.1\}$, $\delta_0 = 0.1$, $\lambda = 10$, $\varepsilon_i = 0.01$, $\mu = 10$, $K_{i1}^o = \text{diag}\{20, 20\}$, $K_{i2}^o = \text{diag}\{100, 100\}$.

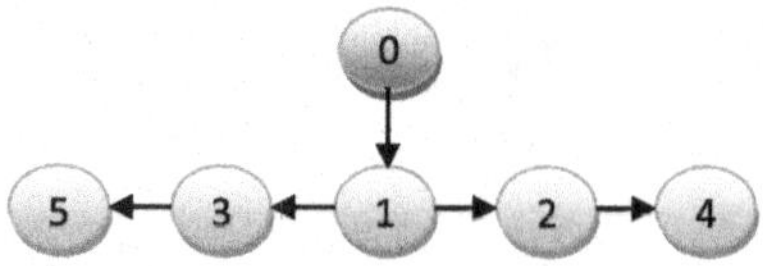

Figure 12.9 Communication graph.

The simulation results are shown in Figs. 12.10–12.13. Fig. 12.10 shows the path-guided distributed time-varying formation control trajectories of the five ASVs. It can be observed that a circular formation shaped by the five ASVs is achieved. Fig. 12.11 depicts the distributed path maneuvering errors of the five ASVs. Fig. 12.12 shows the designed guidance signals for five ASVs. Fig. 12.13 gives the profile of the path variable.

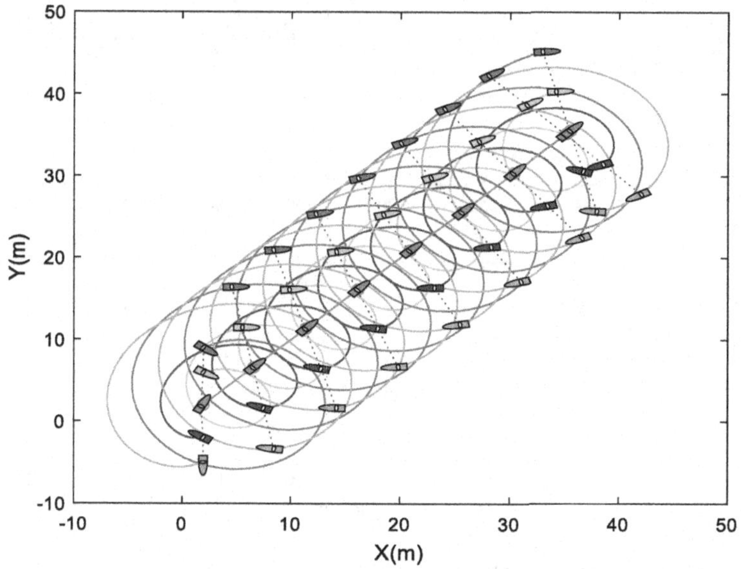

Figure 12.10 Formation pattern using the proposed distributed guidance law.

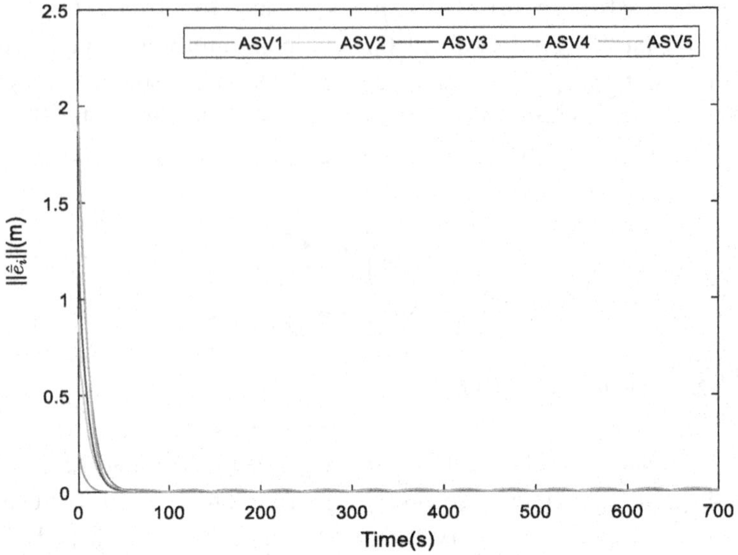

Figure 12.11 Profile of the distributed path maneuvering errors of five underactuated ASVs.

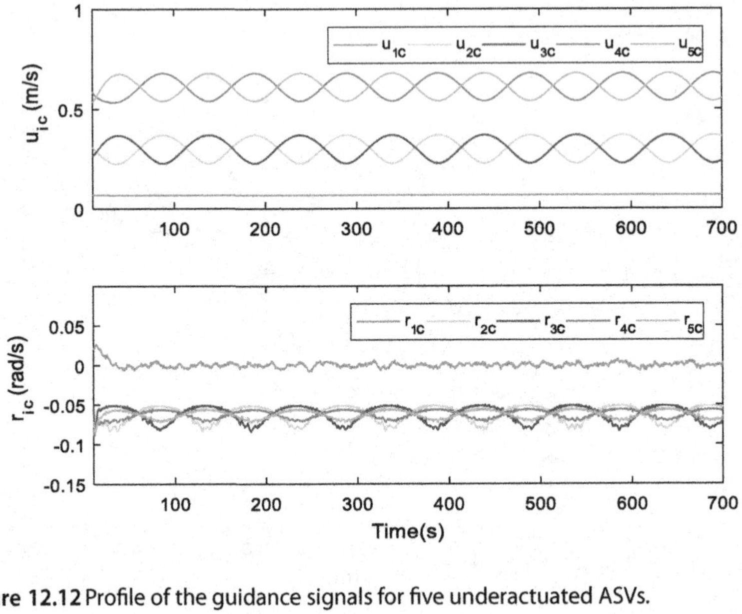

Figure 12.12 Profile of the guidance signals for five underactuated ASVs.

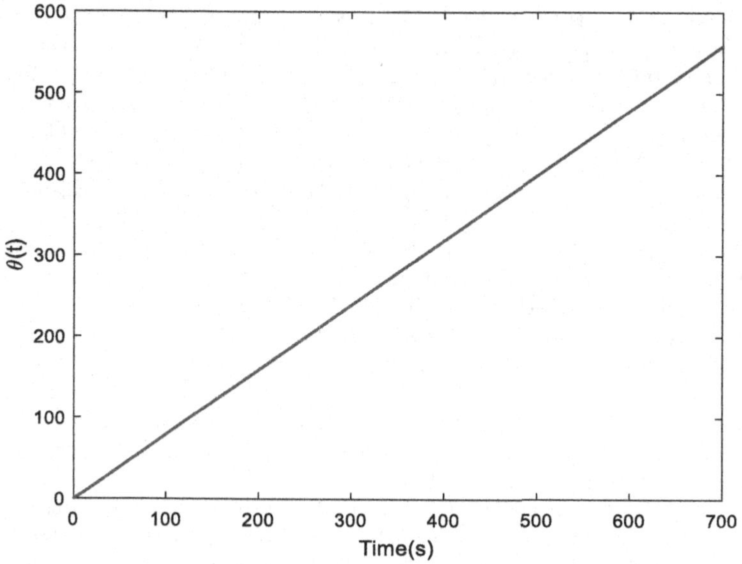

Figure 12.13 Profile of the path update.

12.4. Conclusions

This chapter presents the robust guidance law design for distributed path maneuvering of multiple ASVs with fully-actuated and underactuated configurations guided by a parameterized path. ESO-based guidance laws based on the relative positions and headings of neighbors are proposed. The stability of the closed-loop guidance laws is analyzed based on input-state-stability and cascade analysis. The prescribed guidance laws are bounded regardless of initial positions. By the proposed guidance laws, the linear velocities and yaw rates of neighbors are not needed. The feasibility and effectiveness of the proposed ESO-based guidance laws are illustrated via simulations.

References

[1] Z. Liu, Y. Zhang, X. Yu, C. Yuan, Unmanned surface vehicles: an overview of developments and challenges, Annual Reviews in Control 41 (2016) 71–93.
[2] Y. Shi, C. Shen, H. Fang, H. Li, Advanced control in marine mechatronic systems: a survey, IEEE/ASME Transactions on Mechatronics 22 (3) (2017) 1121–1131.
[3] E. Zereik, M. Bibuli, N. Mišković, P. Ridao, A.M. Pascoal, Challenges and future trends in marine robotics, Annual Reviews in Control 46 (2018) 350–368.
[4] H.R. Karimi, Offshore Mechatronics Systems Engineering, CRC Press, 2018.

[5] H. Ren, H.R. Karimi, R. Lu, Y. Wu, Synchronization of network systems via aperiodic sampled-data control with constant delay and application to unmanned ground vehicles, IEEE Transactions on Industrial Electronics 67 (6) (2020) 4980–4990.
[6] M. Breivik, V.E. Hovstein, T.I. Fossen, Straight-line target tracking for unmanned surface vehicles, Modeling, Identification and Control 29 (4) (2008) 131–149.
[7] M. Bibuli, M. Caccia, L. Lapierre, G. Bruzzone, Guidance of unmanned surface vehicles: experiments in vehicle following, IEEE Robotics & Automation Magazine 19 (3) (2012) 92–102.
[8] Z. Peng, D. Wang, Z. Chen, X. Hu, W. Lan, Adaptive dynamic surface control for formations of autonomous surface vehicles with uncertain dynamics, IEEE Transactions on Control Systems Technology 21 (2) (2013) 513–520.
[9] T. Glotzbach, M. Schneider, P. Otto, Cooperative line of sight target tracking for heterogeneous unmanned marine vehicle teams: from theory to practice, Robotics and Autonomous Systems 67 (2015) 53–60.
[10] S. Gao, Z. Peng, D. Wang, L. Liu, Extended-state-observer-based collision-free guidance law for target tracking of autonomous surface vehicles with unknown target dynamics, Complexity 4154670 (2018).
[11] S. Dai, S. He, C. Wang, Platoon formation control with prescribed performance guarantees for USVs, IEEE Transactions on Industrial Electronics 65 (5) (2018) 4237–4246.
[12] L. Liu, D. Wang, Z. Peng, C.L.P. Chen, T. Li, Bounded neural network control for target tracking of underactuated autonomous surface vehicles in the presence of uncertain target dynamics, IEEE Transactions on Neural Networks and Learning Systems 30 (4) (2019) 1241–1249.
[13] Z. Peng, D. Wang, J. Wang, Cooperative dynamic positioning of multiple marine offshore vessels: a modular design, IEEE/ASME Transactions on Mechatronics 21 (Jun 2016) 1210–1221.
[14] A. Liu, W. Zhang, L. Yu, H. Yan, R. Zhang, Formation control of multiple mobile robots incorporating an extended state observer and distributed model predictive approach, IEEE Transactions on Systems, Man, and Cybernetics: Systems (2020), https://doi.org/10.1109/TSMC.2018.2855444.
[15] B.S. Park, S.J. Yoo, An error transformation approach for connectivity-preserving and collision-avoiding formation tracking of networked uncertain underactuated surface vessels, IEEE Transactions on Cybernetics 49 (8) (2019) 2955–2966.
[16] X. Xiang, C. Liu, H. Su, Q. Zhang, On decentralized adaptive full-order sliding mode control of multiple UAVs, ISA Transactions 71 (2) (2017) 196–205.
[17] H. Qin, H. Chen, Y. Sun, Z. Wu, The distributed adaptive finite-time chattering reduction containment control for multiple ocean bottom flying nodes, International Journal of Fuzzy Systems 21 (2) (2018) 607–619.
[18] H. Qin, H. Chen, Y. Sun, Distributed finite-time fault-tolerant containment control for multiple ocean bottom flying nodes, Journal of the Franklin Institute (2020), https://doi.org/10.1016/j.jfranklin.2019.05.034.
[19] H. Qin, H. Chen, Y. Sun, L. Chen, Distributed finite-time fault-tolerant containment control for multiple ocean bottom flying node systems with error constraints, Ocean Engineering 106341 (2019).
[20] R. Skjetne, T.I. Fossen, P.V. Kokotović, Robust output maneuvering for a class of nonlinear systems, Automatica 40 (3) (2004) 373–383.
[21] I.F. Ihle, M. Arcak, T.I. Fossen, Passivity-based designs for synchronized path-following, Automatica 43 (9) (2007) 1508–1518.

[22] J. Almeida, C. Silvestre, A. Pascoal, Cooperative control of multiple surface vessels in the presence of ocean currents and parametric model uncertainty, International Journal of Robust and Nonlinear Control 20 (14) (2010) 1549–1565.
[23] J. Almeida, C. Silvestre, A. Pascoal, Cooperative control of multiple surface vessels with discrete-time periodic communications, International Journal of Robust and Nonlinear Control 22 (4) (2011) 398–419.
[24] Y. Chen, P. Wei, Coordinated adaptive control for coordinated path-following surface vessels with a time-invariant orbital velocity, IEEE/CAA Journal of Automatica Sinica 1 (4) (2014) 337–346.
[25] L. Liu, D. Wang, Z. Peng, H.H.T. Liu, Saturated coordinated control of multiple underactuated unmanned surface vehicles over a closed curve, Science China Information Sciences 60 (2017) 070203.
[26] L. Liu, D. Wang, Z. Peng, T. Li, Modular adaptive control for LOS-based cooperative path maneuvering of multiple underactuated autonomous surface vehicles, IEEE Transactions on Systems, Man, and Cybernetics: Systems 47 (7) (2017) 1613–1624.
[27] Z. Peng, J. Wang, D. Wang, Containment maneuvering of marine surface vehicles with multiple parameterized paths via spatial-temporal decoupling, IEEE/ASME Transactions on Mechatronics 22 (2) (2017) 1026–1036.
[28] Z. Peng, J. Wang, D. Wang, Distributed containment maneuvering of multiple marine vessels via neurodynamics-based output feedback, IEEE Transactions on Industrial Electronics 64 (5) (2017) 3831–3839.
[29] Z. Peng, J. Wang, D. Wang, Distributed maneuvering of autonomous surface vehicles based on neurodynamic optimization and fuzzy approximation, IEEE Transactions on Control Systems Technology 26 (3) (2018) 1083–1090.
[30] Z. Peng, J. Wang, Q. Han, Path-following control of autonomous underwater vehicles subject to velocity and input constraints via neurodynamic optimization, IEEE Transactions on Industrial Electronics 66 (11) (2019) 8724–8732.
[31] Z. Peng, D. Wang, T. Li, M. Han, Output-feedback cooperative formation maneuvering of autonomous surface vehicles with connectivity preservation and collision avoidance, IEEE Transactions on Cybernetics 50 (6) (2020) 2527–2535.
[32] Z. Peng, N. Gu, Y. Zhang, Y. Liu, D. Wang, L. Liu, Path-guided time-varying formation control with collision avoidance and connectivity preservation of under-actuated autonomous surface vehicles subject to unknown input gains, Ocean Engineering 191 (2019) 106501.
[33] N. Gu, D. Wang, Z. Peng, L. Liu, Adaptive bounded neural network control for coordinated path-following of networked underactuated autonomous surface vehicles under time-varying state-dependent cyber-attack, ISA Transactions (2020), https://doi.org/10.1016/j.isatra.2018.12.051.
[34] N. Gu, Z. Peng, D. Wang, Y. Shi, T. Wang, Antidisturbance coordinated path-following control of robotic autonomous surface vehicles: theory and experiment, IEEE/ASME Transactions on Mechatronics 24 (5) (2019) 2386–2396.
[35] N. Gu, D. Wang, Z. Peng, L. Liu, Distributed containment maneuvering of uncertain under-actuated unmanned surface vehicles guided by multiple virtual leaders with a formation, Ocean Engineering 187 (2019) 105996.
[36] N. Gu, D. Wang, Z. Peng, L. Liu, Observer-based finite-time control for distributed path maneuvering of underactuated unmanned surface vehicles with collision avoidance and connectivity maintenance, IEEE Transactions on Systems, Man, and Cybernetics: Systems (2020), https://doi.org/10.1109/TSMC.2019.2944521.

[37] Y. Zhang, D. Wang, Z. Peng, T. Li, Distributed containment maneuvering of uncertain multiagent systems in MIMO strict-feedback form, IEEE Transactions on Systems, Man, and Cybernetics: Systems (2020), https://doi.org/10.1109/TSMC.2019.2896662.

[38] L. Liu, D. Wang, Z. Peng, T. Li, C.L.P. Chen, Cooperative path following ring-networked under-actuated autonomous surface vehicles: algorithms and experimental results, IEEE Transactions on Cybernetics 50 (4) (2020) 1519–1529.

[39] Z. Peng, Y. Jiang, J. Wang, Event-triggered dynamic surface control of an under-actuated autonomous surface vehicle for target enclosing, IEEE Transactions on Industrial Electronics (2020), https://doi.org/10.1109/TIE.2020.2978713.

[40] Y. Jiang, Z. Peng, D. Wang, C.L.P. Chen, Line-of-sight target enclosing of an under-actuated autonomous surface vehicle with experiment results, IEEE Transactions on Industrial Informatics 16 (2) (2020) 832–841.

[41] L. Liu, D. Wang, Z. Peng, ESO-based line-of-sight guidance law for path following of underactuated marine surface vehicles with exact sideslip compensation, IEEE Journal of Oceanic Engineering 42 (2) (2017) 477–487.

[42] L. Liu, D. Wang, Z. Peng, State recovery and disturbance estimation of unmanned surface vehicles based on nonlinear extended state observers, Ocean Engineering 171 (2019) 625–632.

[43] L. Liu, W. Zhang, D. Wang, Z. Peng, Event-triggered extended state observers design for dynamic positioning vessels subject to unknown sea loads, Ocean Engineering 209 (2020) 107242.

[44] B. Guo, Z. Wu, H. Zhou, Active disturbance rejection control approach to output-feedback stabilization of a class of uncertain nonlinear systems subject to stochastic disturbance, IEEE Transactions on Automation Control 61 (6) (2016) 1613–1618.

[45] H. Khalil, Nonlinear Control, Pearson Education, 2015.

[46] R. Skjetne, T.I. Fossen, P.V. Kokotović, Adaptive maneuvering, with experiments, for a model ship in a marine control laboratory, Automatica 41 (2) (2005) 289–298.

CHAPTER THIRTEEN

Finite-time extended state observer based fault tolerant output feedback control for UAV attitude stabilization under actuator failures and disturbances

Bo Li[a], Wen Liu[a], Ke Qin[a], Bing Xiao[b], Yongsheng Yang[a]
[a]Institute of Logistics Science and Engineering, Shanghai Maritime University, Shanghai, China
[b]School of Automation, Northwestern Polytechnical University, Xi'an, China

Contents

Abstract

This chapter investigates the finite-time fault tolerant attitude stabilization control for Unmanned Aerial Vehicles (UAV) without the angular velocity measurements, in the presence of external disturbances and actuator failures. Firstly, a novel continuous finite-time Extended State Observer is established to observe the unmeasurable attitude angular velocity and synthetic failure simultaneously. Unlike the existing observers, the finite-time methodology and Extended State Observer are utilized, the attitude angular velocity and extended state observation errors are realized to be the uniformly ultimately finite-time bounded stable. Furthermore, based on the nonsingular terminal sliding mode control and supertwisting method, a novel continuous

Fundamental Design and Automation Technologies in Offshore Robotics
https://doi.org/10.1016/B978-0-12-820271-5.00018-3

finite-time attitude controller is developed for fast robust fault tolerant attitude control. The main feature of this work stems from the multiply advanced techniques or methodologies. It enables the finite-time stability of the closed-loop attitude control system and the designed control scheme to be continuous with the property of chattering restraining. Finally, some numerical simulation results are presented to verify the effectiveness and fine performances of the UAV attitude stabilization control system driven by the proposed finite-time robust fault tolerant attitude control scheme.

Keywords

Quadrotor UAV, Attitude stabilization, Extended state observer, Output feedback, Fault tolerant control

13.1. Introduction

The Unmanned Aerial Vehicle (UAV) is a vehicle with the capabilities of autonomous navigation, real-time monitoring, and tasking scheduling by utilizing the ground stations. Especially for the quadrotor UAV, it has superiorities of flexible flight, vertical take-off and landing, autonomous hovering, strong maneuverability, and the ability to execute specified missions in dangerous or/and complex environments, where it is difficult for human or ground vehicles to work well. Recently, quadrotor UAV has attracted much attention and applications in military and civil fields, such as attack and rendezvous, search-and-rescue, fire monitoring, electric power overhaul, aerial photography, urban management, port and shipping management, logistics transportation, and so on [1–5]. For example, automated container terminal and smart port are the advanced form of the port, in which all the parts of the port operations work automatically without any person in the container yard. In view of this, UAVs can be utilized to execute some potential tasks, such as safety inspection to ensure the high security level, 3D modeling and assessment to calculate the quantities of the container or break-bulk cargos, remote fault diagnosis, and state monitoring of the port machinery, etc. Additionally, in the logistics and transportation industry, the "last mile" distribution problem is related to the efficiency and cost of the entire logistics process. UAVs can be advantageous in the logistics transportation together with some other vehicles. As a very critical requirement in the intelligent logistics systems, it is becoming more and more popular. In recent years, some internet e-commerce and logistics companies such as Amazon, DHL, SF Express, and JD have deployed their UAV delivery strategies and started doing research. In the above applications or missions of UAVs, many control problems should be solved well to guarantee the stability, accuracy, and reliability of the control system with constraints.

The quadrotor UAV has the characteristics of underactuation, nonlinearity, and strong coupling, which brings great difficulties to the design of a flight control system. Thus, the control problems of quadrotor UAV have attracted more and more attention from researchers. Particularly, attitude changes in attitude loop will directly affect the position changes in position loop, the study for attitude control problem has more important significance. Many technical results dealing with practical problems and constraints for the quadrotor attitude control system in the presence of external disturbance and/or uncertainties have been developed [6–9]. In the existing control strategies, most of them require the accurate attitude angular velocity information for the full-state feedback control design. However, the above requirement may not always be met in some severe situations, or for some specific microquadrotor without gyro/rate sensors due to either cost saving or implementation constraints. What is worse, the unexpected malfunctions or catastrophic failures of the equipped actuators can bring the further considerable complexity and diversity in the control system design. Therefore, as a key issue for the practical engineering applications of the quadrotor UAV attitude control system, the output-feedback based fast fault tolerant control without angular velocity measurements must be taken into consideration.

In the absence of attitude angular velocity information, the problem of fast and accurate attitude control becomes much more challenging. To this end, several ways have been proposed, such as passivity-based attitude control [10], angular velocity filter [11,12] observer based controller architectures, and so on [13–15]. In [16], a high-order sliding mode observer was designed to reconstruct the perturbation and attitude angular velocity information for quadrotor attitude tracking control system without angular velocity measurements. To handle the unmeasurable velocity states and disturbances, the extended state observer (ESO) based robust integral of the sign of the error (RISE) feedback control was proposed in [14] to enhance the robustness of the control system. But the above observers can only reconstruct the actual angular velocity in infinite time. To achieve the fast convergence and high-precise of the observers, finite-time control techniques have been utilized for the design of observers, such as homogeneous observer, terminal sliding mode observer, and so on [17,18]. The homogeneous observer is an effective way to estimate the unknown angular velocity in finite time in the absence of disturbances or uncertainties, whereas the terminal sliding mode observer can deal with the finite-time observation of the angular velocity as well. But the property of discontinuous high-frequency switching dynamics may pull down the control system performance. In [19], an adaptive model-free state observer was designed to estimate the quaternion derivative in finite time, then, based on an angular velocity calculation algorithm, the unmeasurable angular velocity was achieved. To handle the finite-time attitude control with unknown angular velocity measurements and synthetic uncertainty, a modified adaptive supertwisting sliding mode state observer and a finite-time continuous adaptive disturbance observer were designed in [20]. Nevertheless, the mutual influence between observers was not taken into account for the design of separate observer. In [21], adding power integrator (API) technique based output feedback control law was adopted to make the system reach a neighborhood of zero within finite time. However, the problem of fault tolerant control under actuator failures cannot put into consideration by the above schemes.

To guaranteed the high robustness for the quadrotor attitude control under unexpected actuator's faults, many fault tolerant control strategies have been investigated [22–24]. Previous FTC approaches can be categorized into the following two main types: passive fault tolerant control and active fault tolerant control (AFTC). Compared with the former, AFTC can obtain the multiple or unknown online fault data through feasible and effective Fault Detection and Diagnosis (FDD) mechanism in real time. It has attracted much more attention in many practical applications [25–28]. As an essential mechanism, FDD should be first considered to reconstruct the failures or uncertainties during the design of AFTC. In [29], an adaptive linear observer was designed for occurring faults estimation. A modified finite-time disturbance observer was designed to reconstruct the synthetic uncertainty deriving from actuator failures and disturbances in [30]. In [31], a novel nonsingular terminal sliding mode based finite-time extended state observer was proposed for finite time bounded observation about faults or uncertainties. It should be noted that most attitude control schemes can only dealt with either actuator failures or unknown angular velocity measurements. To handle the gyro and velocity measurement sensor failures in quadrotor UAV attitude system, an analytical redundancy based fault-tolerant filter was formed to improve the state estimation accuracy in [32]. In [33], the supertwisting sliding mode observer was provided to estimates the angular velocity accurately in finite time, and a neural networks based adaptive fault-tolerant controller was proposed to achieve attitude tracking exponentially. However, it employed the passive or robust way to achieve fault tolerant control.

This work investigates a finite-time fault tolerant output feedback control scheme, which incorporate a continuous finite-time ESO with a finite-time robust fault tolerant control law for the quadrotor UAV attitude stabilization control system. The constraints and problems of unavailable angular velocity measurements, external disturbances, actuator failures and misalignments are taken into consideration explicitly. The main contributions are summarized as follows.

(1) By utilizing the finite-time methodology and ESO technique synthetically, a novel continuous finite-time ESO (FTESO) is firstly established to observe the attitude angular velocity and the synthetic failure simultaneously. The observation errors of the attitude angular velocity and extended state are proved to be uniformly ultimately finite-time bounded stable.

(2) With the reconstructed information of attitude angular velocity and the synthetic failures deriving from the proposed FTESO, a novel nonsingular terminal sliding mode and supertwisting based finite-time attitude dynamic feedback control law (FTDFC) is presented. And the finite-time stability of the closed-loop attitude control system is proved by using the Lyapunov method and homogeneous theory.

(3) From both the theoretical analysis and simulations, the fault tolerant output feedback control scheme proposed in this work is effective, simple, and continuous with chattering restraining. These fine properties will improve the engineering significance of the proposed scheme, and the results can also provide some positive and efficient guidance for the engineers.

13.2. Quadrotor dynamics and kinematics

In this paper, the method of Euler angles is applied to describe the 3-degree-of-freedom (DOF) rotational attitude dynamics of a quadrotor UAV, and expressed as [9,34]

$$\begin{cases} \dot{\boldsymbol{\Theta}} = \boldsymbol{G}\boldsymbol{\omega} \\ \boldsymbol{J}\dot{\boldsymbol{\omega}} = -\boldsymbol{\omega}^{\times}\boldsymbol{J}\boldsymbol{\omega} + \boldsymbol{E}\boldsymbol{\tau} + \boldsymbol{d} \end{cases} \tag{13.1}$$

where $\boldsymbol{\Theta} = [\varphi, \theta, \psi]^{\mathrm{T}}$ and $\boldsymbol{\omega} \in \mathbf{R}^{3\times1}$ denote the Euler anger vector and the angular velocity vector expressed in inertial frame, respectively; $\boldsymbol{\tau} = [\tau_1, \tau_2, \tau_3]^{\mathrm{T}} \in \mathbf{R}^{3\times1}$ denotes the attitude control torque acting on the channel of roll, pitch, and yaw; $\boldsymbol{J} \in \mathbf{R}^{3\times 3}$ is the positive-definite symmetric inertia matrix of the quadrotor UAV; $\boldsymbol{E} = \text{diag}\{\varepsilon_{11}, \varepsilon_{22}, \varepsilon_{33}\} \in \mathbf{R}^{3\times 3}$ represents the actuator effectiveness matrix on the three-attitude control channels (roll, pitch, and yaw), in which $0 < \varepsilon_{ii} \leq 1\,(i = 1, 2, 3)$ represents the actuator effectiveness factor on the ith attitude control channel. Note that only healthy condition $\varepsilon_{ii} = 1$ and partial loss of actuation effectiveness $0 < \varepsilon_{ii} < 1$ are considered, but the case of complete failure $\varepsilon_{ii} = 0$ is not considered in this work. In addition, $\boldsymbol{d} \in \mathbf{R}^{3\times 1}$ denotes the bounded external disturbance torque. And $\mathbf{G} \in \mathbf{R}^{3\times 3}$ is defined as

$$\mathbf{G} = \boldsymbol{W}(\theta)\boldsymbol{T}(\phi) \tag{13.2}$$

where

$$\boldsymbol{W}(\theta) = \begin{bmatrix} 1 & 0 & \tan\theta \\ 0 & 1 & 0 \\ 0 & 0 & \sec\theta \end{bmatrix}, \qquad \boldsymbol{T}(\phi) = \begin{bmatrix} 1 & 0 & 0 \\ 0 & \cos\phi & -\sin\phi \\ 0 & \sin\phi & \cos\phi \end{bmatrix}. \tag{13.3}$$

Furthermore, define the skew-symmetric matrix $(\cdot)^{\times}$ as

$$\boldsymbol{x}^{\times} = \begin{bmatrix} 0 & -x_3 & x_2 \\ x_3 & 0 & -x_1 \\ -x_2 & x_1 & 0 \end{bmatrix}, \qquad \boldsymbol{x} = [x_1, x_2, x_3]^{\mathrm{T}} \in \mathbf{R}^{3\times 1}. \tag{13.4}$$

According to (13.1), one can express

$$\boldsymbol{\omega} = \mathbf{G}^{-1}\dot{\boldsymbol{\Theta}}. \tag{13.5}$$

And then the following formula can be obtained:

$$\dot{\boldsymbol{\omega}} = \dot{\mathbf{G}}^{-1}\dot{\boldsymbol{\Theta}} + \mathbf{G}^{-1}\ddot{\boldsymbol{\Theta}}. \tag{13.6}$$

Substituting (13.5) and (13.6) into (13.1), the attitude dynamics can be rewritten as Euler–Lagrange formulation, and described as follows:

$$\boldsymbol{M}\ddot{\boldsymbol{\Theta}} + \boldsymbol{C}\left(\boldsymbol{\Theta}, \dot{\boldsymbol{\Theta}}\right)\dot{\boldsymbol{\Theta}} = \boldsymbol{F}^{\mathrm{T}}\boldsymbol{\tau} + \bar{\boldsymbol{d}} \tag{13.7}$$

where $\boldsymbol{M} = \mathbf{G}^{-\mathrm{T}}\boldsymbol{J}\mathbf{G}^{-1}$, $\boldsymbol{C}\left(\boldsymbol{\Theta}, \dot{\boldsymbol{\Theta}}\right) = \mathbf{G}^{-\mathrm{T}}\boldsymbol{J}\dot{\mathbf{G}}^{-1} - \mathbf{G}^{-\mathrm{T}}\left(\boldsymbol{J}\mathbf{G}^{-1}\dot{\boldsymbol{\Theta}}\right)^{\times}\mathbf{G}^{-1}$, $\boldsymbol{F}^{\mathrm{T}} = \mathbf{G}^{-\mathrm{T}}$, and $\bar{\boldsymbol{d}} = \mathbf{G}^{-\mathrm{T}}\left((\boldsymbol{E} - \boldsymbol{I}_3)\boldsymbol{\tau} + \boldsymbol{d}\right)$.

Without loss of generality, for (13.7), the following common properties [35,36] can be assumed to be satisfied: (1) $\dot{\boldsymbol{M}} - 2\boldsymbol{C}(\boldsymbol{\Theta}, \dot{\boldsymbol{\Theta}})$ is a skew-symmetric matrix, i.e., $\boldsymbol{x}^{\mathrm{T}}\left[\dot{\boldsymbol{M}} - 2\boldsymbol{C}(\boldsymbol{\Theta}, \dot{\boldsymbol{\Theta}})\right]\boldsymbol{x} = 0$ can be obtained for any vector $\boldsymbol{x} \in \mathbf{R}^{3\times 1}$; (2) $\boldsymbol{M}$ satisfies $0 < \beta_{\min}(\boldsymbol{M})\boldsymbol{I}_3 \leq \boldsymbol{M} \leq \beta_{\max}(\boldsymbol{M})\boldsymbol{I}_3 < \infty$, where $\beta_{\min}(\cdot)$ and $\beta_{\max}(\cdot)$ represent the minimum and maximum eigenvalues of the specified matrix $(\cdot)$, respectively; (3) Additionally, the dynamics are linearly parameterizable, and the matrix $\boldsymbol{C}(\boldsymbol{\Theta}, \dot{\boldsymbol{\Theta}})$ is also bounded, which satisfies the inequality $\left\|\boldsymbol{C}(\boldsymbol{\Theta}, \dot{\boldsymbol{\Theta}})\dot{\boldsymbol{\Theta}}\right\| \leq h_3 \left\|\dot{\boldsymbol{\Theta}}\right\|^2$ with $h_3 > 0$; (4) For any matrix $\boldsymbol{A}, \boldsymbol{B} \in \mathbf{R}^n$, if $\mathrm{Rank}(\boldsymbol{A}) = \mathrm{Rank}(\boldsymbol{B}) = n$, $\mathrm{Rank}(\boldsymbol{AB}) = n$ can be ensured.

Assumption 13.1. The output torque magnitudes of quadrotor UAV are assumed to subject to the same constraint value, which is noted as $\boldsymbol{\tau}_{\max}$ for simplicity. In view of the definition of the synthetic uncertainty term $\overline{\boldsymbol{d}}$, it is supposed to be bounded and differentiable. Meanwhile, it is assumed that there always exists a constant d_m that satisfies $\|\overline{\boldsymbol{d}}\| \leq d_m$.

Assumption 13.2. In practical engineering, the unlimited attitude angular velocity will lead to severe chattering, even instability of the attitude control system. Thus, the actual attitude angular velocities are always bounded during the missions, for keeping the stability of the attitude control system, restraining the vibration of the payloads, and so on. In this work, only a small attitude maneuver or variation is considered for the UAV attitude stabilization control mission. Therefore, it is reasonable to assume that there exists a positive constant k_ω, such that $\left\|\dot{\boldsymbol{\Theta}}\right\| \leq k_\omega$ all the time.

Assumption 13.3. There might be a singularity problem in kinematics equations with $\theta = \pm\frac{\pi}{2}$. To avoid this problem, the initial pitch angle θ of the quadrotor UAV should be restricted by $-\frac{\pi}{2} < \theta(0) < \frac{\pi}{2}$; $-\frac{\pi}{2} < \theta(t) < \frac{\pi}{2}$ for all $t > 0$ should be guaranteed by the developed attitude stabilization controller. Then the matrix $\boldsymbol{M}$ can be guaranteed to be invertible.

13.3. Finite-time observer based fault tolerant attitude control scheme design

13.3.1 Preliminaries

In this chapter, some notations are defined as $\mathrm{sig}^{\alpha}(\boldsymbol{x}) = \mathrm{sign}(\boldsymbol{x})\,|\boldsymbol{x}|^{\alpha}$, $|\boldsymbol{x}|^{\alpha} = \left[|\boldsymbol{x}_1|^{\alpha}, |\boldsymbol{x}_2|^{\alpha}, \ldots, |\boldsymbol{x}_n|^{\alpha}\right]^{\mathrm{T}}$, $\alpha \in \mathbf{R}$, and $\mathrm{sign}(\cdot)$ is a sign function with $\mathrm{sign}(0) = 0$.

Consider the following nonlinear system:

$$x = f(x), \quad x(0) = x_0, \quad f(0) = 0, \quad x \in \mathbf{R}^n \tag{13.8}$$

where $f: U \to \mathbf{R}^n$ is a continuous function in an open neighborhood U of the origin. And it is assumed that (13.8) has a unique solution in forward time for all initial conditions.

Definition 13.1 ([31,37,38]). For the given vector x, the dilation operator is defined as $\boldsymbol{\Gamma}_{\varepsilon}^{r} x := (\varepsilon^{r_1} x_1, \varepsilon^{r_2} x_2, \ldots, \varepsilon^{r_n} x_n)$, $\forall \varepsilon > 0$, and the exponential variables $r_i > 0, i = 1, \ldots, n$ are the weights of the coordinates, and $r = (r_1, r_2, \ldots, r_n) \in \mathbf{R}^n$ denotes the vector of weights. A function $V: \mathbf{R}^n \to \mathbf{R}$ (or a vector field $f: \mathbf{R}^n \to \mathbf{R}^n$, or a vector-set field $F(x) \subset \mathbf{R}^n$ is called homogeneous of degree $\delta \in \mathbf{R}$ with the dilation $\boldsymbol{\Gamma}_{\varepsilon}^{r} x$, if the identity $V\left(\boldsymbol{\Gamma}_{\varepsilon}^{r} x\right) = \varepsilon^{\delta} V(x)$ (or $f\left(\boldsymbol{\Gamma}_{\varepsilon}^{r} x\right) = \varepsilon^{\delta} \boldsymbol{\Gamma}_{\varepsilon}^{r} f(x)$, or $F\left(\boldsymbol{\Gamma}_{\varepsilon}^{r} x\right) = \varepsilon^{\delta} \boldsymbol{\Gamma}_{\varepsilon}^{r} F(x)$, respectively). The system is called homogeneous of some degree.

Lemma 13.1 ([37]). *Consider the system in (13.8), and suppose there exists a Lyapunov function $V(x)$ defined on a neighborhood $U \in \mathbf{R}^n$ of the origin, and $\dot{V}(x) + \beta_{11} V(x)^{\alpha_{11}} < 0$ with $x \in U \backslash \{0\}$, $0 < \alpha_{11} < 1$, $\beta_{11} > 0$. Then the system is locally finite-time stable, and $T \leq \frac{1}{\beta_{11}(1-\alpha_{11})} \left| V(x_0) \right|^{1-\alpha_{11}}$ is the time needed to reach $V(x) = 0$, where $V(x_0)$ denotes the initial value of the Lyapunov function $V(x)$.*

Lemma 13.2 ([18,39]). *Consider the above nonlinear system in (13.8), and suppose there exists a Lyapunov function $V(x)$, with $V(x_0)$ denoting its initial value. The trajectory of the system in (13.8) is finite-time uniformly ultimately bounded stable within the region of $Q_{21} = \left\{ x \mid V(x)^{\alpha_{21}-\alpha_{22}} < \frac{\beta_{22}}{\vartheta_{21}} \right\}$, if $\dot{V}(x) \leq -\beta_{21} V(x)^{\alpha_{21}} + \beta_{22} V(x)^{\alpha_{22}}$ for $0 < \alpha_{21} < 1$, $\alpha_{21} > \alpha_{22}$, $\beta_{21} > 0$, $\beta_{22} > 0$, $\vartheta_{21} \in (0, \beta_{21})$. The settling time for the states reaching the stable residual set is subject to the constraint $T_{21} \leq \frac{V(x_0)^{1-\alpha_{21}}}{(\beta_{21}-\vartheta_{21})(1-\alpha_{21})}$.*

13.3.2 Finite-time extended state observer design

13.3.2.1 Structure of the finite-time extended state observer

In this subsection, a novel continuous finite-time ESO (FTESO) will be studied to observe the attitude angular velocity and the synthetic uncertainty/failure simultaneously.

Defining two new auxiliary variables as $x_1 = \boldsymbol{\Theta}$ and $x_2 = \dot{\boldsymbol{\Theta}}$, the attitude control system in (13.7) can be rewritten in the form of

$$\dot{x}_1 = x_2, \tag{13.9a}$$

$$\dot{x}_2 = -M^{-1}C(x_1, x_2)x_2 + M^{-1}F^{\mathrm{T}}\tau + M^{-1}\overline{d}. \tag{13.9b}$$

By using the ESO technique, an extended state variable is defined as $x_3 = M^{-1}\overline{d}$ with $\dot{x}_3 = g(t)$. It should be noted that the function $g(t)$ is assumed to be bounded. In other words, there exists a positive constant $\overline{g} > 0$ such that $|g_i(t)| \leq \overline{g}$, $i = 1, 2, 3$. Then, the quadrotor UAV attitude stabilization control system governed by (13.9) can be extended as follows:

$$\dot{x}_1 = x_2, \tag{13.10a}$$

$$\dot{x}_2 = -M^{-1}C(x_1, x_2)x_2 + M^{-1}F^{\mathrm{T}}\tau + x_3, \tag{13.10b}$$

$$\dot{x}_3 = g(t). \tag{13.10c}$$

Denote $\hat{x}_1$, $\hat{x}_2$, and $\hat{x}_3$ as the observation values of states x_1, x_2, and x_3 in (13.10), and $e_1 = \hat{x}_1 - x_1$, $e_2 = \hat{x}_2 - x_2$, $e_3 = \hat{x}_3 - x_3$ as the observation errors of the attitude, angular velocity, and extended state of the synthetic failure, respectively. Afterwards, a continuous FTESO is proposed as

$$\dot{\hat{x}}_1 = -\beta_1 \mathrm{sig}^{\alpha_1}(e_1) + \hat{x}_2, \tag{13.11a}$$

$$\dot{\hat{x}}_2 = -\beta_2 \mathrm{sig}^{\alpha_2}(e_1) - M^{-1}C(x_1, \hat{x}_2)\hat{x}_2 + M^{-1}F^{\mathrm{T}}\tau + \hat{x}_3, \tag{13.11b}$$

$$\dot{\hat{x}}_3 = -\beta_3 \mathrm{sig}^{\alpha_2}(e_1) \tag{13.11c}$$

where the FTESO gains meet the following constraints: $\beta_1 > 0$, $\beta_2 > 0$, $\beta_3 > 0$, $\frac{1}{2} < \alpha_1 < 1$, and $\alpha_2 = 2\alpha_1 - 1$. In view of the above FTESO in (13.11), it depends on the attitude and the actual actuator output torques from the quadrotor UAV attitude control system. The dynamic observation process of the FTESO has nothing to do with the actual spacecraft attitude angular velocity ω, synthetic failure x_3, and its differential $g(t)$, whereas it requires only the actual action for the finite-time stability of the control system that the extended state x_3 is bounded according to the principles of the ESO technique. Therefore, the assumption of the boundedness of $g(t)$ is reasonable and feasible for the design and analysis of the proposed FTESO.

13.3.2.2 Finite-time ESO convergence analysis

According to the extended attitude control system in (13.10) and the proposed FTESO in (13.11), one can obtain the observation error dynamics as follows:

$$\dot{e}_1 = -\beta_1 \mathrm{sig}^{\alpha_1}(e_1) + e_2, \tag{13.12a}$$

$$\dot{e}_2 = -\beta_2 \mathrm{sig}^{\alpha_2}(e_1) - \boldsymbol{M}^{-1}\boldsymbol{C}(x_1, \hat{x}_2)\hat{x}_2 + \boldsymbol{M}^{-1}\boldsymbol{F}^{\mathrm{T}}\boldsymbol{\tau} + e_3, \tag{13.12b}$$

$$\dot{e}_3 = -\beta_3 \mathrm{sig}^{\alpha_2}(e_1) - \boldsymbol{g}(t). \tag{13.12c}$$

Theorem 13.1. *Consider the attitude stabilization control system in (13.7) under Assumptions 13.1–13.2. If the FTESO is proposed in the form of (13.11), then the observation errors* $\boldsymbol{e} = \left[\boldsymbol{e}_1^{\mathrm{T}}, \boldsymbol{e}_2^{\mathrm{T}}, \boldsymbol{e}_3^{\mathrm{T}}\right]^{\mathrm{T}}$ *will converge to the region*

$$D_1 = \left\{ \boldsymbol{e} \,|\, V_1(\boldsymbol{e}) < \left(\frac{\lambda_4}{\lambda_5}\right)^{\frac{2\alpha_1}{2\alpha_1 - 1}} \right\} \tag{13.13}$$

in finite time. It also implies that the observation error dynamics system (13.12) is locally finite-time uniformly ultimately bounded stable.

Proof of Theorem 13.1. Define

$$\boldsymbol{\zeta}\left(\boldsymbol{x}_1, \hat{\boldsymbol{x}}_2\right) = \boldsymbol{M}^{-1}\boldsymbol{C}\left(\boldsymbol{x}_1, \hat{\boldsymbol{x}}_2\right)\hat{\boldsymbol{x}}_2, \qquad \boldsymbol{\zeta}(\boldsymbol{x}_1, \boldsymbol{x}_2) = \boldsymbol{M}^{-1}\boldsymbol{C}(\boldsymbol{x}_1, \boldsymbol{x}_2)\boldsymbol{x}_2.$$

Then one can obtain $\overline{\boldsymbol{\zeta}}(\boldsymbol{x}_1, \boldsymbol{x}_2) = \boldsymbol{\zeta}(\boldsymbol{x}_1, \boldsymbol{x}_2) - \boldsymbol{\zeta}\left(\boldsymbol{x}_1, \hat{\boldsymbol{x}}_2\right)$. According to the properties of the Euler–Lagrange formulation of the attitude dynamics (13.7), one can get the following equations:

$$\boldsymbol{\zeta}(\boldsymbol{x}_1, \boldsymbol{x}_2) - \boldsymbol{\zeta}\left(\boldsymbol{x}_1, \hat{\boldsymbol{x}}_2\right) = \boldsymbol{M}^{-1}\boldsymbol{C}(\boldsymbol{x}_1, \boldsymbol{x}_2)\boldsymbol{e}_2 - \boldsymbol{M}^{-1}\boldsymbol{C}(\boldsymbol{x}_1, \boldsymbol{e}_2)(\boldsymbol{x}_2 + \boldsymbol{e}_2). \tag{13.14}$$

Then, we have

$$\begin{aligned} \left\|\boldsymbol{\zeta}(\boldsymbol{x}_1, \boldsymbol{x}_2) - \boldsymbol{\zeta}\left(\boldsymbol{x}_1, \hat{\boldsymbol{x}}_2\right)\right\| &\le k_{c1}\|\boldsymbol{x}_2\|\|\boldsymbol{e}_2\| + k_{c2}\left\|\boldsymbol{\xi}_2\right\|(\|\boldsymbol{x}_2\| + \|\boldsymbol{e}_2\|) \\ &\le \left(k_{c1} + k_{c2}\right)k_{\omega}\|\boldsymbol{e}_2\| + h_{31}\|\boldsymbol{e}_2\|^2 \end{aligned} \tag{13.15}$$

where k_{c1} and k_{c2} are existing positive constants.

Define the following auxiliary variable:

$$\overline{\boldsymbol{\varepsilon}} = \left[\left(\mathrm{sig}^{\alpha_1}(\boldsymbol{e}_1)\right)^{\mathrm{T}}, \boldsymbol{e}_2^{\mathrm{T}}, \boldsymbol{e}_3^{\mathrm{T}}\right]^{\mathrm{T}}. \tag{13.16}$$

It should be noted that $\boldsymbol{e}_1$, $\boldsymbol{e}_2$, and $\boldsymbol{e}_3$ will converge to the origin in finite time, if the new state $\overline{\boldsymbol{\varepsilon}}$ is finite-time stable. Differentiating $\overline{\boldsymbol{\varepsilon}}$ in (13.16) yields

$$
\begin{aligned}
\dot{\overline{\boldsymbol{\varepsilon}}} &= \begin{bmatrix} \alpha_1 \mathrm{diag}\left(|e_1|^{\alpha_1-1}\right)\left(-\beta_1 \mathrm{sig}^{\alpha_1}(e_1)+e_2\right) \\ -\beta_2 \mathrm{sig}^{\alpha_2}(e_1)+\overline{\zeta}+e_3 \\ -\beta_3 \mathrm{sig}^{\alpha_2}(e_1)-g \end{bmatrix} \\
&= \begin{bmatrix} \alpha_1 \mathrm{diag}\left(|e_1|^{\alpha_1-1}\right)\left(-\beta_1 \mathrm{sig}^{\alpha_1}(e_1)+e_2\right) \\ -\beta_2 \mathrm{sig}^{\alpha_2}(e_1)+e_3 \\ -\beta_3 \mathrm{sig}^{\alpha_2}(e_1) \end{bmatrix} + \begin{bmatrix} \mathbf{0}_3 \\ \overline{\zeta} \\ \mathbf{0}_3 \end{bmatrix} + \begin{bmatrix} \mathbf{0}_3 \\ \mathbf{0}_3 \\ -g \end{bmatrix} \\
&= \mathrm{diag}\left(\left[|e_1|^{\alpha_1-1}, |e_1|^{\alpha_1-1}, |e_1|^{\alpha_1-1}\right]\right)\boldsymbol{H}\overline{\boldsymbol{\varepsilon}}+\underline{\overline{\zeta}}+\boldsymbol{\Gamma}
\end{aligned}
\tag{13.17}
$$

where $\boldsymbol{H}$ is defined as $\boldsymbol{H}=\left[-\alpha_1\beta_1\boldsymbol{I}_3, \alpha_1\boldsymbol{I}_3, \mathbf{0}_3; -\beta_2\boldsymbol{I}_3, \mathbf{0}_3, \boldsymbol{I}_3; -\beta_3\boldsymbol{I}_3, \mathbf{0}_3, \mathbf{0}_3\right]$, $\underline{\overline{\zeta}}=\left[\mathbf{0}_3; \overline{\zeta}; \mathbf{0}_3\right]$, $\boldsymbol{\Gamma}=\left[\mathbf{0}_3; \mathbf{0}_3; -g\right]$, and $\mathbf{0}_3$ denotes the zero 3×3 matrix. In view of the state-space equation in (13.17), one can get the characteristic function of this system

$$\lambda^3+\alpha_1\beta_1\lambda^2+\alpha_1\beta_2\lambda+\alpha_1\beta_3=0 \tag{13.18}$$

where λ is the characteristic variable of the characteristic function in (13.18). If one designs the gain parameters of the proposed FTESO subject to the constraint $\beta_3<\alpha_1\beta_1\beta_2$, all the eigenvalues of the characteristic function in (13.18) have positive real parts. That is, the system (13.17) is stable according to Routh–Hurwitz criterion. Thus, the coefficient matrix $\boldsymbol{H}$ in (13.17) is a Hurwitz matrix. There exists a symmetric and positive-definite matrix $\boldsymbol{P}=\boldsymbol{P}^{\mathrm{T}}$, such that the following algebraic Lyapunov equation holds:

$$\boldsymbol{H}^{\mathrm{T}}\boldsymbol{P}+\boldsymbol{P}\boldsymbol{H}=-\mathbf{Q} \tag{13.19}$$

where $\mathbf{Q}=\mathbf{Q}^{\mathrm{T}}$ is a Hermitian matrix, which is also a symmetric and positive-definite.

Then, we select a positive-definite candidate Lyapunov function as

$$V_1(\boldsymbol{e})=\overline{\boldsymbol{\varepsilon}}^{\mathrm{T}}\boldsymbol{P}\overline{\boldsymbol{\varepsilon}}. \tag{13.20}$$

Differentiating V_1 in (13.20) with respect to time yields

$$
\begin{aligned}
\dot{V}_1 &= \overline{\boldsymbol{\varepsilon}}^{\mathrm{T}}\left[\mathrm{diag}\left(\left[|e_1|^{\alpha_1-1}, |e_1|^{\alpha_1-1}, |e_1|^{\alpha_1-1}\right]\right)\left(\boldsymbol{H}^{\mathrm{T}}\boldsymbol{P}+\boldsymbol{P}\boldsymbol{H}\right)\right]\overline{\boldsymbol{\varepsilon}} \\
&\quad +2\overline{\boldsymbol{\varepsilon}}^{\mathrm{T}}\boldsymbol{P}\left(\underline{\overline{\zeta}}+\boldsymbol{\Gamma}\right) \\
&\leq -\left(|e_1|_{\max}\right)^{\alpha_1-1}\overline{\boldsymbol{\varepsilon}}^{\mathrm{T}}\mathbf{Q}\overline{\boldsymbol{\varepsilon}}+2\left\|\overline{\boldsymbol{\varepsilon}}\right\|\left\|\boldsymbol{P}\right\|\left\|\underline{\overline{\zeta}}\right\|+2\left\|\overline{\boldsymbol{\varepsilon}}\right\|\left\|\boldsymbol{P}\right\|\left\|\boldsymbol{\Gamma}\right\|
\end{aligned}
\tag{13.21}
$$

where $|e_1|_{\max} = \max\{|e_{11}|, |e_{12}|, |e_{13}|\}$. Given $|e_1|_{\max} \le \|e_1\| \le \|\bar{\varepsilon}\|^{\frac{1}{\alpha_1}}$ and $\alpha_1 \in \left(\frac{1}{2}, 1\right)$, one can achieve that

$$\dot{V}_1 \le -\lambda_{\min}(\mathbf{Q}) \|\bar{\varepsilon}\|^{3-\frac{1}{\alpha_1}} + 2\|\bar{\varepsilon}\| \|\boldsymbol{P}\| \left(\left(k_{c1} + k_{c2}\right) k_{\omega} \|e_2\| + h_{32} \|e_2\|^2\right) + 2\|\bar{\varepsilon}\| \|\boldsymbol{P}\| \|\boldsymbol{\Gamma}\|. \tag{13.22}$$

Since the terms $\bar{\boldsymbol{d}}$ and $\boldsymbol{g}$ are supposed to be limited by $\|\bar{\boldsymbol{d}}\| \le \bar{d}_1$ and $|g_i(t)| \le \bar{g}$, respectively, one gets

$$2\|\bar{\varepsilon}\| \|\boldsymbol{P}\| \|\boldsymbol{\Gamma}\| \le 2\sqrt{3}\bar{g} \|\bar{\varepsilon}\| \|\boldsymbol{P}\| \le 2\sqrt{3}\lambda_{\min}(\boldsymbol{P})^{-\frac{1}{2}} V_1^{\frac{1}{2}} \|\boldsymbol{P}\|$$

by using the inequality $\lambda_{\min}(\boldsymbol{P}) \|\bar{\varepsilon}\|^2 \le V_1 \le \lambda_{\max}(\boldsymbol{P}) \|\bar{\varepsilon}\|^2$. Then, the inequality (13.22) yields

$$\begin{aligned}\dot{V}_1 \le & -\lambda_{\min}(\mathbf{Q}) \lambda_{\max}(\boldsymbol{P})^{\frac{1}{2\alpha_1}-\frac{3}{2}} V_1^{\frac{3}{2}-\frac{1}{2\alpha_1}} + 2\|\boldsymbol{P}\| \|\bar{\varepsilon}\|^2 \left(\left(k_{c1} + k_{c2}\right) k_{\omega} + k_{c2} \|\bar{\varepsilon}\|\right) \\ & + 2\sqrt{3}\bar{g}\lambda_{\min}(\boldsymbol{P})^{-\frac{1}{2}} \|\boldsymbol{P}\| V_1^{\frac{1}{2}} \\ \le & -\lambda_1 V_1^{\frac{3}{2}-\frac{1}{2\alpha_1}} + \lambda_2 V_1 + \lambda_3 V_1^{\frac{3}{2}} + \lambda_4 V_1^{\frac{1}{2}}\end{aligned} \tag{13.23}$$

with $\lambda_1 = \lambda_{\min}(\mathbf{Q}) \lambda_{\max}(\boldsymbol{P})^{\frac{1}{2\alpha_1}-\frac{3}{2}}$, $\lambda_2 = 2\left(k_{c1} + k_{c2}\right) k_{\omega} \|\boldsymbol{P}\| \lambda_{\min}(\boldsymbol{P})^{-1}$, $\lambda_3 = 2k_{c2} \|\boldsymbol{P}\| \lambda_{\min}(\boldsymbol{P})^{-\frac{3}{2}}$, and $\lambda_4 = 2\sqrt{3}\bar{g}\lambda_{\min}(\boldsymbol{P})^{-\frac{1}{2}} \|\boldsymbol{P}\|$. Define two new variables as $\bar{\lambda}_2 = \lambda_2 V_1(e_0)^{-\frac{1}{2}+\frac{1}{2\alpha_1}}$, $\bar{\lambda}_3 = \lambda_3 V_1(e_0)^{\frac{1}{2\alpha_1}}$, and an attraction region $\Omega_1 = \left\{e \mid \lambda_2 V_1^{-\frac{1}{2}+\frac{1}{2\alpha_1}} + \lambda_3 V_1^{\frac{1}{2\alpha_1}} < \lambda_1\right\}$. This specified attraction region sets a limit to the initial values of the states, which should be subject to the set Ω_1 for the stability of the system. For all the $\boldsymbol{e} \in \Omega_1$, one has $\dot{V}_1 < 0$. It implies that V_1 is a decreasing progressively function with respect to time, then one has $V_1(e_0) \ge V_1(\boldsymbol{e})$ and $\bar{\lambda}_2 + \bar{\lambda}_3 < \lambda_1$. Utilizing the above definitions and properties, inequality (13.23) yields

$$\begin{aligned}\dot{V}_1 \le & -\left(\lambda_1 - \bar{\lambda}_2 - \bar{\lambda}_3\right) V_1^{\frac{3}{2}-\frac{1}{2\alpha_1}} + \lambda_2 V_1 - \bar{\lambda}_2 V_1^{\frac{3}{2}-\frac{1}{2\alpha_1}} + \lambda_3 V_1^{\frac{3}{2}} \\ & - \bar{\lambda}_3 V_1^{\frac{3}{2}-\frac{1}{2\alpha_1}} + \lambda_4 V_1^{\frac{1}{2}} \\ \le & -\left(\lambda_1 - \bar{\lambda}_2 - \bar{\lambda}_3\right) V_1^{\frac{3}{2}-\frac{1}{2\alpha_1}} + \lambda_4 V_1^{\frac{1}{2}} \triangleq -\lambda_5 V_1^{\frac{3}{2}-\frac{1}{2\alpha_1}} + \lambda_4 V_1^{\frac{1}{2}}\end{aligned} \tag{13.24}$$

with $\lambda_5 = \lambda_1 - \bar{\lambda}_2 - \bar{\lambda}_3$. According to Lemma 13.2, the trajectory of the proposed FTESO in (13.11) is finite-time uniformly ultimately bounded stable. It also implies that the state observation errors $\boldsymbol{e}_1$, $\boldsymbol{e}_2$, and $\boldsymbol{e}_3$ will be convergent to a small region of the origin in finite time T_1. Furthermore, the stabilization time T_1 is

$$T_1 \leq \frac{2\alpha_1 V_1(\boldsymbol{e}_0)^{\frac{1}{2\alpha_1}-\frac{1}{2}}}{(\lambda_5 - \gamma_5)(1-\alpha_1)} \tag{13.25}$$

with $\gamma_5 \in (0, \beta_5)$. And the stable region is

$$D_1 = \left\{ \boldsymbol{e} \mid V_1(\boldsymbol{e}) < \left(\frac{\lambda_4}{\lambda_5}\right)^{\frac{2\alpha_1}{2\alpha_1 - 1}} \right\}.$$

Then, the observation errors of the attitude angular velocity $\boldsymbol{e}_2$ and synthetic failure/uncertainty $\boldsymbol{e}_3$ are proved to be bounded theoretically. The boundaries of $\boldsymbol{e}_2$ and $\boldsymbol{e}_3$ can be denoted as $\overline{\boldsymbol{\sigma}}_2 = [\overline{\sigma}_{21}, \overline{\sigma}_{22}, \overline{\sigma}_{23}]^{\mathrm{T}}$ and $\overline{\boldsymbol{\sigma}}_3 = [\overline{\sigma}_{31}, \overline{\sigma}_{32}, \overline{\sigma}_{33}]^{\mathrm{T}}$, respectively. Thus, the observation errors $\boldsymbol{e}$ of the proposed FTESO will be convergent to the small residual set D_1 in finite time T_1. □

Remark 13.1. It should be noted that the finite-time stability of the proposed FTESO is achieved under the condition that the observation errors $\boldsymbol{e}$ belong to the attraction region Ω_1. The size of Ω_1 depends on the selection of the observer gains β_1, β_2, β_3, and α_1. And the attraction region of observation error system could be enlarged as much as desired by increasing β_1, β_2 or β_3, or by decreasing α_1.

13.3.3 Finite-time attitude dynamic feedback control algorithm

In order to achieve fast convergence performance of the control system, terminal sliding mode control has attracted significant interest [40]. However, it has two noteworthy disadvantages, that is, the singularity problem and chattering phenomenon. To this end, a supertwisting algorithm based continuous homogeneous sliding mode control scheme was proposed for a perturbed second-order system in [41]. It has ensured the finite-time convergence of the system states and chattering reducing performance. And that, it is worth mentioning that the supertwisting algorithm based sliding mode control laws can actuate continuous control signals to restrain the chattering phenomenon and obtain the finite-time convergence [42]. Inspired by these approaches, a novel continuous finite-time control law is investigated by using the reconstructed angular velocity and system failure information deriving from the proposed FTESO. The proposed control law is considered as a combination of supertwisting and nonsingular terminal sliding mode control, with the superiorities of chattering attenuating and nonsingularity.

Consider the attitude dynamics in the form of the second-order plant as (13.9) by denoting $\boldsymbol{x}_1 = \boldsymbol{\Theta}$ and $\boldsymbol{x}_2 = \dot{\boldsymbol{\Theta}}$. The attitude control system can be rewritten as

$$\dot{\boldsymbol{x}}_1 = \boldsymbol{x}_2, \tag{13.26a}$$

$$\dot{\boldsymbol{x}}_2 = -\boldsymbol{M}^{-1}\boldsymbol{C}(\boldsymbol{x}_1, \boldsymbol{x}_2)\boldsymbol{x}_2 + \overline{\boldsymbol{B}}\boldsymbol{\tau} + \boldsymbol{x}_3, \tag{13.26b}$$

with $\overline{\boldsymbol{B}} = \boldsymbol{M}^{-1}\boldsymbol{F}^{\mathrm{T}}$ and $\boldsymbol{x}_3 = \boldsymbol{M}^{-1}\overline{\boldsymbol{d}}$. Then, a nonsingular terminal sliding mode surface (NTSMS) [40] can be defined as

$$\boldsymbol{S}_L = \boldsymbol{x}_1 + \frac{\alpha_{31}}{L^{\frac{1}{2}}} \mathrm{sig}^{\frac{3}{2}}(\hat{\boldsymbol{x}}_2) \tag{13.27}$$

where $\boldsymbol{S}_L = [s_{L1}, s_{L2}, s_{L3}]^{\mathrm{T}}$, $L > 0$ and $\alpha_{31} > 0$ are positive gains to be designed. It should be noted that (13.27) is introduced as a real sliding mode surface. When $L = 1$, we define $\boldsymbol{S}_L$ simply as $\boldsymbol{S}$. For the attitude control system in (13.26) and the NTSMS in (13.27), a novel simple FTDFC is developed in the form of

$$\boldsymbol{\tau} = \overline{\boldsymbol{B}}^{-1}\left(-k_1 L^{\frac{2}{3}} \mathrm{sig}^{\frac{1}{3}}(\boldsymbol{S}_L) + \boldsymbol{M}^{-1}\boldsymbol{C}(\boldsymbol{x}_1, \hat{\boldsymbol{x}}_2)\hat{\boldsymbol{x}}_2 - \hat{\boldsymbol{x}}_3 + \boldsymbol{z}_1\right), \tag{13.28}$$

$$\dot{\boldsymbol{z}}_1 = -k_2 L\, \mathrm{sign}(\boldsymbol{S}_L) \tag{13.29}$$

where $\hat{\boldsymbol{x}}_2$ and $\hat{\boldsymbol{x}}_3$ are the observation values of the unknown attitude angular velocity and synthetic failure deriving from the proposed FTESO in Sect. 13.3.2, whereas k_1 and k_2 are positive gains to be designed. Submitting the FTDFC in (13.28)–(13.29) into the attitude dynamics in (13.26), it follows that

$$\dot{\boldsymbol{x}}_1 = \boldsymbol{x}_2, \tag{13.30a}$$

$$\dot{\boldsymbol{x}}_2 = -k_1 L^{\frac{2}{3}} \mathrm{sig}^{\frac{1}{3}}(\boldsymbol{S}_L) - \tilde{\boldsymbol{\zeta}} - \boldsymbol{e}_3 + \boldsymbol{z}_1. \tag{13.30b}$$

Defining $\boldsymbol{\mu} = -\tilde{\boldsymbol{\zeta}} - \boldsymbol{e}_3$ and $\boldsymbol{z}_3 \triangleq -\tilde{\boldsymbol{\zeta}} - \boldsymbol{e}_3 + \boldsymbol{z}_1$, the closed-loop attitude control system in (13.26)–(13.29) yields the following third order plant:

$$\dot{\boldsymbol{x}}_1 = \boldsymbol{x}_2, \tag{13.31a}$$
$$\dot{\boldsymbol{x}}_2 = -k_1 L^{\frac{2}{3}} \operatorname{sig}^{\frac{1}{3}}(\boldsymbol{S}_L) + \boldsymbol{z}_3, \tag{13.31b}$$
$$\dot{\boldsymbol{z}}_3 = -k_2 L \operatorname{sign}(\boldsymbol{S}_L) + \dot{\mu} \tag{13.31c}$$

where $\boldsymbol{x}_1$, $\boldsymbol{x}_2$, and $\boldsymbol{z}_3$ are the states of the above closed-loop plant. According to the results in [41], the solutions of the above third order system will be found in the sense of Filippov. It is easily obtained that the Filippov differential inclusion corresponding to the closed-loop third-order attitude stabilization control system in (13.31) is homogeneous of degree (scaled to) $\delta = -1$ with weights $(r_1 = 3, r_2 = 2, r_3 = 1)$. Then, the other key contribution of this paper is presented as the following theorem.

Theorem 13.2. *Consider the simplified attitude stabilization control system in (13.9) with the assumption that $|\dot{\mu}_i| \leq \overline{\mu}$, $i = 1, 2, 3$, $\overline{\mu} > 0$. The attitudes are globally and uniformly finite-time stable, if the NTSMS based attitude control scheme FTDFC is developed in the form of (13.28)–(13.29).*

To prove and analyze the above Theorem 13.2, the following definition will be introduced and used [41,43].

Definition 13.2. The origin $x = 0$ of a differential inclusion $\dot{x} \in F(x)$ (a differential equation $\dot{x} = f(x)$) is called globally uniformly finite-time stable, if it is Lyapunov stable and there exists $T > 0$ such that the trajectory with initial condition $\|x_0\| < \varsigma$ $(\varsigma > 0)$ will converge to $x = 0$ at time T.

Proof of Theorem 13.2. Firstly, select the candidate Lyapunov function as follows:

$$V_{2i} = \frac{2}{5}\alpha_{31} |x_{2i}|^{\frac{5}{2}} + \beta_{31} |x_{1i}|^{\frac{5}{3}} + x_{1i}x_{2i} - \frac{1}{k_1^3} x_{2i} z_{3i}^3 + k_3 |z_{3i}|^5, \quad i = 1, 2, 3 \tag{13.32}$$

where $k_3 > 0$ and $\beta_{31} > 0$. For the above candidate Lyapunov function V_{2i}, one can obtain easily that V_{2i} is homogeneous with the degree $\delta_{V_{2i}} = 5$, and also continuously differentiable. Utilizing the Young's inequality, V_{2i} can be calculated to be positive definite and bounded, if the control gains are selected as $\beta_{31} > \frac{3}{5}\left(\frac{4}{\alpha_{31}}\right)^{\frac{2}{3}}$ and $k_1^5 k_3 > \frac{3}{5}\left(\frac{4}{\alpha_{31}}\right)^{\frac{2}{3}}$.

Taking the derivative of V_{2i} with respect to time, it follows that

$$\begin{aligned}\dot{V}_{2i} = &-\left(\alpha_{31}\text{sig}^{\frac{3}{2}}\left(s_{Li}\right) - \frac{1}{k_1^3}z_{3i}^3\right)\left(k_1\text{sig}^{\frac{1}{3}}\left(s_{Li}\right) - z_{3i}\right) \\ &+ \left(\frac{5}{3}\beta_{31}\text{sig}^{\frac{2}{3}}\left(x_{1i}\right) + x_{2i}\right)x_{2i} \\ &- \left|z_{3i}\right|^2\left(5k_3\left|z_{3i}\right|^2 - \frac{3}{k_1^3}x_{2i}\right)\left(k_2\,\text{sign}\left(s_{Li}\right) + \frac{\dot{\mu}_i}{L}\right).\end{aligned} \tag{13.33}$$

Utilizing Lemma 2 and Procedure 1 in [41], one can get that $\dot{V}_{2i}$ is continuous and negative definite. And that $\dot{V}_{2i}$ is homogeneous with the degree $\delta_{\dot{V}_{2i}} = 4$ according to the theory of homogeneous functions. Then, one can see that

$$\dot{V}_{2i} \leq -\eta_i V_{2i}^{\frac{4}{5}} \tag{13.34}$$

where η_i is a positive constant. Thus, the attitude stabilization control system in (13.9) is uniformly globally asymptotically stable according to the basic Lyapunov theory for differential inclusions. Furthermore, it is globally uniformly finite-time stable since the closed-loop system in (13.31) is homogeneous of degree $\delta = -1$. Certainly, the results in Theorem 13.2 could be achieved by using Lemma 13.1 as well. □

Remark 13.2. The control scheme proposed in (13.28)–(13.29) is inspired by [41]. This work applies the control methods in [41] to the quadrotor UAV attitude stabilization problem even though there exist model uncertainty, actuator failures, and disturbances. As for the specific analysis of Theorem 13.2, it is similar to [41] and the detailed analysis is omitted here for space limitation. Nevertheless, it should be noted that the assumption $|\dot{\mu}_i| \leq \overline{\mu}$ is reasonable. According to Theorem 13.1, one can easily obtain that $\boldsymbol{e}_2$, $\dot{\boldsymbol{e}}_2$, $\boldsymbol{e}_3$, and $\dot{\boldsymbol{e}}_3$ are all continuous and bounded under the assumption that $\boldsymbol{g}(t)$ is bounded. And $\tilde{\boldsymbol{\xi}}(\boldsymbol{x}_1, \boldsymbol{x}_2) = \boldsymbol{M}^{-1}\boldsymbol{C}(\boldsymbol{x}_1, \boldsymbol{x}_2)\boldsymbol{e}_2 - \boldsymbol{M}^{-1}\boldsymbol{C}(\boldsymbol{x}_1, \boldsymbol{e}_2)(\boldsymbol{x}_2 + \boldsymbol{e}_2)$ is also bounded deriving from (13.15), and so is $\boldsymbol{\mu} = -\tilde{\boldsymbol{\xi}} - \boldsymbol{e}_3$. Thus, it is also reasonable to suppose that there exists a positive and sufficiently large constant $\overline{\mu}$ such that $|\dot{\mu}_i| \leq \overline{\mu}$ on the basis of Assumption 13.2. Although the assumption may appear restrictive, it readily provides some help for the finite-time stability analysis of the proposed control law. And also this bound $\overline{\mu}$ could be an arbitrary positive value, because it has nothing to do with the dynamic process of the attitude control. Therefore, Theorem 13.2 will be applicable with the only requirement that $|\dot{\mu}_i|$ is bounded.

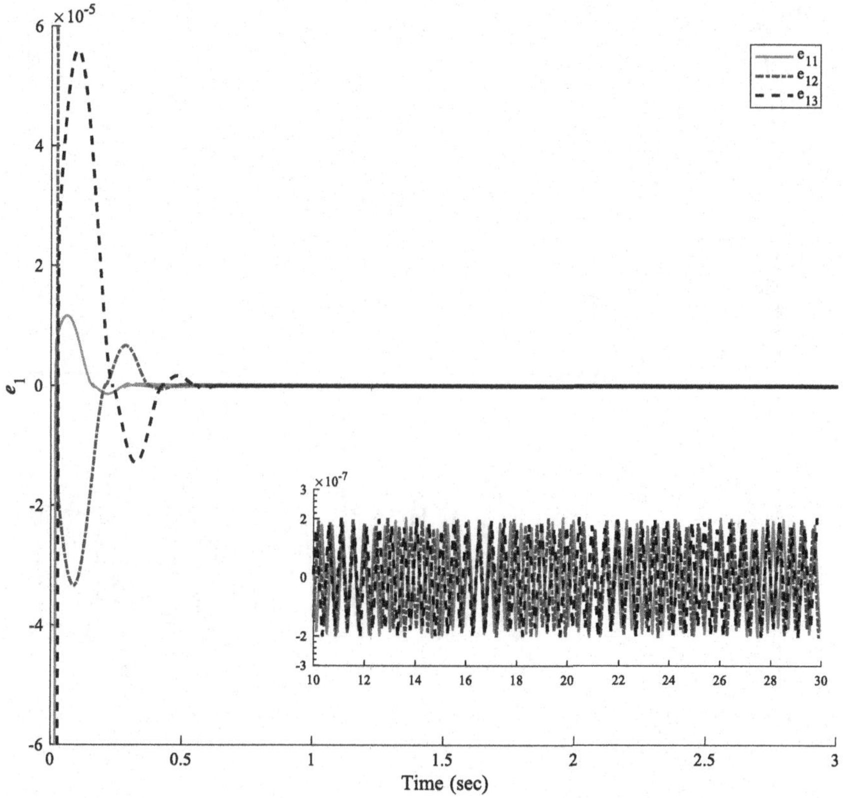

Figure 13.1 Time response of the observation error e_1 without actuator failures.

13.4. Simulation and analysis

To illustrate the effectiveness and fine performances of the proposed FTESO based FTDFC in (13.11) and (13.28)–(13.29) (FTESO+FTDFC), some numerical simulation results for a quadrotor UAV with unavailable angular velocity measurements and actuator failures will be shown in this section. All the numerical simulations are carried out by the software MATLAB®/SIMULINK. For the numerical simulations, the inertia matrix of a quadrotor UAV is selected as $\boldsymbol{J} = \mathrm{diag}\{1.25, 1.25, 2.5\}$ $(\mathrm{kg} \cdot \mathrm{m}^2)$. The initial attitude angle is set as $\boldsymbol{\Theta}(0) = [0.1, -0.2, 0.3]^{\mathrm{T}}$ (rad) and the initial angular velocity is set as $\boldsymbol{\omega}(0) = [0, 0, 0]^{\mathrm{T}}$ (rad/s). Additionally, the disturbances acting on the quadrotor UAV are supposed in the form of

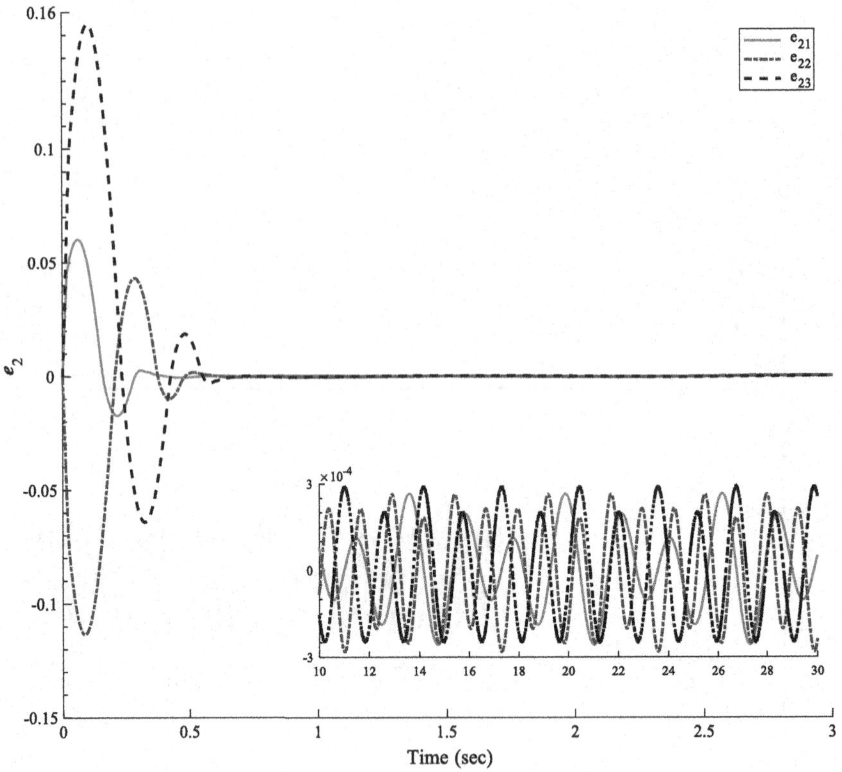

Figure 13.2 Time response of the observation error e_2 without actuator failures.

$$\boldsymbol{d} = 0.01 \cdot \begin{bmatrix} 3\cos(2t) + 4\sin(3t) - 5 \\ -1.5\sin(2t) + 3\cos(5t) + 5 \\ 3\sin(2t) - 8\sin(4t) + 5 \end{bmatrix} \ (\mathrm{N} \cdot \mathrm{m}).$$

Consider a practical engineering application where the maximum output torque of the quadrotor UAV is set as $|\boldsymbol{\tau}_{\max}| = [0.4, 0.4, 0.4]^{\mathrm{T}}$ $(\mathrm{N} \cdot \mathrm{m})$. To achieve the control performance, the chosen control gains of the proposed schemes are described in Table 13.1.

13.4.1 Observation performance of the FTESO without actuator failures

In this subsection, the observation performance of the proposed FTESO in (13.11) is presented and analyzed in the presence of unavailable attitude angular velocity and external disturbances, without actuator failures and

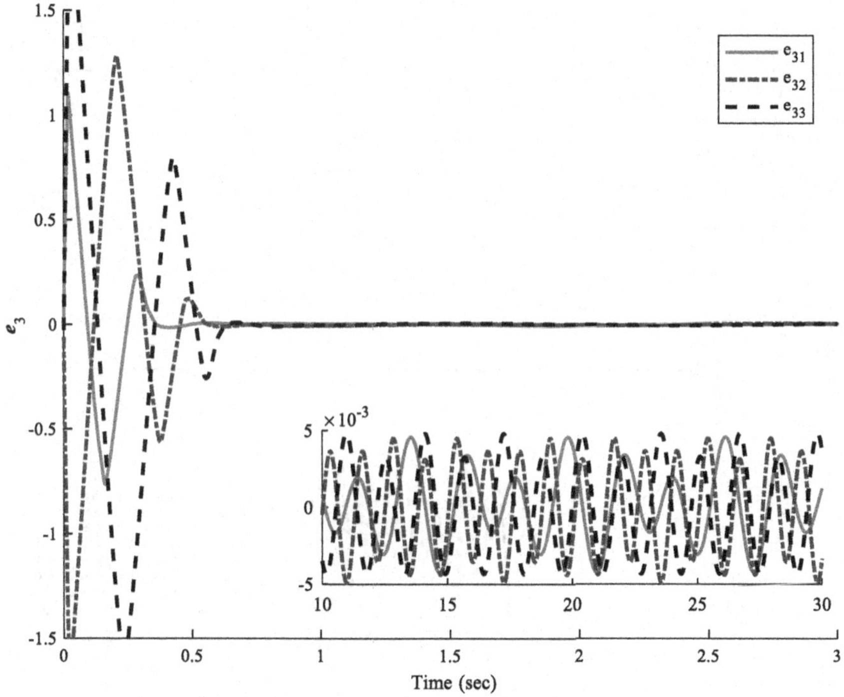

Figure 13.3 Time response of the observation error e_3 without actuator failures.

Table 13.1 Control parameters selected for the numerical simulations.

Control scheme	Control gains
FTESO	$\alpha_1 = 0.6$, $\beta_1 = 55$, $\beta_2 = 4.5$, $\beta_3 = 141$
FTDFC	$\alpha_{31} = 6.7$, $L = 1$, $k_1 = 0.1$, $k_2 = 0.003$

uncertainties. In view of the FTESO proposed in this work in (13.11), the FTESO can achieve the accurate observation only with the help of the UAV attitude information. As is shown in Fig. 13.1, the time responses of the attitude observation error $\boldsymbol{e}_1$ are described. The attitude tracking errors can converge to a small bounded region $|e_{1i}| \leq 2.5 \times 10^{-5}$ $(i = 1, 2, 3)$ in finite time, with the settling time less than 1 s. As observed in Figs. 13.2 and 13.3, the observation errors of the extended system's states $\boldsymbol{x}_2$ and $\boldsymbol{x}_3$

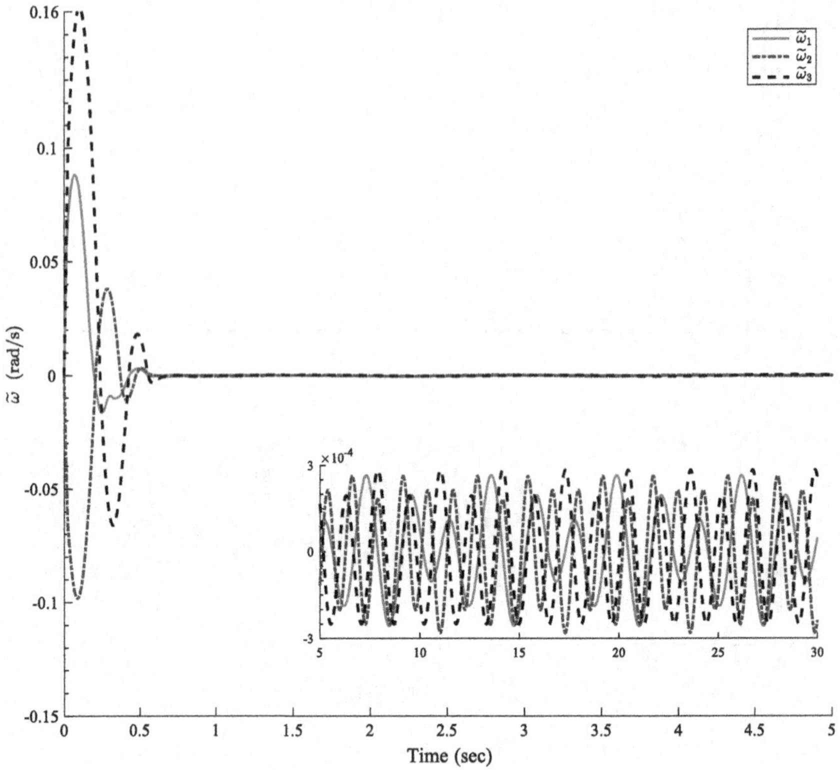

Figure 13.4 Observation errors $\tilde{\omega}$ without actuator failures.

in (13.10) are convergent and stable in finite time within 1 s, with the high accuracies of $|e_{2i}| \leq 3 \times 10^{-4}$ and $|e_{3i}| \leq 5 \times 10^{-3}$, respectively. According to the above figures, it is easy to obtain that the proposed FTESO in this chapter can successfully estimate the unknown system state and the synthetic failure or lumped disturbance in finite time.

According to the equation in (13.11b), the unavailable angular velocity can be reconstructed by the output of the FTESO $\hat{\boldsymbol{x}}_2$. From (13.1), the observation error of the angular velocity can be defined as $\tilde{\boldsymbol{\omega}} = \boldsymbol{M}^{-1}\boldsymbol{e}_2$. As is shown in Fig. 13.4, the observation angular velocity observation error $\tilde{\boldsymbol{\omega}}$ is demonstrated. The observation performance of the proposed FTESO can be perfect after 1 s with the accuracy of $|\tilde{\boldsymbol{\omega}}_i| \leq 3 \times 10^{-4}$ (rad/s). Thus, the FTESO is effective for the estimation of the angular velocity and without synthetic failures.

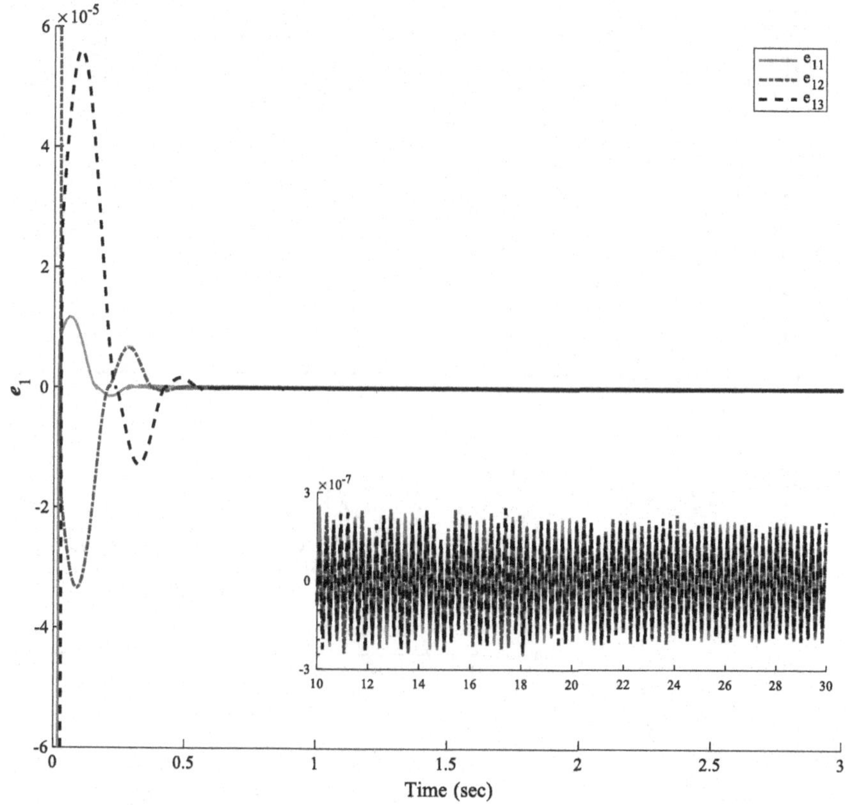

Figure 13.5 Time response of the observation error e_1 under actuator failures.

13.4.2 Attitude control performance of the FTDFC under actuator failures

In this subsection, the actuator failures and uncertainties will be taken into consideration additionally in the simulations to further verify the effectiveness of the proposed schemes in this work. The failure scenarios of the actuators are mathematically described in Table 13.2.

Table 13.2 Actuator control effectiveness.

ε_{ii}	t			
	0–10 s	**10–18 s**	**18–24 s**	**24–30 s**
ε_{11}	1	0.40	0.65	1
ε_{22}	1	0.40	0.70	1
ε_{33}	1	0.40	0.65	1

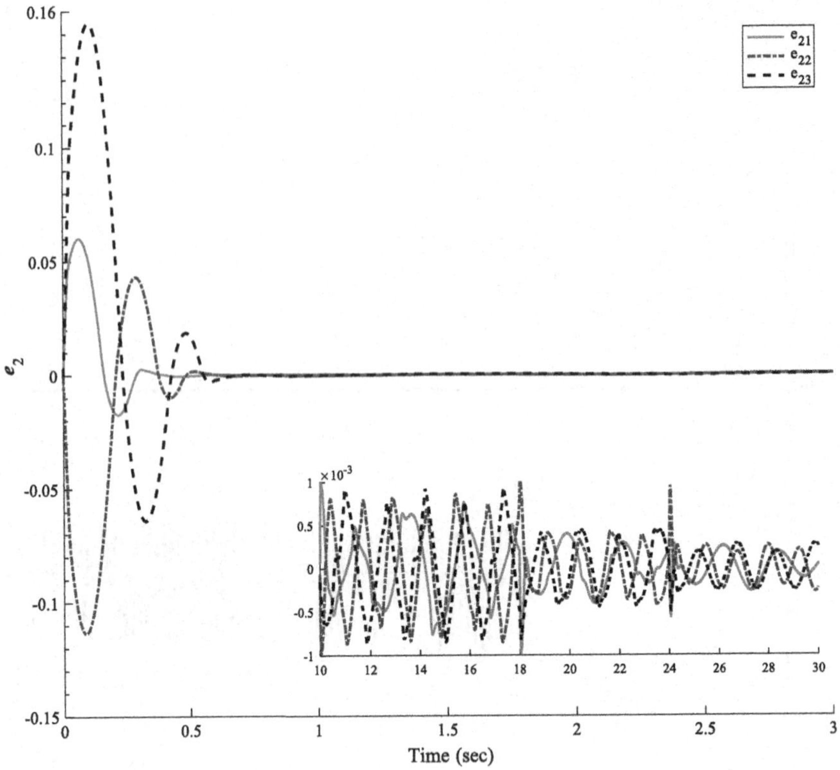

Figure 13.6 Time response of the observation error e_2 under actuator failures.

The time responses of observation errors $\boldsymbol{e}_1$, $\boldsymbol{e}_2$, and $\boldsymbol{e}_3$ obtained from the proposed FTESO in (13.11) under the actuator failures in Table 13.2 are presented in Figs. 13.5–13.7. It is showed that the reconstruction errors can converge to a small neighborhood of the origin under the actuator failures and uncertainties within 1 s. And the overall stabilization accuracy for each reconstruction error can reach $|e_{1i}| \leq 3 \times 10^{-7}$, $|e_{2i}| \leq 1 \times 10^{-3}$, and $|e_{3i}| \leq 4 \times 10^{-2}$, respectively. It should be noted that $\boldsymbol{e}_2$, $\boldsymbol{e}_3$ can change correspondingly under the different actuator failures and uncertainties, and the more serious variation of the actuator failures, the greater the corresponding observation error. Thus, the proposed FTESO can achieve the high accuracy observation of the attitude angle and synthetic failures in finite time under actuator failures and uncertainties. With the reconstructed information from the FTESO, the time responses of the attitude angular $\boldsymbol{\Theta}$ are illustrated in Fig. 13.8. The attitude angle $\boldsymbol{\Theta}$ can reach a stable state

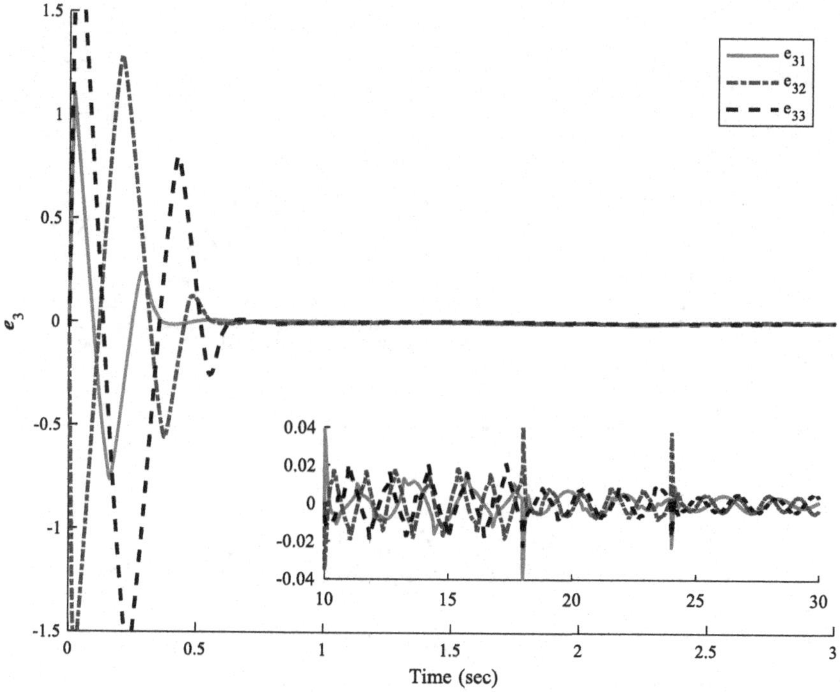

Figure 13.7 Time response of the observation error e_3 under actuator failures.

within 7 s with the steady-state accuracy of $|\boldsymbol{\Theta}| \leq 3 \times 10^{-3}$ under actuator failures and uncertainties. Then, the steady-state accuracy of $|\boldsymbol{\Theta}| \leq 3 \times 10^{-4}$ can be reached in the absence of actuator failures. The proposed observer-based finite-time attitude stabilization control scheme can obtain fine performance. In Fig. 13.9, a comparison of the actual actuator outputs $\boldsymbol{\tau}$ of FTESO+FTDFC and FTESO+PD are illustrated. Considering the limitation of the actual actuator outputs, compared with the FTESO+PD control scheme, the FTESO+FTDFC control scheme can stabilize the control system in a shorter time and provide more accurate control torque to achieve higher steady-state accuracy.

13.5. Conclusions

In this chapter, a novel FTESO is firstly developed to reconstruct the attitude angular velocity and the synthetic failure simultaneously for the quadrotor UAV attitude control system, which is the key feature of

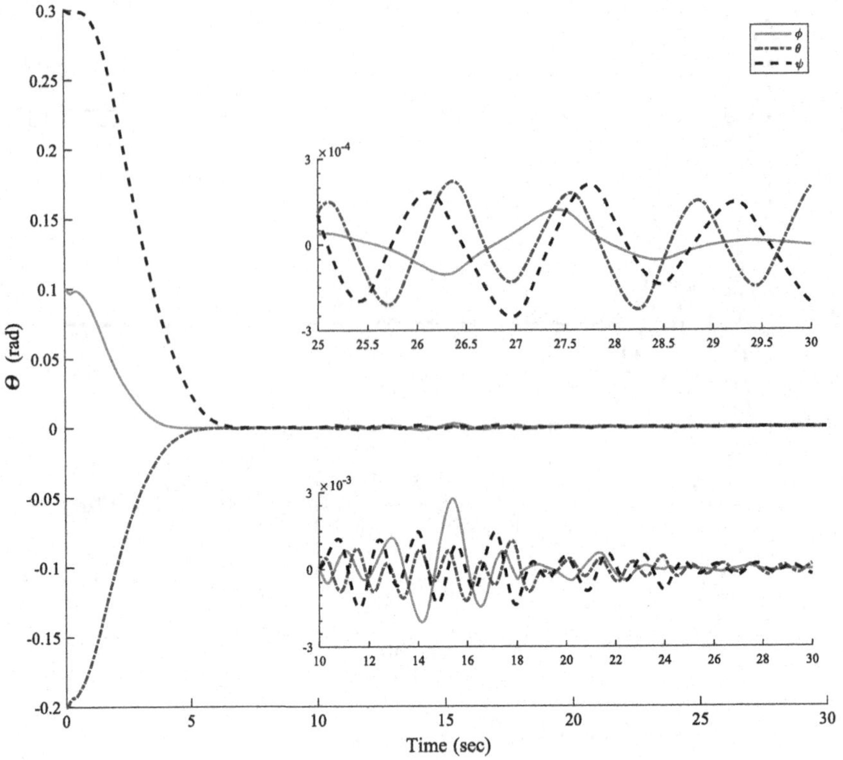

Figure 13.8 Time response of UAV attitude Θ under actuator failures.

the proposed fault tolerant control scheme. With the reconstructed information, a supertwisting and nonsingular terminal sliding mode based attitude control law is designed to guarantee the finite-time stability of the closed-loop attitude control system. From both the theoretical analysis and simulation results, the proposed fault tolerant output feedback control scheme is continuous with the property of chattering rejection. However, the reconstructed information of the extended state or the synthetic failure from the FTESO is the observation value of the lumped term including different external disturbances and actuator failures. The issue of isolating and identifying the individual failure of the actuators or system is one of the authors' future work. In addition, it is also worth doing some research to solve the cooperative control problems for multi-UAVs or multi-UAVs and AGVs in the future.

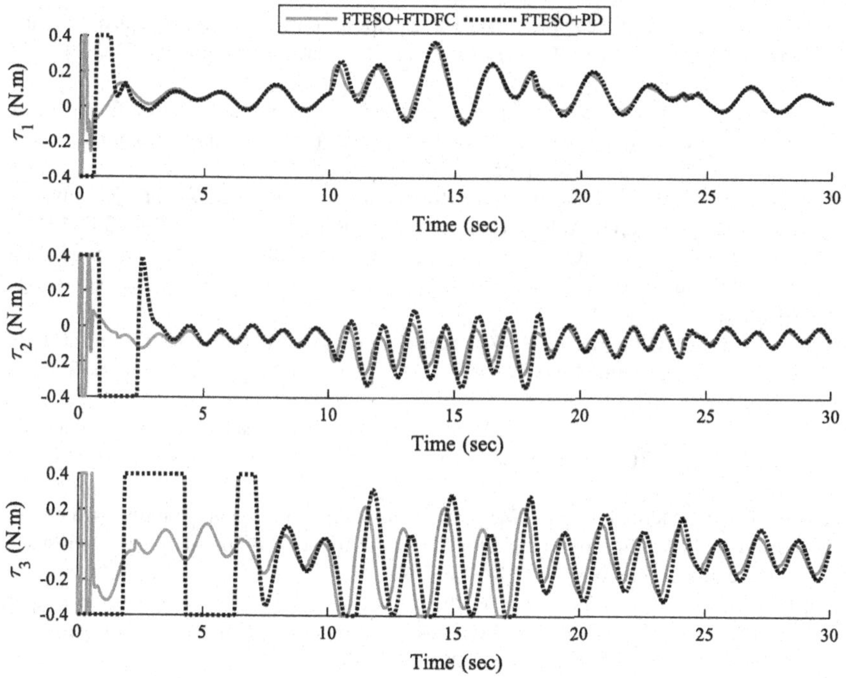

Figure 13.9 Time response of the actual actuator control torques τ under actuator failures.

References

[1] S. Hassan, A. Ali, Y. Rafic, A survey on quadrotors: configurations, modeling and identification, control, collision avoidance, fault diagnosis and tolerant control, IEEE Aerospace and Electronic Systems Magazine 33 (7) (2018) 14–33.

[2] H. Enke, J. Letschnik, Investigation of Ka-band satcom link performance for teleoperated search and rescue applications, IEEE Aerospace and Electronic Systems Magazine 34 (1) (2019) 28–38.

[3] H. Byun, Park S. Thrust, Control loop design for electric-powered UAV, International Journal of Aeronautical and Space Sciences 19 (1) (2018) 100–110.

[4] A. Gurtner, D.G. Greer, R. Glassock, L. Mejias, R.A. Walker, W.W. Boles, Investigation of fish-eye lenses for small-UAV aerial photography, IEEE Transactions on Geoscience and Remote Sensing 47 (3) (2009) 709–721.

[5] S.Z. Wang, B. Xian, S. Yang, Anti-swing controller design for an unmanned aerial vehicle with a slung-load, Acta Automatica Sinica 44 (10) (2018) 1771–1780.

[6] B.L. Tian, L.H. Liu, H.C. Lu, Z.Y. Zuo, Q. Zong, Y.P. Zhang, Multivariable finite time attitude control for quadrotor UAV: theory and experimentation, IEEE Transactions on Industrial Electronics 65 (3) (2018) 2567–2577.

[7] D.D. Wang, Q. Zong, B.L. Tian, S.K. Shao, X.Y. Zhang, X.Y. Zhao, Neural network disturbance observer-based distributed finite-time formation tracking control for multiple unmanned helicopters, ISA Transactions 73 (2018) 208–226.

[8] B. Xiao, S. Yin, A new disturbance attenuation control scheme for quadrotor unmanned aerial vehicles, IEEE Transactions on Industrial Informatics 13 (6) (2017) 2922–2932.
[9] K. Lee, J. Back, I. Choy, Nonlinear disturbance observer based robust attitude tracking controller for quadrotor UAVs, International Journal of Control, Automation, and Systems 12 (6) (2014) 1266–1275.
[10] C.S. Ha, Z. Zuo, F.B. Choi, et al., Passivity-based adaptive backstepping control of quadrotor-type UAVs, Robotics and Autonomous Systems 62 (9) (2014) 1305–1315.
[11] L.Q. Dou, C. Tao, M. Qi, Disturbance observer-based distributed formation tracking control for multiple quadrotor, Journal of Tianjin University: Science and Technology 51 (8) (2018) 817–824.
[12] J. Gośliński, W. Giernacki, A. Królikowski, A nonlinear filter for efficient attitude estimation of unmanned aerial vehicle (UAV), Journal of Intelligent & Robotic Systems 95 (3) (2019) 1079–1095.
[13] T.H. Wu, T. Lee, Angular velocity observer for attitude tracking on SO(3) with the separation property, International Journal of Control, Automation, and Systems 14 (5) (2016) 1289–1298.
[14] X. Shao, Q. Meng, J. Liu, H. Wang, RISE and disturbance compensation based trajectory tracking control for a quadrotor UAV without velocity measurements, Aerospace Science and Technology 74 (2018) 145–159.
[15] H.M. Guzey, T. Dierks, S. Jagannathan, L. Acar, Modified consensus-based output feedback control of quadrotor UAV formations using neural networks, Journal of Intelligent & Robotic Systems 84 (1) (2018) 283–300.
[16] N. Wang, Q. Deng, G.M. Xie, X.X. Pan, Hybrid finite-time trajectory tracking control of a quadrotor, ISA Transactions 90 (2019) 278–286.
[17] Q.L. Hu, B.Y. Jiang, Continuous finite-time attitude control for rigid spacecraft based on angular velocity observer, IEEE Transactions on Aerospace and Electronic Systems 54 (3) (2018) 1082–1092.
[18] Q.L. Hu, B.Y. Jiang, Y.M. Zhang, Observer-based output feedback attitude stabilization for spacecraft with finite-time convergence, IEEE Transactions on Control Systems Technology 27 (2) (2019) 781–789.
[19] S.K. Shao, Q. Zong, B.L. Tian, F. Wang, Finite-time sliding mode attitude control for rigid spacecraft without angular velocity measurement, Journal of the Franklin Institute 354 (12) (2017) 4656–4674.
[20] Q. Zong, S.K. Shao, X.Y. Zhang, Finite-time output feedback attitude tracking control for rigid spacecraft, Journal of Harbin Institute of Technology 49 (9) (2017) 136–147.
[21] L. Yuan, G. Ma, C. Li, B.Y. Jiang, Finite-time attitude tracking control for spacecraft without angular velocity measurements, Journal of Systems Engineering and Electronics 28 (6) (2017) 1174–1185.
[22] M.S. Qian, B. Jiang, H.HT. Liu, Dynamic surface active fault tolerant control design for the attitude control systems of UAV with actuator fault, International Journal of Control, Automation, and Systems 14 (3) (2016) 723–732.
[23] W. Hao, B. Xian, Nonlinear adaptive fault-tolerant control for a quadrotor UAV based on immersion and invariance methodology, Nonlinear Dynamics 90 (4) (2017) 2813–2826.
[24] Q. Shen, C. Yue, C.H. Goh, B.L. Wu, D.W. Wang, Rigid-body attitude tracking control under actuator faults and angular velocity constraints, IEEE/ASME Transactions on Mechatronics 23 (3) (2018) 1338–1349.

[25] Y.M. Zhang, A. Chamseddine, C.A. Rabbath, B.W. Gordon, C.Y. Su, S. Rakheja, et al., Development of advanced FDD and FTC techniques with application to an unmanned quadrotor helicopter testbed, Journal of the Franklin Institute 350 (9) (2013) 2396–2422.
[26] Z. Yu, Y. Qu, Y.M. Zhang, Fault-tolerant containment control of multiple unmanned aerial vehicles based on distributed sliding-mode observer, Journal of Intelligent & Robotic Systems 93 (1) (2019) 163–177.
[27] B. Xiao, S. Yin, H.J. Gao, Reconfigurable tolerant control of uncertain mechanical systems with actuator faults: a sliding mode observer-based approach, IEEE Transactions on Control Systems Technology 26 (4) (2018) 1249–1258.
[28] X. Wang, S. Sun, E.J. van Kampen, Q. Chu, Quadrotor fault tolerant incremental sliding mode control driven by sliding mode disturbance observers, Aerospace Science and Technology 87 (2019) 417–430.
[29] F. Chen, R. Jiang, K. Zhang, B. Jiang, T. Gang, Robust backstepping sliding mode control and observer-based fault estimation for a quadrotor UAV, IEEE Transactions on Industrial Electronics 63 (8) (2016) 5044–5056.
[30] B. Li, Q.L. Hu, Y.S. Yang, A.P. Octavian, Finite-time disturbance observer based integral sliding mode control for attitude stabilisation under actuator failure, IET Control Theory & Applications 13 (1) (2019) 50–58.
[31] B. Li, Q.L. Hu, Y.S. Yang, Continuous finite-time extended state observer based fault tolerant control for attitude stabilization, Aerospace Science and Technology 84 (2019) 204–213.
[32] S.C. Liu, P. Lyu, J.Z. Lai, C. Yuan, B.Q. Wang, A fault-tolerant attitude estimation method for quadrotors based on analytical redundancy, Aerospace Science and Technology 93 (2019) 105290.
[33] X. Wang, C.P. Tan, F. Wu, J. Wang, Fault-tolerant attitude control for rigid spacecraft without angular velocity measurements, IEEE Transactions on Cybernetics (2019) 1–14.
[34] Y. Song, L. He, D. Zhang, J. Qian, J. Fu, Neuroadaptive fault-tolerant control of quadrotor UAVs: a more affordable solution, IEEE Transactions on Neural Networks and Learning Systems 30 (7) (2019) 1975–1983.
[35] K. Zhang, M.A. Demetriou, Adaptive attitude synchronization control of spacecraft formation with adaptive synchronization gains, Journal of Guidance, Control, and Dynamics 37 (5) (2014) 1644–1651.
[36] B. Xiao, S. Yin, Exponential tracking control of robotic manipulators with uncertain kinematics and dynamics, IEEE Transactions on Industrial Informatics 15 (2) (2019) 689–698.
[37] S.P. Bhat, D.S. Bernstein, Geometric homogeneity with applications to finite time stability, Mathematics of Control, Signals, and Systems 17 (2) (2005) 101–127.
[38] E. Cruz-Zavala, J.A. Moreno, Homogeneous high order sliding mode design: a Lyapunov approach, Automatica 80 (2017) 232–238.
[39] Q.L. Hu, B.Y. Jiang, Continuous finite-time attitude control for rigid spacecraft based on angular velocity observer, IEEE Transactions on Aerospace and Electronic Systems 54 (3) (2018) 1082–1092.
[40] Y. Feng, X.H. Yu, F.L. Han, On nonsingular terminal sliding-mode control of nonlinear systems, Automatica 49 (6) (2013) 1715–1722.
[41] S. Kamal, J.A. Moreno, A. Chalanga, B. Bandyopadhyay, L.M. Fridman, Continuous terminal sliding-mode controller, Automatica 69 (C) (2016) 308–314.

[42] L. Fridman, Sliding mode enforcement after: main results and some open problems, in: L. Fridman, J. Moreno, R. Iriarte (Eds.), Sliding Modes after the First Decade of the 21st Century, in: LNCIS, Springer, Berlin, 1990, pp. 3–57.
[43] A. Levant, Homogeneity approach to high-order sliding mode design, Automatica 41 (5) (2005) 823–830.

Index

A

D

F

G

H

O

P

Printed in the United States
By Bookmasters